Interactive Task Learning

Humans, Robots, and Agents Acquiring
New Tasks through Natural Interactions

Strüngmann Forum Reports

Julia R. Lupp, series editor

The Ernst Strüngmann Forum is made possible through
the generous support of the Ernst Strüngmann Foundation,
inaugurated by Dr. Andreas and Dr. Thomas Strüngmann.

This Forum was supported by the
Deutsche Forschungsgemeinschaft

Interactive Task Learning

Humans, Robots, and Agents Acquiring New Tasks through Natural Interactions

Edited by

Kevin A. Gluck and John E. Laird

Program Advisory Committee:

Kenneth M. Ford, Kevin A. Gluck, John E. Laird, Elena Lieven,
Julia R. Lupp, Luc Steels, and Niels A. Taatgen

The MIT Press

Cambridge, Massachusetts
London, England

Series Editor: J. R. Lupp
Editorial Assistance: A. Ducey-Gessner, M. Turner, C. Stephen
Photographs: N. Miguletz
Lektorat: BerlinScienceWorks

The book was set in TimesNewRoman and Arial.
Printed and bound in the United States of America.

Library of Congress Cataloging-in-Publication Data is available

Ernst Strüngmann Forum (26th: 2017 : Frankfurt am Main, Germany)

ISBN: 978-0-262-03882-9

10 9 8 7 6 5 4 3 2 1

Contents

Instruction

Learning New Tasks

Ethical Considerations

In Memoriam

During the preparation of this volume, our valued colleague Charles "Chuck" Rich lost his battle against pancreatic cancer. Chuck's distinguished career involved pioneering contributions to interactive task learning long before that term was coined. At MIT, Mitsubishi Electric Corporation, and finally Worcester Polytechnic Institute, he researched, created, and taught about collaborative, conversational, natural interaction between humans and learning agents and robots. We were thrilled when Chuck accepted our invitation to author a background paper on task knowledge (see Chapter 5). His commitment and passion for this field of research were evident in carrying through with the writing, even after his devastating diagnosis. The love and dedication of his research partner and wife, Candace Sidner, have also been evident in her support through the final editorial processes. It is with deep appreciation that we thank you, Chuck and Candace, for your contributions to this Forum and book, and for providing some of the shoulders on which interactive task learning stands.

Preface

Science is a highly specialized enterprise—one that enables areas of enquiry to be minutely pursued, establishes working paradigms and normative standards, and supports rigor in experimental research. Some issues, however, do not fall neatly into the purview of any one particular area of science and in these cases, specialization can actually hinder conceptualization and/or limit the generation of problem-solving approaches.

The Ernst Strüngmann Forum was established in 2006 to address these types of issues. Utilizing a specific approach, it addresses problems that emerge from ongoing research: topics that transcend classic disciplinary boundaries, emanating from areas where conceptualization may have stagnated and the way forward is uncertain. This approach facilitates open discourse and encourages divergent perspectives, both viewed as integral to the expansion of knowledge. Consensus is never a goal, explicitly or implicitly. Put simply, topics are placed on the proverbial discussion table and examined collectively from multiple vantage points: existing gaps in knowledge are exposed, key questions formulated, and ways to "fill" such gaps put forward.

Topic proposals are accepted from leading researchers and reviewed by the Scientific Advisory Board of the Ernst Strüngmann Forum. Once approved, the Ernst Strüngmann Forum convenes a steering committee to transform the proposal into a scientific framework that will support the focal meeting, or Forum, which is best imagined as a week-long intellectual retreat. At a Forum, formal presentations are taboo. Instead, invitees engage in various forms of dialogue that maximize intellectual output. To prepare for this discourse, invited "background papers," written in advance, introduce key issues and expose areas for discussion. At a Forum, the central topic is then addressed from different perspectives by four working groups, each comprised of 10–11 researchers. Groups work autonomously yet build on the interactions with each other. To ensure that emerging insights do not get lost, a designated group member generates a draft report during the Forum.

The central topic of the 26th Forum—interactive task learning—was brought to our attention by Kevin A. Gluck and John E. Laird. The overarching idea was to convene a diverse group of experts to examine the processes by which new tasks are acquired through natural interaction between humans, humans and agents, and agents. It was felt that the fractionated state of relevant scientific and technical disciplines had thus far hindered progress in this area—one fundamental to both artificial intelligence and cognitive psychology. Thus, creating collaboration within the broad research community as well as delineating research challenges and future trajectories were set as primary goals.

Joining us on the Program Advisory Committee to realize these goals were Ken Ford, Elena Lieven, Luc Steels, and Niels Taatgen. From April 22–24, 2016, the committee met to refine the scientific framework of the proposal and identify participants to the Forum, which was held in Frankfurt am Main from May 21–26, 2017.

This volume synthesizes the ideas and perspectives that emerged from the entire process and is comprised of two types of contributions:

- Key questions and aspects of interactive task learning are presented. Most of these chapters served as background to the Forum while others emerged out of the discussion. Both have been peer reviewed and edited to provide up-to-date information.

- In Chapters 3, 7, 11, and 15, the working groups from the Forum provide a synthesis of their multifaceted discussions. Edited to ensure accessibility, these chapters should not be understood as proceedings or consensus documents. Their intent is to summarize perspectives, expose diverging opinions as well as remaining open questions, and highlight areas for future enquiry.

Every Forum creates its own unique dynamics and puts demands on all who participate. Each invitee played an active role at this Forum and for their efforts, I wish to thank everyone. I extend a special word of appreciation to the Program Advisory Committee, to the authors and reviewers of the background papers, as well as to the moderators of the individual working groups: Elena Lieven, Niels Taatgen, John Laird, and Kevin Gluck. For their efforts in drafting and finalizing the reports, special recognition goes to the rapporteurs of the working groups: Andrea Thomaz, Robert Wray III, Dario Salvucci, and Julie Shah. Finally, I wish to extend my sincere appreciation to Kevin Gluck and John Laird, whose commitment and cooperation were integral at each stage.

To conduct its work, the Ernst Strüngmann Forum relies on institutional stability and an environment that encourages free thought. The generous support of the Ernst Strüngmann Foundation, established by Dr. Andreas and Dr. Thomas Strüngmann in honor of their father, enables the Ernst Strüngmann Forum to pursue its work in the service of science. Additional partnerships include the Scientific Advisory Board, which ensures the scientific independence of the Forum; the German Science Foundation, for its supplemental financial support; and the Frankfurt Institute for Advanced Studies, which shares its vibrant setting with the Forum.

Breaking new intellectual ground is never easy. Yet, when the edges of the unknown begin to appear and the gaps in understanding are identified, the act of formulating strategies to fill these gaps becomes a most invigorating activity. On behalf of everyone involved, I hope this volume will spur further discussion and research into the field of interactive task learning.

Julia R. Lupp, Director
Ernst Strüngmann Forum
Frankfurt Institute for Advanced Studies (FIAS)
Ruth-Moufang-Str. 1, 60438 Frankfurt am Main, Germany
https://esforum.de/

List of Contributors

Belpaeme, Tony Centre for Robotics and Neural Systems, Plymouth University, Plymouth, PL4 8AA, U.K.

Beuls, Katrien Artificial Intelligence Laboratory, Vrije Universiteit Brussel, 1050 Brussels, Belgium

Cakmak, Maya Paul G. Allen School of Computer Science and Engineering, University of Washington, Seattle, WA 98195, U.S.A.

Chai, Joyce Y. Department of Computer Science and Engineering, Michigan State University, East Lansing, MI 48864, U.S.A.

Chang, Franklin Kobe City University for Foreign Studies, Kobe, Japan

Denga, Ropafadzo Rensselaer Polytechnic Institute, Troy, NY 12180, U.S.A.

Destefano, Marc Google LLC, Mountain View, CA 94043, U.S.A.

d'Inverno, Mark Department of Computing, Goldsmiths, University of London, London SE14 6NW, U.K.

Forbus, Kenneth D. EECS, Northwestern University, Evanston, IL 60208, U.S.A.

Garrod, Simon Institute of Neuroscience and Psychology, University of Glasgow, Glasgow G12 8QB, U.K.

Gluck, Kevin A. Cognitive Models and Agents Branch, Air Force Research Laboratory, Wright-Patterson AFB, OH 45433, U.S.A.

Gray, Wayne D. Cognitive Science, Rensselaer Polytechnic Institute, Troy, NY 12180, U.S.A.

Kirk, James CSE, University of Michigan, Ann Arbor, MI 48109-2121, U.S.A.

Koedinger, Kenneth R. Human–Computer Interaction and Psychology Institute, Carnegie Mellon University, Pittsburgh, PA 15213, U.S.A.

Kordjamshidi, Parisa Computer Science, Tulane University, New Orleans, LA 70118, U.S.A., and the Florida Institute for Human and Machine Cognition, Pensacola, FL, U.S.A.

Laird, John E. CSE, University of Michigan, Ann Arbor, MI 48109-2121, U.S.A.

Lebiere, Christian Psychology, Carnegie Mellon University, Pittsburgh, PA 15213, U.S.A.

Levinson, Stephen C. Language and Cognition, Max Planck Institute for Psycholinguistics, 6525 XD Nijmegen, The Netherlands

Lieven, Elena LuCiD Child Study Centre, School of Health Sciences, University of Manchester, Manchester M13 9PL, U.K.

Lindstedt, John K. Department of Psychological Sciences, Rice University, Houston, TX 77005, U.S.A.

Mininger, Aaron CSE, University of Michigan, Ann Arbor, MI 48109-2121, U.S.A.

Mitchell, Tom M. Machine Learning Department, School of Computer Science, Carnegie Mellon University, Pittsburgh, PA 15213, U.S.A.

Mohan, Shiwali Interaction Analytics Lab, Palo Alto Research Center, Palo Alto, CA 94304, U.S.A.

Paiva, Ana INESC-ID and IST, Instituto Superior Técnico, University of Lisbon, 2744–016 Porto Salvo, Portugal

Pastra, Katerina Cognitive Systems Research Institute, 11525 Athens, Greece

Pirolli, Peter Florida Institute for Human and Machine Cognition, Pensacola, FL 32502, U.S.A.

Rahman, Roussel Cognitive Science, Rensselaer Polytechnic Institute, Troy, NY 12180, U.S.A.

Rich, Charles Department of Computer Science, Worcester Polytechnic Institute, Worcester, MA 01609, U.S.A.

Rohlfing, Katharina J. Faculty of Arts and Humanities, Paderborn University, 33098 Paderborn, Germany

Rosenbloom, Paul S. Department of Computer Science, Institute for Creative Technologies, University of Southern California, Playa Vista, CA 90094, U.S.A.

Russwinkel, Nele Cognitive Modeling in Dynamic Human-Machine Systems, TU Berlin, 10587 Berlin, Germany

Salvucci, Dario D. Department of Computer Science, Drexel University, Philadelphia, PA 19104, U.S.A.

Sangster, Matthew-Donald D. Rensselaer Polytechnic Institute, Troy, NY 12180, U.S.A.

Scheutz, Matthias Computer Science, Tufts University, Medford, MA 02155, U.S.A.

Shah, Julie A. Aeronautics and Astronautics, Massachusetts Institute of Technology, Cambridge, MA 02139, U.S.A.

Sidner, Candace L. Department of Computer Science, Worcester Polytechnic Institute, Worcester, MA 01609, U.S.A.

Sibert, Catherine Rensselaer Polytechnic Institute, Troy, NY 12180, U.S.A.

Spranger, Michael Fundamental Research Laboratory, Sony Computer Science Laboratories Inc., 141-0022 Tokyo, Japan

Steels, Luc Institut de Biologia Evolutiva, (UPF-CSIC), 08003 Barcelona, Spain

Stevenson, Suzanne Department of Computer Science, University of Toronto, Toronto, ON, M5S 3G4, Canada

Stewart, Terrence C. Centre for Theoretical Neuroscience, University of Waterloo, Waterloo, ON, Canada N2L 3G1, Canada

Still, Arthur Department of Psychology, University of Durham, Durham DH1 3LE, U.K.

Stocco, Andrea Department of Psychology and Institute for Learning and Brain Sciences, University of Washington, Seattle, WA 98195, U.S.A.

Taatgen, Niels A. Artificial Intelligence, University of Groningen, AG Groningen, The Netherlands

Thomaz, Andrea L. Electrical and Computer Engineering, University of Texas at Austin, Austin, TX 78701, U.S.A.

Trafton, J. Gregory U.S. Naval Research Laboratory, Washington, D.C. 20375, U.S.A.

van der Maas, Han L. J. Psychology, University of Amsterdam, 1018WS Amsterdam, The Netherlands

Van Eecke, Paul Artificial Intelligence Laboratory, Vrije Universiteit Brussel, 1050 Brussels, Belgium

VanLehn, Kurt School of Computing, Informatics and Decision Science Engineering, Arizona State University, Tempe, AZ 85284, U.S.A.

Vollmer, Anna-Lisa Applied Informatics Group, CITEC Bielefeld University, 33619 Bielefeld, Germany

Wiles, Janet School of Information Technology and Electrical Engineering, The University of Queensland, St. Lucia QLD 4072, Australia

Wray, Robert E., III Soar Technology, Inc., Chapel Hill, NC 27516, U.S.A.

Yee-King, Matthew Department of Computing, Goldsmiths, University of London, London SE14 6NW, U.K.

1

Looking Forward to Interactive Task Learning

Kevin A. Gluck and John E. Laird

Human learning has been the subject of extensive research in multiple areas of science. People are always learning, from whatever sources of knowledge are available. Language and other forms of natural communication enable us to master novel tasks quickly; once we do, we often share the resultant knowledge with others. In just a few minutes, we can grasp how to play a new game, use a new device (e.g., smart phones, industrial machinery), or assist a disabled family member in meeting specific challenges. Importantly, as we learn and hone performance on a task, we adapt in real time to emergent needs—sometimes figuring things out for ourselves, sometimes interacting with others to gain efficiencies or address any problems we encounter.

Contemporary artificial agents, by contrast, are bound to the specific tasks for which they were originally programmed. Even systems designed to acquire knowledge and expertise can learn only a single task at a time (e.g., Chess, Go, video games), becoming idiot savants with amazing capabilities, but without any abilities beyond that narrow specialization. Without doubt, advances in artificial intelligence, cognitive science, and robotics point to future systems with sufficient cognitive and physical capabilities to perform a wide variety of diverse tasks. But how will they learn tasks that arise unexpectedly—tasks that cannot be anticipated and therefore preprogrammed or trained for? How can agents pursue a task when there is insufficient prior knowledge or time for exploration to guide learning?

Interactive task learning (ITL) attempts to answer those questions by providing a conceptual framework for agents to learn not only how to perform tasks better, but also to learn new tasks from scratch through natural, real-time interactions with others. ITL involves interactions between an agent (human or machine), its world, and, crucially, other agents in the world. ITL is a bidirectional process between teacher and learner (both of which can be humans or machines) that results in collaboration and knowledge creation.

The catalyzing idea behind ITL is as follows: for artificial systems to learn from and teach us entirely new things at any given moment, we must advance beyond the traditional approach of creating specialized AI agents for single, predetermined niche purposes and instead incorporate a rich set of natural interaction and learning mechanisms into our systems. This involves two crucial requisites. First, the way in which artificial systems learn and teach new tasks must be natural for people, not constrained by traditional programming and digital forensics. Second, learning and performing multiple tasks can only be bounded by physical and informational limits, not by design, implementation, or optimization for single task performance.

The concept of ITL is controversial, in that it disrupts the status quo and involves diverse challenges. ITL requires theoretical and practical advances in the integration of a broad range of capabilities associated with cognition, including extracting task-relevant meaning from perception, task-relevant action, grounded language processing, dialogue and interaction management, integrated knowledge-rich relational reasoning, problem solving, learning, and metacognition. This integration contrasts with the general trend toward increasing fragmentation and focus on narrow capabilities and problems. Beyond these challenges, ITL creates an opportunity to rethink the fundamental nature of our most advanced and capable artifacts. How can we move beyond artifacts that are designed for a single use or purpose, to ones that can be dynamically adapted to our changing needs, increasing the rate of our progress and the quality of our lives as individuals and societies? Isolated efforts to develop more intelligent agents and robots are already underway in areas such as healthcare, in-home assistance, education, and transportation. We propose the missing link among them is the unifying vision of ITL.

Our optimism that the time is right for a coordinated and concerted push toward ITL is grounded in our assessment that despite shortcomings, gaps, and challenges, the research community has made progress on important component capabilities and their integration. To shape R&D investments in a way that advances ITL in artificial systems requires the identification of broad organizing themes. *Pace*, *persistence*, and *partnering* are core characteristics that constitute research challenges around which we can rally our science and technology investments.

Pace, Persistence, and Partnering

The human capacity for rapid, nearly instantaneous learning of entirely new tasks on the basis of brief communications and one, two, or a few demonstrations sets a *pace* requirement for ITL. The pace of the interaction, the pace of the teaching and learning, and the pace of task completion must all occur on timescales aligned with and amenable to real-time human experience.

Humans are engines of creation. From the imaginative play of early childhood, to the generative nature of language, to scientific discovery and technological innovation, we are constantly creating new constructs, concepts, and capabilities. Our *persistence* throughout these activities requires us to both assimilate and accommodate newly gained knowledge and existing knowledge. As we work to create intelligent artifacts that interact, we must recognize that it will never be possible to anticipate and represent, in advance, all the knowledge and skill that may be required in the future. ITL agents must be able to learn and adapt continually with robust success, over a long period of time in environments that are dynamic, nonstationary, and boundlessly novel.

Human beings help each other. It is what we do. We organize in ways that support joint objectives and goals. Our most valued relationships are with family, friends, partners, and teammates. These relationships develop over time out of shared experiences in which we demonstrate an ability and willingness to be there for each other in times of need. We are at our best when we take the initiative to assist or compensate without being asked, simply because we know it will be helpful. By contrast, contemporary machine artifacts do none of this. They function merely as tools, responding as designed, reactive but not proactive. They are unable to engage in true *partnering*. ITL systems need to be more like partners or teammates, and not merely tools.

Each of these core characteristics has received some attention from isolated subsets of the research community. To achieve ITL in future agents, we must find a way to integrate these characteristics into systems. This will not be easy, for at the core of each characteristic and their integration is the challenge of understanding.

The Challenge of Understanding

Perhaps the most important limitation of our contemporary intelligent machines is that they are not capable of understanding with the depth and breadth found in humans. Many impressive accomplishments have been achieved in the cognitive and computational sciences in recent decades. Most of those are best known to isolated subcommunities of researchers toiling away on issues with great scientific merit. A precious few have captured the imagination of the public due to high profile events, demonstrations, and competitions. Algorithmic advances, blazing fast processors, and massive amounts of training data make it possible to show that silicon-based computation can classify objects, learn well-defined games, and answer some types of questions as well as or better than people can. Less well hyped is the characteristic fragility of these systems. When they are wrong, they are often wrong in ways that are surprising and confusing to people. This is because people understand the questions, images, and activities within the broad context of not just a single task, but within the myriad of tasks, experiences, and relationships they develop over time in ways that the algorithms do not.

At least as troubling as the lack of understanding in our most advanced artificially intelligent machine learners is the fact that we often don't understand them. This is certainly true for the general public, who tend to ascribe assortments of sophisticated humanlike intellectual capacities to computational systems where it is not warranted. It is often also true for the developers of some of our most impressive learning machines. Among those working at the leading edge of science and technology, the issue is not one of unjustified anthropomorphism. Rather, it is the reality of human cognitive limitations running up against complex, hybrid computational systems. The emphasis on powerful learning mechanisms scaled for use on big data sources has abandoned transparency and left even the innovators of these capabilities scratching their heads and asking, "Why and how is it doing that? What did it learn?" Generally those questions can be answered by engaging in some committed digital forensics, but the time and energy required for those analyses far exceed what would be tolerable in the context of ITL. Queryability, explainability, and transparency must be baked into these systems in order to foster natural, efficient understanding.

Finally, in a recursive descent into scientific and technological challenges, as a research community we must face the reality that the root cause of our machines' poor understanding and of our poor understanding of complex learning machines is the fact that we simply do not understand the concept of understanding. There is, in effect, no scientific consensus about what understanding actually is, despite an abundance of work by philosophers, psychologists, neuroscientists, and computer scientists. Indicative of this absence of agreement is a great deal of ambiguity regarding how to assess understanding. This should come as no surprise, given the inconsistent and haphazard manner in which we, as individuals, evaluate the understanding of other people in our daily lives. We tend to assume a great deal of understanding in the minds of others. Sometimes those assumptions are valid and supported by social cues, prior experience, or knowledge of the other, which makes these assumptions defensible. Other times they are simply efficient conveniences. Rarely do we bother to rigorously evaluate the extent to which another person understands.

Up to now, we have been able to overlook our ignorance regarding the fundamental nature of understanding, our poor understanding of complex learning systems, and the absence of understanding in our machines. We have been satisfied with the traditional approach of implementing systems for pre-determined niche purposes. However, the vision for ITL in artificial systems creates a forcing function to address these issues. The development, improvement, and evaluation of understanding in humans, robots, and agents is critical to the creation of ITL.

Moving Forward: A Multidisciplinary Challenge

Clearly, we are enthusiastic about the potential societal benefits that ITL systems could bring. Nonetheless, we appreciate that the challenges are daunting. To even begin, experts from multiple areas of science and technology must be able to communicate, find common ground, and implement novel capabilities across disciplinary divides.

With the support of the Ernst Strüngmann Forum we sought to initiate a dialogue among experts from robotics, cognitive modeling, computer science, artificial intelligence, and developmental and comparative psychology. This discourse aimed to analyze how humans and artificial agents acquire new tasks through natural interactions as well as to define ITL from various perspectives, in an effort to establish a foundational reference and organizing framework. The results of this multifaceted dialogue are captured in this volume. Organized around the following primary topics, each contribution explores key aspects of ITL:

1. *Knowledge*: In Chapter 3, Robert Wray III et al. discuss the functional roles of knowledge in ITL, examine central challenges that must be overcome, and pose research questions to direct future research. From a formal, computational perspective, Christian Lebiere (Chapter 4) presents different forms of knowledge and skills involved in ITL. Through an examination of the collaborative interactions inherent in learning and teaching, Charles Rich (Chapter 5) analyzes the abstract form, nature, and organization of task knowledge. Concluding this section, Niels Taatgen (Chapter 6) explores what is needed to construct a cognitive architecture capable of supporting flexible knowledge and skills.

2. *Interaction*: In Chapter 7, Andrea Thomaz et al. consider which qualities of human interaction and learning will be most effective and natural to incorporate into an ITL agent; central to this is the alignment of common ground between a teacher agent and a learner agent. In his analysis of natural forms of purposeful interaction among humans, Stephen Levinson (Chapter 8) delineates the basic organization of interactive language use and discusses the challenge of incorporating the predictive nature of human comprehension into an ITL agent. In Chapter 9, Joyce Chai et al. outline the different types of knowledge that can be transferred between agents and discuss the perception, action, and coordination capabilities that enable teaching–learning interactions; in addition, they consider challenges and research opportunities associated with enabling *natural* interaction in artificial agents. To conclude this section, Wayne Gray et al. (Chapter 10) explore how experimental psychology, machine learning, and advanced statistical analyses can be used to understand the complexity of interactive performance in complex tasks involving single or multiple interactive agents in dynamic environments.

3. *Instruction*: In Chapter 11, Julie Shah et al. present frameworks, models, and methods for task instruction, broadly connecting structural and adaptive improvements to instruction, historical developments in programming, and the extraordinary challenge that fluid, flexible,

co-constructive task instruction and learning places on the vision for ITL. In Chapter 12, Kurt VanLehn looks at prototypical human tutoring behavior, analyzing what exceptional tutors sometimes do (but most tutors do not) and comparing the effectiveness of human versus computer tutors. In Chapter 13, Katrin Beuls et al. examine what type of general architecture is needed to construct artificial agents that can assume the role of teacher (by carrying out teaching strategies) or the role of learner (by carrying out learning strategies that benefit from these teaching strategies); they argue that a meta-layer is necessary to understand and implement strategies and point to operational examples in the domain of second language teaching. In Chapter 14, Arthur Still et al. explore the concept of creativity and its relationship to the development of education theory, focusing on what is necessary to inform teaching practice and development of education technology.

4. *Learning*: Summarizing their discussions at the Forum, Dario Salvucci et al. explore in Chapter 15 the learning of task knowledge through interaction, the capabilities that facilitate learning, aspects of interaction that relate closely to learning, as well as evaluation dimensions and metrics for ITL systems. Based on knowledge of preexisting capabilities that appear early in human development, Franklin Chang (Chapter 16) introduces a world-state prediction model—one that can learn detailed physical regularities in the environment and develop representations for predicting the actions and goals of animate agents—to suggest that prediction and prediction error are capabilities that could improve ITL systems. In Chapter 17, using an existing agent, Rosie, to illustrate how an ITL agent can learn many tasks in a variety of domains, John Laird et al. present characteristics of the learning problem and examine how these influence underlying learning algorithms; learning approaches are discussed that respond to the unique challenges of ITL.

Throughout this process, the open exchange of ideas and perspectives—hallmarks of the Ernst Strüngmann Forum—was bolstered by our own inclination toward asking lots of questions, especially the hard ones, and to accept disagreement, countervailing opinions, and inevitable failures on the path of progress. As one might imagine, many questions surfaced and, where appropriate, ways of pursuing these have been highlighted. Two priorities that emerged, however, have been given special attention. The first, a common reference frame to guide future discussion, is presented by Tom Mitchell et al. in Chapter 2. The second is an appreciation that we must commence with ethical considerations now, even as we debate the nature, viability, and path toward ITL. There will be valid security and privacy concerns, and it is certain that people with malicious intent will attempt to repurpose ITL for harm. Thus, now is the time to think about and take action on these matters. To that end, in Chapter 18, Matthias Scheutz examines different ethical aspects of ITL.

Despite all the challenges, we believe ITL offers great potential for humanity and hope this volume will inspire the international research community to pursue the necessary science and technology. We look forward to working with a global community of researchers to realize this vision.

Acknowledgments

We thank the Ernst Strüngmann Foundation for its extraordinary commitment to the exploration of multidisciplinary scientific challenges, and especially for approving and funding this Forum on Interactive Task Learning. However, organizations are only as good as the people within them, and it is the collective contributions of the individuals leading the Ernst Strungmann Forum that make it the world-class, impactful experience that it is. Our sincere appreciation to Julia Lupp, its Director, for her deep immersion and commitment, unwavering support, and impressive patience, and to Aimée Ducey-Gessner, Marina Turner, and Catherine Stephen for their excellent professional support throughout this process.

Participating in an Ernst Strungmann Forum is a significant investment of time and intellectual energy. We thank all of the participants for setting aside their many other existing commitments to join us in this endeavor. Special appreciation is due to our Program Advisory Committee (Ken Ford, Elena Lieven, Julia Lupp, Luc Steels, and Niels Taatgen), who worked with us to shape new ideas and rough intentions into a more complete and concrete plan for the Forum. We must also recognize the impressive work of our rapporteurs (Dario Salvucci, Julie Shah, Andrea Thomaz, and Bob Wray) who toiled diligently throughout and following the Forum to summarize, represent, and organize diverse discussion points into the group reports introducing each section of the book.

Finally, although we were as comprehensive and inclusive as possible, there is no way to include everyone who is doing important and relevant work in a single event such as this. Thus we thank our many additional colleagues and collaborators from the cognitive, computing, social, and psychological sciences who are chipping away at the barriers, making progress on the challenges, and choosing to travel the path toward ITL. You are inspiring, and we look forward to learning with you in future interactions.

2

Framing the Problem of Interactive Task Learning

Tom M. Mitchell, Simon Garrod, John E. Laird,
Stephen C. Levinson, and Kenneth R. Koedinger

Abstract

To support a precise discussion of interactive task learning, the problem setting in which teachers and learners interact in a shared world must be clearly defined and understood. This chapter provides a formalism to enable discussion of the different types of interactive learning: from teaching a robot to grasp a novel object, to instructing a mobile phone how to reach a friend in an emergency. It provides a way to speak precisely about notions such as shared knowledge between teachers and learners, presents working definitions of the internal structures of the agent, and describes the relationships between the task environment and the communication channel. It focuses on the *problem* of interactive task learning, not its solution, as a backdrop to further discourse in this volume.

Formulating the Problem

In interactive task learning (ITL), we assume that a set of agents jointly occupies a shared world. Each agent, A, has a set of sensors, $S = \{s_1, s_2, \ldots\}$, it uses to sense the world, and a set of effectors, $EF = \{ef_1, ef_2, \ldots\}$, it can use to alter the world through actions. For example, a robot might have sensors such as a camera and microphone and effectors such as wheels and arms. Similarly, a mobile phone might have both physical sensors and effectors, such as a microphone to sense the world and a speaker to affect it, as well as cyber-sensors, such as a web browser to observe the internet, and cyber-effectors, such as the ability to send text messages.

In general, we assume that these agents exist in a shared world, the state of which at time t is denoted by $w(t)$. The sensors and effectors of each agent operate on this world, producing a time series of its observations and actions. In general, the world may contain any number of agents with differing capabilities.

Communication

To capture communication among agents (e.g., speech, nonverbal gestures), we assume there is a communication channel between each pair of agents. Given that all communications must go through the physical world in some form, we capture the communication channel in our model by designating a subset of effectors, EF_C (e.g., speaking, pointing), and sensors, S_C (e.g., listening and interpreting), that implement this communication channel. We think of these communication sensors and effectors as operating on a set of objects (e.g., words, symbols, fingers that point, underlines that highlight) that *refer* (among other things) to other objects in the world. However, communication differs from other forms of sensing and acting in at least two key ways:

1. Unlike the use of sensors to observe properties of nonhuman physical objects, the interpretation of sensor inputs in communication (e.g., input speech) is based on *shared conventions* between agents about the meanings conveyed by different communication actions and communication objects.
2. The communication channel can convey meanings that go beyond what can be perceived by other sensors of the physical world, including information about mental states such as "I believe Mary thinks the train is late."

Determining whether another agent is using its effectors in an attempt to communicate rather to perform some instrumental action in the world is part of any interactive task for a robot. This is discussed below under the section, "Task Environment and the Communications Channel."

Learning

Now let us consider some learning agent A. To define a learning problem precisely, we say that agent A learns to improve its performance, P, at task, T, through experience, E. In fact, we assume every learning problem can be defined in terms of some triple <P, T, E>. To illustrate, an agent might face a learning problem in which the task is to play the game Go (T), the training experience (E) consists of playing 1000 practice games against itself, and the goal of learning is to improve performance (P) measured by how frequently it can defeat a second agent in a 100-game tournament. A related learning problem might have the same task, T, and performance metric, P, but differ in the type of training experience available (e.g., learning from advice received by a teacher while playing games). As a further example, a robot might face the task of learning to set the dining table (T), from watching videos of a person performing that task (E), where performance (P) is measured by the precision of the final table setting minus the number of dishes dropped along the way.

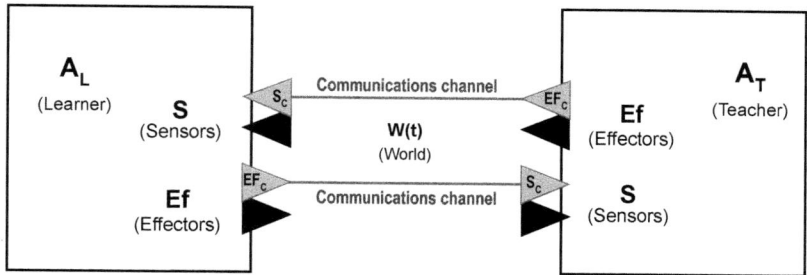

Figure 2.1   Simple world with learning agent A_L and teaching agent A_T. Black sensors and effectors sense and alter the agents' shared physical world. Gray sensors and effectors indicate the communication channel between agents.

It is important to keep in mind that the performance task (e.g., playing Go) is distinct from the learning problem itself (e.g., *learning* to play Go).

This definition of learning covers cases where the learning agent acts alone in the world to generate its own training experience (e.g., experimenting in the world via its sensors and effectors), as well as cases where the training experience involves other agents acting as teachers or collaborators to help shape the training experience. Figure 2.1 illustrates a prototypical situation of interactive learning between a learning agent, A_L, and a teaching agent, A_T.

Working Definition of ITL

We define ITL to be any process by which an agent (A) improves its performance (P) on some task (T) through experience (E), when E consists of a series of sensing, effecting, and communicating interactions between A, its world, and crucially other agents in the world.

There are many nuances to the ITL concept, as presented in subsequent chapters in this volume. These discussions introduce additional structure into the above problem formulation, including assumptions about how a teacher might impart different types of knowledge to a learner, the roles of demonstration and direct teaching by instruction, the internal architecture of the student and teacher, as well as the need for building common ground between the teacher and learner.

Internal Structures of an Agent

Having described the problem faced by a learning agent, we now turn to a discussion of the internal structures of the agent. The determination of what constitutes a proper internal structure for an agent to succeed at ITL is the focus of ongoing research, and although consensus has yet to emerge, many proposed approaches share certain assumptions; namely, that an agent possesses internal structures such as knowledge and goals. Here we introduce a simple vocabulary to support discussion of the knowledge and goals held by agents, as well as the mental states of others.

Knowledge

We assume that each Agent A possesses a set of beliefs which constitute its knowledge, K(A), about its world and agents. Knowledge can be about absolutely anything. It can include, for example, knowledge about the world (e.g., "there is a chair in front of Agent A"), other agents (e.g., "Agent A_2 is also in the room"), capabilities of other agents (e.g., "Agent A_2 is a good Go player"), as well as knowledge about the knowledge of other agents (e.g., "Agent A_2 knows that Agent A_1 knows there is a chair in the room").

Some of an agent's knowledge may be correct whereas other knowledge may be incorrect, and the agent might or might not know which is correct. Much of learning involves acquiring new knowledge and correcting or refining current knowledge.

Agents may possess (or act as though they possess) knowledge of which they are unaware, though their ability to communicate this knowledge depends on mental access to this knowledge in some declarative form. For example, a human agent may know how to recognize their mother yet be unable to communicate this knowledge to another agent in declarative form.

Goals

Agent A may have goals G(A) at any given time. These may include, for instance, goals applicable to the external world (e.g., set the dining table), to the agent's internal mental world (e.g., learn how to set dining tables), as well as to the mental world of another agent (e.g., help Agent A_L learn to set tables). Goals can have subgoals which contribute to an agent achieving the overall goal.

Common Ground

Shared knowledge and goals of agents are essential to communication, in general, and to ITL in particular. Communication and teaching can be viewed as processes used to establish and refine a set of shared goals and knowledge about the task at hand—the common ground between agents. When discussing common ground, it is helpful to define two related but distinct notions:

- The set of shared knowledge between agents A_1 and A_2 is defined as the intersection of knowledge between these agents: $IK(A_1, A_2)$.
- Shared goals is the set of goals held by both agents A_1 and A_2: $IG(A_1, A_2)$.

Again, it is important to note that agents A_1 and A_2 may themselves be unable to know perfectly which of their knowledge and goals are shared by the other agent. In fact, the ITL process is often driven by the attempts of these agents to better know what they already share, and the attempt to transfer knowledge between teacher A_T and learner A_L, thereby growing the shared knowledge, $IK(A_T, A_L)$.

The intersection of knowledge is, however, not sufficient to *ground* communication. For example, two people may both have the knowledge that one of two cars in a lot is electric, but they may not realize that the other person knows this as well; thus, when differentiating between the two cars in conversation, they will not refer to the car as "the electric car" but rather draw upon other criteria (e.g., color, model) to make the distinction. This demonstrates the need for a more specific notion than simply the intersection of knowledge (see Smith 1982). For successful communication, we need a set of assertions, S, such that A_1 knows S, A_2 knows S, A_1 knows that A_2 knows S, and A_2 knows that A_1 knows S. In this case, we will say that A_1 and A_2 *mutually know* S and that S is in the set of mutual knowledge $MK(A_1, A_2)$. This is the notion of *common ground* needed for establishing communication.

There is no foolproof way for agents to infer what is contained in common ground. So when human agents interact, they might try to discover a shared basis for common ground before treating anything as being in it. For instance, if two agents see a cup in front of them and both observe that the other also sees the cup, this may act as a shared basis for that cup to be considered in common ground. However, if it is not apparent that the cup is salient to both agents, one agent might attempt to *ground* the information by holding, for example, the cup up or pointing to it (Clark 1996:98). Interactions between human and nonhuman agents (e.g., robots) present challenging problems of finding the appropriate shared basis for inferring that something is in common ground.

In summary, common ground is always specific to two or more agents at a point in an interaction. It can be thought of as nested: all the mutual beliefs any human could be presumed to have, all the mutual beliefs members of a speech community or nation may be presumed to have, all the beliefs presumably shared by our special interest groups, or friendship network, partners, and so forth. Above all, common ground contains the history of previous interactions of not only what we have established as mutual beliefs, but also the particular referential tokens we agreed to use to refer to them (e.g., the blue car is the electric car) through a *referential pact* (Clark 1996).

Task Environment and the Communication Channel

Tasks are performed in the world, but that same world also provides the channel for communication between humans. Spoken language is the most obvious communication channel between humans but there are others as well (e.g., gestures, documents). It is therefore worthwhile to distinguish the parts of the world used in task performance from those used in referential communication about task performance.

We define *task environment* as the set of objects in the world used or created in performing or attempting to perform tasks. When a learner agent has learned a task, it can manipulate these objects to achieve the task goals. A teacher agent may also manipulate these objects, for example, to demonstrate task steps.

Recall that above we defined a communication channel between agents in terms of sensors and effectors that operate on objects used to refer to the task environment, such as words or fingers that point. Depending on the nature of the task, objects can be in either the task environment or the communication channel, or both. For example, robot fingers may be used to point (communication) and to manipulate physical parts during a repair task (task environment). Similarly, words are typically part of the communication channel, but they can be situated in the task environment if the task domain is learning a second language. If an object is used or created while performing a task, then that object is part of the task environment. If an object is used to refer to an object in the task environment, then we say it is a "referring object" and is part of the communication channel. Because there is no clear *a priori* division between the sensors and effectors used for communication versus those used for accomplishing tasks, (e.g., a person may indicate agreement/disagreement with a word, a nod, eye gaze, a gesture, movement of an object, or even the length of pause before a response), one of the great challenges in social robotics is to read such personalized signals on the fly, as humans do. An important challenge for ITL is to understand how communicative signals are bootstrapped through other actions.

Some instructional strategies are executed primarily in the task environment, such as demonstration or corrective action (e.g., the teacher moves the fork to the left of the plate after the robot has placed it on the right). Other instructional strategies are executed primarily in the communication channel (e.g., the teacher says "put the fork to the left of the plate"). Some instructional strategies use both, whereby the teacher performs actions in the task environment while using the communication channel (see Chai et al., this volume).

In everyday terms, teaching which moves in the task environment includes "showing" what should be done or "correcting" what the learner has done incorrectly. Similarly, teaching that moves in the communication channel includes "telling" what should be done or providing "feedback" on what the learner has done incorrectly. When a teacher is showing and telling, performing corrective actions and giving feedback, that teacher is simultaneously using both the task environment and the communications channel. Whether "showing" or "telling," interaction is critical because of inherent ambiguities in communication through both the task environment and communications channel.

Conclusion

This chapter has laid out a framing of the problem for ITL and proposed a vocabulary that will be used in the remaining chapters when discussing agents, their knowledge and goals, as well as their interactions.

Knowledge

3

Functional Knowledge Requirements for Interactive Task Learning

Robert E. Wray III, Niels A. Taatgen, Christian Lebiere,
Katerina Pastra, Peter Pirolli, Paul S. Rosenbloom,
Matthias Scheutz, Terrence C. Stewart, and Janet Wiles

Abstract

What knowledge needs to be learned to acquire a novel task? What background knowledge does an agent need to use newly acquired knowledge effectively? This chapter considers the functional roles of knowledge in task learning. These roles of knowledge span interaction with other entities and the environment and core functional capabilities of the reasoning system itself (i.e., architecture). Perspectives are offered on the definition of "task" and the relationship between task and knowledge. In addition, three specific challenges central to the role of knowledge in interactive task learning (ITL) are examined: the identification of architectural primitives (basic functional and representational building blocks) needed for ITL, requirements for enabling shared understanding ("common ground") between learner and instructor, and conditions that support projection and anticipation of future states. In conclusion, specific research questions are put forth to address these challenges and advance ITL as a field of inquiry.

Introduction

What knowledge needs to be learned to acquire a novel task? What background knowledge does an agent need to use newly acquired knowledge effectively?

Group photos (top left to bottom right) Robert Wray, Niels Taatgen, Katerina Pastra, Peter Pirolli, Janet Wiles, Paul Rosenbloom, Terry Stewart, Matthias Scheutz, Christian Lebiere, Janet Wiles, Robert Wray, Niels Taatgen, Katerina Pastra, Paul Rosenbloom, Terry Stewart, Janet Wiles, Peter Pirolli, Christian Lebiere, Robert Wray, Matthias Scheutz, Niels Taatgen

Answering these questions requires some consideration of the functional roles of knowledge in task learning. These roles include interaction with other entities, interaction with the environment, and the core functional capabilities of the reasoning system itself (i.e., the agent architecture).

This chapter offers an introductory consideration of these functional issues. We begin by defining tasks, elaborating on the relationship of task and knowledge, and then suggesting a formulation of the interactive task learning (ITL) challenge that emphasizes the role of knowledge. We focus primarily on a human teaching an artificial agent a new task but also consider other configurations of learner and instructor (for a general treatment of the topic of task instruction, see Shah et al., this volume). Thereafter we consider three goals or challenges which we regard as fundamental for understanding the functional role of knowledge in ITL:

1. *Identify the basic building blocks of ITL functionality, both computational and representational.* Can the field come to understand which building blocks are required? Is it reasonable or feasible to assume that such building blocks can be found?

2. *Enable the development of a common, shared understanding during interaction* (i.e., common ground) (Clark and Brennan 1991). Several chapters in this volume (e.g., Mitchell et al., Levinson, Chai et al., and Thomaz et al.) emphasize the importance of achieving common ground during interaction. Here, we explore the computational implications of this requirement, both in terms of "background knowledge" (general knowledge of the world that learner and instructor bring to the interaction) and dynamic shared understanding within the instructional interaction.

3. *Support rapid, pervasive anticipation and prediction of future states.* Both interaction and learning benefit functionally from the ability to anticipate (or to "predict") future states. What are the expected benefits of prediction capabilities for ITL? What are the requirements for prediction within the context of ITL systems?

ITL represents a scientific challenge that can only be addressed by multiple researchers, from diverse areas of expertise, over many years. How might we better identify common goals and work together to achieve them? First, we must acknowledge and capture some of the diverse perspectives on the goals of ITL, before considering community perspectives on the endeavor and the potential for shared tools. We recommend that common, shared learning tasks would aid improved communication and sharing of results, and we outline the notion of challenge problems for the community, including one specific example. To this end, we summarize our analysis and present a number of high-level research challenges for the community of ITL researchers.

The Nature of Tasks for Interactive Task Learning

The breadth and depth of tasks that one could ask a human or agent to perform are considerable. This raises the question of whether there is a general definition of "task" that can encompass this breadth and depth, or whether "task" is a cluster concept where instances only bear a family resemblance (in deference to Wittgenstein). Cluster concepts have clear (noncontroversial) instances, clear non-instances, and a large set of border cases where even experts will not agree on whether they ought to belong to a given class. For the concept of "task," composing and sending out a set of customized emails about an upcoming event is a clear instance. Eye-blinking is a clear non-instance. Here, we attempt to define "task" in two different ways. First, we outline a conceptual definition of "task," with the caveat that the proposed definition is incomplete and may raise objections; still, we hope that it will help the research community understand which kinds of new learner behaviors are targeted for an ITL system. Second, we provide a list of specific and diverse examples of tasks.

What Is a Task?

Colloquially, modern human life is replete with tasks. We run errands to buy groceries and fuel. We cook dinner, set the table, and wash the dishes. We schedule meetings, take notes, write summaries. We dance, play a piano, and sing. We play football, shoot basketball, run for fitness. We play cards and games. We tend lawns and gardens, monitor for plumbing leaks, and repair those leaks when we find them. We move boxes and cartons, plan recreational activities, and balance a banking account. We call or text a friend. We help our young children dress and eat. We surprise our partners with some token of our affection.

Because so much of human activity can be characterized as tasks, we can ask: What is not a task? Is recognizing a face a task? Is daydreaming a task? Is earning an undergraduate degree a task? Is ocean surfing a task?

Table 3.1 lists the primary characteristics that we feel are central to the notion of "task" in ITL. This characterization is not definitive. It is meant to convey the direction of current focus in research, as well as the ambition to move to tasks that are more complex, more meaningful to the public (outside our research laboratories), and more broadly inclusive of what it may mean to learn a new task or to instruct another agent in learning a task.

The properties of tasks listed in Table 3.1 allow us to consider the questions introduced earlier. Is daydreaming a task? The answer would be "no," because there is no (deliberate) goal for daydreaming; that is, daydreaming may have functional benefits in some cases, but the essence of daydreaming is typically a wandering away from more purposeful behavior or thought. Similarly, facial recognition, by itself, is not a task. Facial recognition in humans can be considered a single-step process at the cognitive level that occurs very quickly

Table 3.1   Characteristics of tasks for interactive task learning.

Dimension	Description
Task goal(s)	A state of the world that the agent should achieve, maintain (homeostatic goal), or perform as a result of the task. Task goals can be seemingly simple (pick up a block) or complex (continually monitor a space station for leaks); they can also condition or modify the performance of the task (move with control and grace).
Multiple steps	A task generally requires that the agent perform a series of separable actions under its own control. In some cases, repeated task performance might lead to more succinct representations of behavior (Laird et al. 1986; Mitchell et al. 1986; Taatgen and Lee 2003) and could result in single-step execution, even if the task is learned as a multistep process.
Temporal bounds	Human-scale tasks are generally ones that are performed on a timescale of minutes to hours.
Instructible tasks	A human must be able to articulate/express how the task should be executed, verbally or nonverbally (e.g., instructions might include demonstrations). This dimension does not imply that a human is necessarily able to perform the task.

(Bruce and Young 1986). It would be difficult for a human to articulate how one should recognize faces in comparison to other objects. However, there are very simple tasks that could include facial recognition as a component: finding and choosing smiling faces in an image would be a task according to the characteristics enumerated in Table 3.1. Earning an undergraduate degree is not a task, according to the listed characteristics, because of the lengthy timescale required to achieve this task. However, there are many individual tasks that would be important to the pursuit of a degree, such as taking class notes or performing algebraic manipulations while solving an equation in differential calculus.

Some tasks can be described as a skill that involves, for example, dexterity (e.g., bipedal walking, drumming, or swinging a tennis racquet to hit a ball). Such tasks are largely continuous in nature and hard to analyze in terms of symbolic complexity. Such skills/tasks can play the role of primitives in more complex tasks, as they have an instrumental role in structuring other tasks. However, the distinction between skills and tasks of definable symbolic complexity is fuzzy. In essence, there is a paradox between task and skill: the more familiar or adept a learner becomes with a task, the less prominent the corresponding task structure knowledge becomes in the learner's memory when executing the task: the task then becomes a "skill" for the learner. For example, when a young adult is learning to drive, the many steps associated with successfully controlling a vehicle are explicitly in mind and consciously attended to during driving. A person who has been driving for many years, however, will likely not regard these individual steps at all. Unless the learner needs to teach or describe the task to someone else, the symbolic task structure used during the learning phase is a piece of task knowledge that is not usually activated in routine task performance. Thus, the "complex" skill can now play the role of the primitive in a task of higher complexity. "Driving to the market" can be considered a single step in the execution of the task to buy groceries at the market.

According to the conception of "task" proposed here, surfing is a task (comprised of individual skills): It has a goal (riding a wave). There are multiple steps in the process of surfing (e.g., identifying a "good" wave, moving from prone to standing, guiding the board along the wave, maintaining balance). A single instance of surfing takes place over a few minutes. Surfing instructors can make a living by teaching others to surf.

The characteristics in Table 3.1 not only help us characterize the concept "task," they also call out what may need to be learned when an agent and instructor undertake ITL. A surfing instructor can describe the goals of surfing, help the learner break down wave-riding into a series of largely decomposable steps, and guide the learner in practice. Although the timescale of performance in surfing is only a few minutes, learning to surf may require many days or weeks of practice, some of which is guided by the instructor. The important observation is that the timescale of learning to perform a task at a high level of proficiency (e.g., consider surfing or playing chess) may be several orders of magnitude larger than the time required to perform the task once.

Practitioners within this emergent research domain have diverse and sometimes divergent perspectives on the nature of ITL and the goals and priorities of research within the field. Mitchell et al. (this volume) provide a high-level definition with which others (including some of the authors of this chapter) may not fully agree. For example, some of us assert that a key characteristic of ITL is that the instructor agent has a goal to increase (make more efficient) the speed of learning on the part of the learner agent. This more strongly emphasizes the role of the instructor in ITL than the definition offered by Mitchell et al. This disagreement about edge cases (i.e., whether they should or should not count as instances of some concept) is the hallmark of a "cluster concept" outlined above.

Although disagreement remains about what constitutes ITL at its boundaries, there is also broad agreement about its core. The notion that tasks have goals is a guiding idea to classify activities that may or may not be tasks (e.g., recognizing particular faces). Moreover, in general, we have chosen in this chapter to exclude task-learning scenarios where there is little ongoing interaction between the teacher and the learner (e.g., single instructions or instructional videos): we regard "interaction" as an exchange between two agents, rather than one-way communication (additionally, regardless of who these agents are, simply interacting with the environment is not sufficient). These definitions exclude learning contexts where the learner acquires competence in performing a task from unsupervised interactions with a task environment, which is a feature of some recent advances in machine learning, such as the ability of Deep Mind to learn to play Atari video games (Mnih et al. 2015).

Finally, we acknowledge that a discussion of the functional knowledge requirements for ITL results necessarily in imprecision, as is inherent to all central concepts involved in ITL. "Task," "learning," and "interaction" as well as concepts involved in spelling out those concepts (e.g., "knowledge") are themselves cluster concepts.

Examples of Domains and Tasks

To explore the notion of "task" for ITL further, we outline specific examples of ITL which the field could and should attempt to address. We deliberately include examples of tasks that are "atypical" to those currently at the center of most ITL research. Our intent is to suggest that the span of tasks which ITL should be able to address is broader than the array of tasks the field is currently addressing. In some cases, suggested tasks will introduce new requirements for ITL capabilities. For each learning task, we also introduce a domain of usage in which the specific task to be learned could be introduced, with the goal of highlighting why the learner's performance of that task could not simply be "programmed" into the original performance system.

Function Composition in User Interfaces

Personal assistants that can automate the tedium of administrative tasks—scheduling meetings (Allen et al. 2007), making travel arrangements (Knoblock 2004), completing "paperwork"—have been a long-time goal of researchers in both artificial intelligence (AI) and human–computer interfaces (Lehman and Carbonell 1989). Such personal assistants may have powerful built-in functionality, but to be maximally useful in everyday contexts, they must be able to learn to perform specific tasks required by the individual user. Today's existing personal assistants (e.g., Apple Siri, Google Now, Amazon Alexa) are limited to preprogrammed tasks (e.g., enter calendar entry). The devices in which these assistants are embedded, however, have the needed infrastructure already in place (e.g., voice commands, application programming interfaces to apps) on most target platforms to extend their capabilities toward task learning and instruction.

ITL could be the means by which a user instructs a personal electronic device (e.g., computer, tablet, smart phone) to perform automatically increasingly complex sequences of actions to achieve high-level user goals (e.g., fill out an expense report from stack of receipts). Such functionality is already being explored in everyday tasks, such as ordering coffee (Azaria et al. 2016). Such task learning offers tremendous potential benefits, from increased work productivity to enabling access to support services for those with cognitive impairments.

The interaction for function composition could be as follows: Users give a verbal description of a target task as they perform it, associating the various components and arguments with the actions performed and values entered. Alternatively, the system starts by being able to perform basic actions (e.g., typical application menu commands, keyboard entry) from voice commands and is instructed step-by-step how to perform more complex ones.

Call Centers and Field Linguistics: Learning to Interact Socially in New Cultural Contexts

Increasing the efficiency of call centers can provide significant cost savings (Gray et al. 1993). Customer satisfaction often declines when automation replaces human operators in call centers, leading to caller frustration and anger. Rather than canned, automated responses, it would be preferable to have call center agents that can adapt to the customers' needs, to the affect that they have on the customer, and to the social context of the caller.

The rudiments for this type of capability might be able to be programmed in advance. However, since cues and dynamics of interaction differ according to social and cultural context, it would be advantageous to have an ITL system that could conduct basic diagnostics and remediation, adapting when necessary to the specific cultural (and even subcultural) cues so as to customize the response to the caller. This approach would benefit the call center by enabling greater scalability (one core system customizable to many different sociocultural contexts) and mitigate the problem of cultural and gender bias that has been observed in some AI systems (Caliskan et al. 2017).

The social interaction learning task would involve an instructor and a learner (an agent), tasked with learning to respond appropriately to specific cultural cues of future callers. The learner would need to learn such things as turn-taking and interruption strategies, theory of mind of the entity with which it is interacting, and how to respond appropriately to affective states or displays. One of the advantages of ITL over preprogramming is that the agent would receive ongoing feedback and thus improve its ability to recognize and to adapt to specific caller cues (a kind of "on the job" training). It is not intended for the learner to learn how to understand or generate the content of the communications, other than that it may need to learn to extract cues from the content for aid in making social decisions.

Another use case for such social learning could be robots and virtual agents being developed to support field linguists. The task of a field linguist begins with segmentation and discovery of the sound system of a new language. What are the distinctions that make a difference to meaning? These can include perceptual distinctions that are not initially in the linguist's repertoire. The linguist needs to discover how sounds are combined in flexible ways into wordlike units, centered on word stems. The structure is inherently combinatorial, with the task being the discovery of alternates for roles and fillers at multiple levels, including identification of word stems and the possible suffixes, affixes, and infixes. The knowledge representation in a language comprises multiple levels, including phonemes (sounds units), morphemes (units of meaning), words, phrases, sentences, and syntax as well as embodied actions such as pointing and gaze.

Evaluation of competence is revealed through transcription of corpora. Much of the task of the field linguist involves the transcription of recorded language material. For ITL, sets of corpora could form a challenge domain with a range of tasks for human and agent learners, which include construction of learning examples and other tasks that reveal the sources of variability in how information is communicated, multilevel structure, and discovery of slot-and-filler syntax in multiscale structures.

More generally, although there are thousands of human natural languages in use today, only several hundred are supported through current information technology tools. An agent designed to act as a surrogate or aid for language study by a field linguist would also benefit from being able to be quickly customized to specific cultural and social patterns of interaction. While it may be possible to encode an expert system that could support the dynamics and nuance of interaction for a particular language, this approach is infeasible due to the funding and technical expertise that would be required. A mature ITL solution that allowed an expert in the language to teach its aid rapidly and without technical know-how would scale to the diverse needs of the thousands of human languages.

Robotic Assistants: Fast Customization for Emerging Tasks

Robotic systems offer significant capability for performing tasks that are repetitive, dangerous, or beyond human capability. While robotic systems may have much inherent capability "out of the box," their function will likely need to be modified or extended, either to amortize the cost of the robot under changing task performance requirements or because the specific behavior in a performance environment can only be finalized or specialized *in situ*. Consider the following three examples of ITL in the robotic domain:

1. *Industrial manufacturing*: For any assembly task, we would like to instruct the robot how to manipulate parts to assemble the prescribed structure (e.g., a piece of furniture, a car, an airplane wing): how to pick up parts; how to insert them into, connect, or mount them onto other parts; how to place and reorient parts, and so on. An ITL robot could be instructed or receive demonstrations to support changing assembly requirements.

2. *Disaster site monitoring*, including routine tasks such as counting containers (e.g., containers of nuclear waste), recording their ID tags, and performing gamma measurements on some of them. The International

Atomic Energy Agency is currently seeking to identify small robot-ized rolling platforms capable of assisting the human inspector by per-forming the following tasks: moving autonomously across a storage area, counting items of a specific geometry, recording their ID tags, and carrying specific instrument payloads. The agency has issued an open challenge for a robot that is partially autonomous (i.e., naviga-tion), with the remaining functions to be teleoperated. This basic ro-botic platform could be enhanced with ITL to enable the robot to learn specific measures needed at a particular site (e.g., where to go to take measurements, how to navigate the inside of structures) and to provide a more autonomous monitoring capability.

3. *NASA's space robotics challenge* identified three tasks aimed at sim-ulating what a robot may be required to do while assisting a NASA mission to Mars, whether in a preparatory capacity before astronauts arrive, or alongside astronauts: (a) aligning a communications array; (b) repairing a broken solar array; and (c) identifying and repairing a habitat leak. All three tasks could be instructed using natural language, including the details of what an object looks like. Currently, this chal-lenge does not require robots to be instructible.

Studying Human Task Learning to Advance ITL

Here we consider four examples of human task learning that might be apt subjects for advancing ITL: playing card games, tying a knot, dancing the Argentine tango, and playing BrainQuest. These examples of human task learning appear relatively straightforward. They can be taught by humans to other humans without requiring the instructor to have a high level of task pro-ficiency or training in instruction. Currently, these tasks present challenges (to lesser and greater degrees) to existing artificial (nonhuman) agents. Thus, they are useful in identifying directions that future research may wish to take.

Playing card games. An agent needs to play a new card game, which can include a software simulation with high-level sensors and actuators (e.g., mov-ing cards, flipping them, reading them). Some aspects of the new card game may rely on preexisting concepts (e.g., tricks, piles, cards-in-your-hand, dis-cards) and actions (e.g., play a card, draw a card) whereas others may be new. As the agent is introduced to a new game, it may need to construct both novel task operators and new terms that describe or summarize task states (e.g., a "good hand"). Once the agent understands the basic rules, it might also be given other instructions, such as suggested strategies or indications as to what "good" moves are. Thus, learning a new card game may involve the use of existing knowledge to support the acquisition of new task knowledge. These are core requirements in ITL (discussed further below).

Tying a knot. This task can be used to test a number of aspects of ITL due to the range of task complexity. One could begin with an easy knot (an overhand knot) and proceed to increasingly complex knots, such as bowline or highly complex decorative knots. In addition, knot tying involves a hierarchy of task primitives: one type of knot learned previously is part of the tying of a more complex one and actually forms a necessary step of it (i.e., there is dependency). Currently, robots exist that have learned to tie different types of knots based on direct demonstration or manipulation (Schulman et al. 2016). These robots offer direct comparisons of the benefits of interactive instruction and generality of the resulting agent capabilities.

Knot tying presents challenges for verbalization (verbal instruction) and thus provides an opportunity to test different instruction modalities and their effects on learning (e.g., use of analogies in language, use of visual demonstration alone, or combinations of verbalization and demonstration). A basic requirement for instruction is the establishment of common ground (e.g., the "tail of the rope" vs. the "standing edge," the notion of a "loop"); term creation is needed. Knot tying is a highly interactive task: the learner needs immediate feedback while performing the task, and it is crucial to know at which part of the process things went wrong. Knot tying would also stress the temporal constraints and social aspects of the interaction: the learner may easily become frustrated and the teacher may find it challenging to communicate the same information in different, more effective ways.

Knot tying requires relatively little background knowledge. It is an activity that involves learners of any age, including children; thus one can test this task developmentally using age range or level of expertise (expert vs. novices). It is an integral part of many different activities in everyday life, such as tying one's shoelaces, getting dressed in a suit, and connecting two items with rope (e.g., in sailing, climbing, fishing, scouting, rescue activities, knitting/creative decorations, medical practice/surgery). Today's robots may lack the necessary dexterity to support near-term, direct reproduction of human instruction. However, simulations of knot tying could be developed to explore what would be required for a robot to master this task.

Argentine tango. Human adults can learn new tasks in a variety of ways, from taking a class at a community college, to online videos or engaging a teacher or coach directly. Learning to perform a new physical skill is an example of one of these tasks. As a particularly challenging example, consider the task of a physically embodied agent learning to dance the tango with a human (or other agent) as its partner. This task requires being taught the basic moves and sequences, and then performing those moves with a partner. When dancing with a partner, there will likely be a lot of nonverbal instruction (or perhaps reinforcement learning) to get the more subtle aspects of the movements that are dependent on the partner. However, the robot could also be verbally instructed about the sorts of nonverbal cues that the leading partner will present to indicate the desired actions of the following partner. One interesting requirement for performing the tango is the aesthetics of the dance: a combination of grace, fluidity, and power characterize a good tango. Thus, learning this task requires performing the dance with these requirements in addition to the more basic expectations of rhythmic steps and stances.

BrainQuest. Children learn new tasks from teachers, parents, and other children. For ITL research, various developmental stages of children offer the opportunity to focus on tasks that require less sophisticated and rich sources of knowledge, thus providing greater opportunities for a cognitive modeling environment to allow the architecture to "show through." As an example, consider BrainQuest, a game that consists of cards with a question on it, typically accompanied by a picture. There are cards for 3- to 4-year-old children, for 5- to 6-year olds, as well as older children. Each card introduces a little task in itself. Take, for example, a picture of a mother duck with five ducklings and a chicken with six chicks. The question posed on the card is: "Which mother bird has six babies?" Another example is a picture of a lion, a dog, and a cat with the question: "I bark and I like to go for walks. Which one am I?" There is overlap in knowledge with other tasks, but the task is almost never exactly the same.

Examples of Desired Features for ITL Research Tasks

In the course of considering a range of tasks for ITL, we identified a number of task features that an agent should learn. These features are not all in agreement with one another. Thus, in our listing of desiderata for ITL research tasks, we summarize some of the rationales that one might make for including a particular task feature:

- *Interactive tasks*: Tasks that require interaction in performance (as well as in learning) are of special interest to some researchers. Interactive task domains (e.g., conversational personal agents such as the help center, robots that support child learning through games and play, tango dancing) represent domains that are currently very difficult to engineer effectively (for a discussion of language interaction requirements, see Levinson, this volume). Using ITL to teach effective interactions (as in the help center and tango dancing domains) would benefit the larger AI community because of the difficulty of providing effective interactions in today's systems.

- *Generalization and transfer*: Researchers in AI and cognitive science, especially cognitive architecture (Laird et al. 2017), are interested in exploring the generality of an agent system and the ability to perform successfully across a wide range of tasks (Anderson et al. 2004; Newell 1990). Card playing (as highlighted above) provides a good test of generality within a limited sphere, as would the task domains proposed for general game playing (Genesereth and Thielscher 2014).

- *"Complex enough" domains*: Toy domains are attractive because the field of ITL is relatively immature and there are many potential barriers to entry. However, toy domains may not offer sufficient complexity, and the lack of complexity may not always be apparent when the domain is chosen. Thus, there is a tension between wanting to work on domains that matter (below) and domains that are not overwhelming. One potential challenge for the field would be to provide systematic characterization of the complexity of the task learning and performance domains, to enable more informed and deliberate selection of research domains.

- *Domains that "matter"*: Developing an ITL research system will require a significant investment and be conducted over many years. Because of this, it may be beneficial to focus on tasks which "matter"; that is, domains (and tasks within them) that offer direct benefit to the larger world, if the research led to field-testable prototypes and field-capable systems. A good example is the assistant for a field linguist, as summarized above, which could play an important role in helping to preserve endangered languages. Another example is ITL agents and robots for disaster response assistance, which is a domain that requires extraordinary flexibility and *in situ* adaptivity. It is important to note that what "matters" is subjective and that researchers may differ in their assessments. Domains that matter need not necessarily be highly complex domains that require huge investments. Stakeholders within each respective domain's community may help clarify important requirements for ITL research.

Computational, Representational, and Task Primitives for ITL

In AI, an agent's behavior and capability is typically assumed to be a function of its architecture operating on its store of knowledge while interacting in an environment (Russell and Norvig 1995). In this context, "architecture" refers to a fixed set of computational operations and representational elements used to build an intelligent agent (Newell 1990).

Cognitive science and AI have produced many different types of agent architectures, with various representational and processing assumptions. Thus, the same type of functionality (e.g., ITL) can be realized in different ways in different architectures. For instance, while most cognitive architectures may require explicit task representations in declarative or procedural memories that can be assembled through learning processes, robotic architectures might represent the task only as an action policy that implicitly represents its tasks in the form of state-action pairs (i.e., mappings from state descriptions to actions that will allow the agent to reach new states, resulting over time in the agent's performance of the task). Rich (this volume) and Scheutz et al. (2013) discuss these distinctions further.

Agent architectures are blueprints for computers: machines capable of universal computation. Computation must be grounded in primitives that are executable. For standard computer architectures, computational primitives form an instruction set that enables execution of complex programs. The composition of representational primitives (bits and bytes) and computational primitives (move, store, add) provide sufficient mechanisms for the creation of complex and sophisticated computer applications. Some combinations of these primitives are sufficiently useful such that they can be organized into a stable and reusable higher level of abstraction above the most primitive level. An operating system defines a set of primitives (files, typed numbers, strings, system calls such as "open," "load," and "execute") which can then be used to compose software programs that depend only on the operating system level of abstraction (e.g., a word processor, a mobile phone application). For a cognitive or robotic architecture, primitives include at least the mechanisms that enable the representation and processing of knowledge and skills. For an artificial neural architecture, it includes the operations that compute outputs from inputs given information in the form of inputs and weights.

What elements comprise an effective architecture (abstract computational level) for pursuing ITL? Does it even make sense to pursue such a goal? Given the variety of architectures, it is important to take a "least commitment perspective" when discussing architectural requirements for ITL. As yet, there is no agreed upon set of common functional components across architectures for different types of agents (virtual and robot) or task domains (e.g., solving mathematical problems vs. performing assembly tasks) although there is long-standing and ongoing interest in such identification and unification (Laird et al. 2017; Newell 1990; Sloman and Scheutz 2002). In this section, we consider the architectural implications of ITL while not committing to specific architectural assumptions; instead, we focus on the notion of *knowledge and process abstraction* as well as the role of *primitives*. We consider four distinct issues related to computational abstractions for ITL:

- Do cognitive architectures offer a useful level of architectural abstraction to pursue ITL?
- Is a fixed level of abstraction for ITL a reasonable goal?
- How might the mixed modalities of interaction within ITL shape requirements for an architectural abstraction?
- How might diverse conceptions of "task" inform the goals of architectural abstraction?

Cognitive Architectures as Potential ITL Architectures

Cognitive architectures provide computational abstractions that are, to varying degrees, designed to realize a human-mind-like virtual machine. For ITL, we need to know whether these existing computational architectures are defined at a useful level of abstraction to support ITL. How do the primitives defined for some particular architecture map to the functional requirements for ITL?

Are existing primitives too low level, too cumbersome, and too tedious to support efficient pursuit of ITL? If so, one direction of research could be to define an ITL capability (a higher-level architecture or virtual machine) that "implements" ITL for some architecture. For example, in cognitive architectures such as ACT-R and Soar, the representation of a task would be distributed across multiple memories (semantic, procedural, and episodic). In an ITL virtual machine, a "task" might be a primitive representation that was then decomposed into the specific and distributed representations of the underlying architecture.

Conversely, it could be that primitives (or at least some primitives) of today's architectures are defined too coarsely to support ITL, thus making it difficult to realize ITL within a given architecture. In standard computer architectures, it may sometimes be necessary to decompose the primitives of an architecture into finer-grained elements (e.g., the primitives of an architecture at a lower level of abstraction). In most cases, an applications developer will work at the operating system level of abstraction, but may need to shift to the assembly-language level in some cases. The instruction set level of architecture may need to be described at the level of electrical current flows. Any computational architecture is grounded in lower-level computational or physical processes. Thus, we need to ask whether ITL requires a reconsideration of many of the "standard" or assumed primitives of computational cognitive architectures.

In Search of Reusable Levels of Abstraction for Task Learning

Within any computational architecture, computation ultimately builds from the execution of its computational primitives, upon which more complex programs can be built. For example, for a cognitive robotic architecture, action primitives would likely specify basic movement behaviors that enable the emergence of more complex ones; such primitives may be explicitly represented (the case in most symbolic representation architectures) or may implicitly emerge from computation (as in neural architectures). Different architectures commit to different computational and representational primitives. As described by Taatgen (this volume), the primitives of a high-level cognitive or robotics architecture can be created from the composition of primitive elements of another architecture defined at a lower level, analogous to the various architectures one finds defined for standard computation, as outlined in the previous section.

This perspective raises questions as to what sufficient and useful primitives are for ITL, and whether these primitives can be mapped to the primitives of existing computational architectures (as outlined above). Should concepts such as "task," "task step or action," or "task state" map directly onto computational primitives for an ITL architecture? The identification and exploitation of the primitives of task representation and task learning is a basic research goal for computational and behavioral disciplines alike. These task primitives are essential in task learning. For example, if task primitives are defined at an appropriate level of abstraction sufficient for representation and computation, task representations will more readily scale to tasks of arbitrary complexity and can be applied creatively when encountering unknown, uncertain, and noisy or ambiguous conditions.

Decomposition of processes, knowledge, and tasks into core nondecomposable units cannot be arbitrary if scalability and effectiveness in learning is of interest. However, the fundamental principles that may govern such decomposition are largely unexplored (Barto et al. 2013).

We assume that task primitives are not necessarily mapped to architectural primitives: the knowledge required to represent and to execute a task may need to be represented at different levels within an ITL system and the level at which task knowledge is represented impacts the composition of the knowledge and potential for reuse (for further discussion, see Taatgen, this volume). ITL increases the challenge because the representation of the task itself changes with task learning. For example, we may teach the learner a sequence of actions ("raise your arm; move your hand a short distance back and forth for a few seconds") and then give that sequence of actions a particular name ("wave") so that it can be referred to in the future by the more abstract label. Importantly, however, there may be times when the learner needs to consider how some task action is composed and further decompose it. For instance, when teaching the learner to reach for an object located above the learner, we may want to refer to a movement previously labeled "raising your arm," but be more specific about which joints to use and their target angles.

Flexible use of various levels of abstraction extends to all parts of the system. The above example focuses on actions, but the same can be said of objects as well. Sometimes we may want to refer to a crowd of people, an individual person, a face, an eye, a circle, or combinations of these. Sometimes it is sufficient to refer to something as "the number 2" whereas at other times we need to be more specific ("a 2 with a loopy bottom") or less specific ("a number"). Furthermore, the representation of time must be flexible as well, as we discuss further below. Actions may take place over milliseconds, seconds, minutes, hours, or longer periods of time. Actions, objects, and durations can also be combined at various levels of abstraction.

Flexible and adaptive abstraction affects communication between the learner and the tutor (these different levels of abstraction must be able to be mutually identified and understood), as well as the internal cognitive architectures of these systems. For example, the motor control system may be given high-level commands ("wave your right hand") or low-level commands ("move this joint to this angle at this speed"). Perceptual categorization may provide objects at many hierarchical levels (e.g., "object," "person," "Jill," "head," "face," "eye"). There is some lowest level of nondecomposable task atoms, but because we do not know what these lowest level primitives are, we cannot be assured that any particular architecture's computational primitives are fixed at a sufficiently low-enough level to support the flexibility required by ITL.

Experimental research in verbal categorization has shown that some verbally expressed categories of entities (and potentially actions, features, and processes of any kind) do exemplify properties that render them ideally informative for sensorimotor similarity-based generalization; for example, basic level concepts in prototype theory (Rosch 1973) conform to this property. Thus, language may point to useful levels of generalization from which the identification of task primitives may be founded to support both decomposition and synthesis. This basic level of generalization (which comprises categories that are neither too general nor too specific) could provide a common abstraction level for tasks of any type and complexity, serving as a representational ground across ITL systems. The fact that tutor–learner communication is primarily (though not exclusively) verbal makes the role of natural language in providing a basic, common abstraction representation level even more significant, and a promising direction for research.

The Role of Modality-Specific Abstraction

Abstraction of task knowledge and processes may be expressed through one or more modalities, each modality being more appropriate for different types and levels of abstraction according to its strengths and limitations. Consider the different modalities that may be used in teaching a robot to grasp and manipulate an object. Verbal representation of task knowledge is symbolic and high level: "Please place the fork next to the plate." On the other hand, sensorimotor representation of task knowledge may be more usefully represented at a subsymbolic level, such as a demonstration of the movement involved in placing the fork as described. Language is particularly suitable for expressing the goal of a task (i.e., the local or end goal; the final location of the fork). A sensorimotor representation would be limited to a description of the achieved state. These visuomotor modalities, however, are clearly more suitable for capturing the actual movements and physical relationships needed to achieve the goal of moving the fork to the desired location. Thus, linguistic and visuomotor primitives for task knowledge and execution need to be coordinated and aligned with one another during all phases of ITL, and these coordination requirements may impose additional constraints on the specific architectural abstraction needed for ITL.

Task Representation

Task descriptions in past and current research tend to be rigid and overly specific. Accomplishing a given task is usually precisely specified, with clear criteria for successful completion. Often there are equally straightforward procedures for accomplishing the task. While this may seem appropriate, and indeed it is useful in the short term, it tends to prevent cumulative progress in the medium and long term. The very task specificity that allows an ITL problem to be solved precisely prevents it from being directly reused in other, related situations. Those situations tend to have a slightly different representation or contextualization, leading the original task solution to be unusable unless it is explicitly modified to accommodate the new specification.

While this specificity is similar to that of traditional programming paradigms, where the approach has been quite successful, it is quite unlike what Herbert Simon referred to as ill-defined problems characterizing natural intelligence, where their very nature requires the flexibility and adaptivity of human cognition (Simon 1996). Newell advanced cognitive architectures as the solution to the "20 questions" problem (Newell 1973b), which resulted from models of distinct cognitive tasks being developed in incompatible frameworks, thus preventing increasingly integrated models from being developed. However, while cognitive architectures have enabled the generation of compatible models of various tasks, they have not generally resulted in composition for increasingly complex cognitive capacities. In Soar, for example, natural language understanding and dynamic plan execution were demonstrated in an integrated agent (Lehman et al. 1995). That demonstration, however, did not result in these capabilities being routinely composable in future Soar models. Similar examples abound across cognitive architectures. This observation is not a criticism of these architectures but rather a caution for ITL: demonstration of significant integrated systems capability is not sufficient to define reusable and useful higher-level abstractions for ITL.

Co-Construction and Common Ground: Shared Contexts for ITL

In this section we explore what is required to achieve common ground (Clark and Brennan 1991) during learning interactions and what architectural primitives are required to support it. Our discussion here is not conclusive. Although we have some general agreement that architectures require mechanisms to support the achievement of common ground, conclusive demonstration of the necessity and sufficiency of those primitives requires development and testing by the emerging ITL community.

Common Ground in ITL

Agents, as well as their human instructors, must support functionality for establishing common ground in ITL. Borrowing from the notion of common ground in communication theory (Clark and Brennan 1991), this means that the shared context and mutual knowledge available to human and AI agents in a collaborative instructional situation have to be updated, maintained, and often repaired. This assumes that we reinterpret ITL as a form of joint action carried out in a coordinated manner by both the human and the AI agent.

This means coordination of the content (e.g., about the to-be-learned task) as well as the process by which the ITL activity progresses. In the ITL situation, common ground makes it possible for a learner and an instructor to coordinate on what the instructor means and what the learner understands the instructor to mean. As a joint activity, ITL implies bilateral processes: instructors must monitor the actions of the learner in context, and the AI learner must somehow signal their current state of learning.

Levinson (this volume) suggests that the functional requirements for common ground in human communication include:

- *Error checking at every level.* Participants continually monitor their understanding. Monitoring must be fast and accomplished at many levels (e.g., simultaneous monitoring of syntax and semantics). The fast execution of error checking and repair within the tempo of the conversation reflects predictive processes that anticipate errors or potential misunderstandings prior to their occurrence.
- *Continual turn-taking.* Participants engage in dynamic and complex turn-taking interactions that require little explicitly acknowledged cues of negotiation about the turn-taking. Such turn-taking includes shared contextual understanding of when interruption may or may not occur.
- *Mirroring of terminology.* As participants engage in dialogue and "co-construct" a shared understanding, they tend to begin using the same terms. There is reduced use of synonyms under conditions of language/capability mismatch.

Levinson argues that it may be impossible, with the current state of scientific and technical knowledge, to produce ITL systems that can reproduce the richness, subtlety, and timeliness of human–human communication. Instead, we may need to adopt engineering shortcuts that take advantage of the human tendency to anthropomorphize as a means of mitigating this issue.

In the field of human–robot interaction, the theory of common ground has been adopted widely and built upon, for example, in the approach of coactive design (Johnson et al. 2014). The idea is that coordinated tasks involving humans and robots require managing the interdependencies and constraints among the agents' activities. The common ground in such collaborative human–robot activity includes the relevant capacities (i.e., the total set of capabilities, knowledge, and resources) of the interdependent agents needed to perform the joint activity. Capacities are defined completely by the interaction of the agent and its environment—an idea similar to Newell's definition of knowledge-level descriptions (Newell 1982). To support common ground functionally in human–robot activity, coactive design (Johnson et al. 2014) promotes the development of interagent interfaces that support observability, predictability, and directability. These principles can be reinterpreted slightly to the ITL situation: *Observability* means the learner makes pertinent aspects of its own state and knowledge of the instructional situation, target task, and environment observable to the instructor. *Predictability* means that the learner's actions are predictable so that the instructor can act appropriately. *Directability* is the ability of the instructor to direct the behavior of the learner, which in the specific situation of ITL means being able to direct the AI agent to engage in learning actions and processes.

"Global" and "Local" Common Ground

Despite different languages and cultural backgrounds, humans share a high level of commonality, at least in comparison to a human and a machine. One can think of common ground as the set of referents and knowledge shared between the learner and the teacher. These are internal representations that are "grounded" in "common" with one another. Clearly, these may not be perfectly identical; however, successful communication relies on them being sufficiently "aligned." Chai et al. (this volume) discuss the importance of background knowledge and comment that the lack of common ground is both a significant limitation of existing ITL robotics systems and a relatively underexplored research area.

We recommend the introduction of minor terminological distinctions to highlight these differences for future communication within the community. Specifically, we suggest a distinction between "local" and "global" common ground. Global common ground is shared background knowledge (or "common knowledge") which agents may have before the interaction starts. Some of this background knowledge may not be shared, and part of the interaction between teacher and learner may involve detecting and fixing misalignments in background knowledge. However, this sort of knowledge seems to be fairly distinct from the common ground that is local to the instruction of the task itself, for example, identifying which object one agent is referring to with "this cup" and knowledge about the actions and events taking place within the interaction.

Functional Requirements for Reaching Local Common Ground

Computational cognitive architectures have largely focused on identifying the components of thought and mind for individual cognition (Anderson et al. 2004; Kieras and Meyer 1997; Newell 1990). A "standard model" of the components comprising a cognitive architecture has been recently proposed (Laird et al. 2017). It attempts to identify common functional elements of these architectures, such as different kinds of memories (episodic, procedural, semantic) and the computational operations that occur within and between these memories to achieve cognition.

An open question is whether the computational decompositions reflected in these individual architectures (and in the emerging standard model) are sufficient for ITL and for the ability of ITL systems to achieve common ground with human instructors. In other words, what are the architectural implications of ITL? If there are specific architectural capabilities that are required for ITL, can the community identify clear requirements, invariants, or "laws" that should be encapsulated by the architecture? How can these cognitive architectures support the functionality described by Levinson (this volume) or coactive design (Johnson et al. 2014)?

We have not yet reached firm conclusions regarding the sufficiency of existing architectures. However, we have identified the following architectural requirements for ITL, focusing especially on co-construction or achieving of shared understanding during ITL:

- The architecture should directly support the *acquisition of new task knowledge*. This requirement implies the acquisition of new procedural representations (e.g., routines, functions). It also includes many other representations as well, such as the ability to learn new procedures from the composition of previously learned procedures, to articulate (at some level) an internal model of one's procedural understanding, and to assess one's confidence of understanding and ability with task procedures. Rich (this volume) refers to these latter elements as *metaknowledge* of the task procedure. This "knowledge of what I can do" appears critical to co-construct shared understanding effectively as the learner grows in ability and knowledge of the task.

- Architectures should support *concept refinement*. Task concepts are not static in human learning and they should not be assumed to be in ITL, if artificial task-learning systems need to achieve shared understanding with human instructors. Introductory assumptions may need to be modified or extended (e.g., imagine introducing the "castle move" in chess sometime after the normal movements of the king and rook have been explained). Task actions or concepts, which may have been treated as unitary in the introduction of the task, may need to be further decomposed, resulting in the breakdown of apparent task primitives into more fine-grained primitives. Architectures must then support fast and flexible reformulation of these task concepts to support ongoing interaction and learning.

- Architectures should support *recognition and rapid response to realignment*. As discussed above, error checking is a pervasive component of the dynamic, shared understanding that occurs in human interaction. The pervasiveness of such capability strongly suggests that this would be an innate (or architectural) capability in humans. However, many existing architectures do not explicitly embed language processing at the architectural level. This raises an open question of what architectural functions in such architectures could give rise to such fast and pervasive low-level language processing capability.

Finding Common Ground with Nonhuman Intelligences

It is easy to fall into the trap of believing that the challenge of ITL is to provide AI agents with humanlike cognitive architectures so that such agents are instructible in a natural humanlike way. If we can just make robots perceive, act, think, learn, and talk like humans then our problems are solved. Although some agents may end up doing tasks in the same physical world as we humans, many will not. Current commercial conversational agents live primarily in a world of application software, with limited communication channels to human users and limited awareness of the physical environment. As roboticists are quick to tell us, humanoid robots do not "see" the world or execute their motor actions in the same way as humans. Agents in virtual environments typically rely on representations of space (and negative space) based on implementation representations rather than the spatial representation a human viewer might perceive.

To take an even more extreme example, imagine a long-range, long-lived, continuously flying, solar-powered drone that seeks to discover huge collections of plastics in the ocean. It may perceive the world through some combination of LIDAR (light detection and ranging), hyperspectral imaging, GPS, etc. and move about using rotors, wing foils, etc., utilizing flight controls and perceptual processing outside of human expertise and experience. It may be continuously learning using deep learning, reinforcement learning, or some hybrid combination of machine learning techniques that are difficult for humans to interpret. So the AI drone's underlying architecture to address its task environment may consist of fundamentally different perceptions, actions, and representations than humans. However, if we include ITL with humans as part of the drone's task environment, then that creates the need to interface the drone's internal representations with those of humans.

It may also be possible to jettison entirely the idea of having a dedicated task learning architecture that maps and executes onto the underlying cognitive architecture of the agent. The agent could support some general functionality for establishing common ground in the ITL environment. This would require the agent to be able to interact with instructors in ways that align common elements of task capabilities and knowledge to communicate about the task environment, those capabilities, and relevant knowledge.

Mental Models and Simulation for ITL

An ITL system will need to know how its environment will change as a result of its actions as well as through the actions of others. Such understanding is necessary to reason about the potential effects or outcomes of various choices, or to anticipate future states based on the actions of others. This sort of prediction has a long history in AI and cognitive science, but ITL systems may place unique demands on such a system. In particular, following novel instructions will, by necessity, require the agent to predict the future *under conditions it has never previously experienced.* A successful ITL agent will require an internal model of the world that is reliable enough to extrapolate to new conditions and still provide good predictions.

We use the term *mental simulation* to refer to the process of predicting some future state(s). Other common terms include prediction, projection, look-ahead, envisionment, simulation, and model-based reasoning. In all these cases, prediction involves imagining future states and actions. In general, systems capable of such mental simulation will have some form of *mental model*, which we take to be the internal representation that supports this process.

The Necessity of Mental Simulation

To clarify why we believe that mental simulation is a required core component for ITL, we first note that humans do this automatically in ITL situations. For example, imagine hearing verbal instructions for tying a knot (as discussed earlier). It is likely that you may imagine or attempt to visualize the manual operations described during the delivery of those instructions. The question of why this occurs then arises. As discussed below, the knowledge to predict the outcome of an action allows an agent to use instructions with incomplete sequences of actions, repair mistakes, and perform goal-directed planning.

The requirement to consider things that are not currently true in the world extends farther than just visualization based on a set of instructions. When humans perform such tasks, they also make use of hypotheticals, counterfactuals, and so on. That is, they can spontaneously consider various possible futures, not just the one that is likely to be arrived at based on following a novel set of instructions. This may be thought of as a kind of analogy over experience, which enables both predication and abduction, as well as generating more than one alternative when multiple prior experiences are retrieved (Forbus and Gentner 1997). Requirements for ITL may also implicate a continuous rather than a discrete future/timeline.

Importance of Context

Mental simulation is more useful if the learner can relate instructions to the world and reason about the consequences of carrying out the instructions. Instructions, however, are often given as a list of actions to be accomplished, whether in recipes, checklists in aviation, instructions for electronic devices, or instructions on how to connect to the office printer. The disadvantage of these types of instructions is that the purpose of each step is not always clear, making it much harder to simulate outcomes mentally. Humans are sometimes able to infer this knowledge, but without this knowledge, it is very hard to modify a procedure or to repurpose the knowledge for other tasks.

We find clear support for this statement in research on the flight management system (FMS) used in commercial aviation (Taatgen et al. 2008). An FMS is an onboard computer that can almost autonomously fly an airplane. To enter routes, change routes, and program the FMS for other tasks, pilots must learn list-based procedures. If pilots do not know what the purpose of the individual steps are in these procedures, they have difficulty memorizing them and generalizing the knowledge to novel situations, even though this is often required in everyday aviation. Taatgen et al. (2008) developed an alternative instruction: subjects were taught the context within which a particular step had to be understood, and the consequences of that step (i.e., a pre- and a postcondition). Compared to the traditional list instruction, subjects in the context condition learned much faster and were able to solve novel problems that required them to step outside of the bounds of the original procedures. These instructions can be supplemented with context, allowing pilots to perform mental simulation based on a richer mental model. Context allows pilots to reason about the outcomes of their actions, enabling them to compensate for missing knowledge and derive new sequences of actions for novel problems.

Imagination and Counterfactuals

Given that human task-based instructions are often underspecified and incomplete (e.g., details are left out, steps are skipped, some task knowledge is assumed), the learner is often required to fill in gaps which can be achieved using a variety of mechanisms, including imagining concrete task-based settings and applying instructions in the imagined environment, or constructing counterfactual scenarios to determine the extent to which an instruction is applicable. Counterfactuals obtained by making changes to the current state, which turn representations of the actual context into a hypothetical context, can serve multiple purposes in the learner's making sense of an instruction:

- They help the learner understand why the instructor gave the instruction in a particular way.
- They permit the learner to correct wrong assumptions about the nature of the task.
- They focus the learner's attention on relevant task aspects.
- They can suggest questions the learner might ask the instructor to clarify instructions.

In addition, counterfactuals may allow the learner to generalize and transfer knowledge to other cases (Wilson et al. 2016).

Machine learning for classification algorithms often requires careful calibration of the training set to balance stimulus dimensions and categories. Otherwise, undesirable biases such as frequency bias can result in skewed outcomes. Real-time learning mechanisms in human cognition, and by extension ITL, do not allow for such batch techniques to be applied. Counterfactual reasoning can be a flexible and efficient way to achieve similar goals (reducing biases that derive from initial sampling) in a more naturalistic way. For instance, consider a selection problem where one choice yields a deterministic payoff while the other follows a probabilistic distribution between higher and lower values, averaging to a somewhat higher payoff than the deterministic choice (i.e., the well-known two-armed bandit problem). If one chooses according to expected values established from past history of each choice, as typical in instance-based learning models of decision making (e.g., Thomson et al. 2015), a run of poor luck with the probabilistic choice will lead to a lower expectation, resulting in the deterministic payoff being chosen. Since no information is typically available for foregone payoffs, this suboptimal policy can persist indefinitely.

However, counterfactual reasoning can alleviate the tendency to fall into local minima. Engaging in an explicit consideration of alternative choice can lead to a plausible instance similar to the one actually experienced. That plausible, imagined instance can then result in a rebalancing of the choice distribution, future choices that take the alternative outcome, and, over time, normalizing of choice expectations. Counterfactual reasoning, though, requires at least simple causal models of the domain for it to be effective. Otherwise it can just as well result in a confirmation of the original choice, if the decision maker gives in to confirmation bias and similar self-confirming tendencies.

Discrete versus Continuous Prediction

In all the above considerations about the use of mental simulation for prediction, we have remained agnostic as to the type of prediction. Some mental models used for prediction are based on some sort of discrete time step. For instance, a mental simulation of navigating through a house might consist of having a current belief about what room one is in, and then imagining the effect of a discrete action, such as "go to the living room," to be a discrete change in location (perhaps via a series of navigation steps). While this is a common case, it does not seem to be the same type of mental simulation that is involved in the tying of a knot. For this case, mental simulation seems to be continuous in time, as are the imagined movements of one's hands and fingers.

Exactly how such continuous mental models are supported is not clear to us. To capture the timing and dynamics of prediction, it certainly appears that dynamical systems models and various kinds of neural networks may be more useful than symbolic or rule-based systems. However, the more important question may be how these systems are learned. In a situation based on discrete time steps, predicting the future is often thought of as learning the function that maps from the current state and action onto the next state. To support this in a continuous time domain, some sort of discretization may have to be used (e.g., predicting the state one second from now), but this imposes a particular time step that may not be appropriate. Research into qualitative representations provides a quantization of continuous behavior that is broadly compatible with human event decompositions in perception and language (Forbus 2011), and hence may be useful for ITL.

One way to address this may be to consider the approach observed in the human hippocampus (Rubin et al. 2014). Here, there seem to be two different ways of representing the future. One is to represent a sequence over time by actually having the mental model (in this case neural activity) change over time in a similar way (although perhaps much faster than the original action). This sort of neural replay uses time to represent time (although at a different scale). The second method is to represent the episode over time as a single static pattern. This might be thought of as using space to represent time, with the temporal sequence being collapsed into a single event. Particular parts within that sequence may be classified, resulting in something like a chunk (with slots for the different temporally ordered events within that sequence). We note, however, that different levels of discretization of the temporal sequence are possible.

The Research Community and Functional Knowledge Requirements

As the community seeks more definitive understanding in this new interdisciplinary field, there will be many different viewpoints on the goals, research practices, fundamental assumptions, and understanding of what is needed to move ITL research and development forward, and how we might work on common goals. Nonetheless, it is still possible to share methods, tools, and goals (in the form of challenge problems), as we outline below.

Options for Shared Methodologies, Tools, and Infrastructure

Shared methods and tools could be very helpful in removing barriers to entry in ITL research. Figure 3.1 elaborates on the conceptual framing of the ITL system from Mitchell et al. (this volume). It highlights seven areas where sharing could be beneficial and productive (indicated in the elliptical labels). While a comprehensive review of specific sharable components is outside the scope of this chapter, we do identify a few examples. We also recommend that the community establish a shared portal (e.g., website, Wiki, repository) where sharable components can be cataloged and stored for community use.

Architectures and Architectural Components

Architectures and architectural components may be part of the learning system, the instructional system (in the case of a synthetic agent as instructor), or both. Examples of architectures include cognitive architectures, such as ACT-R and Soar, as well as robotic frameworks, such as the Robotic Operating System.

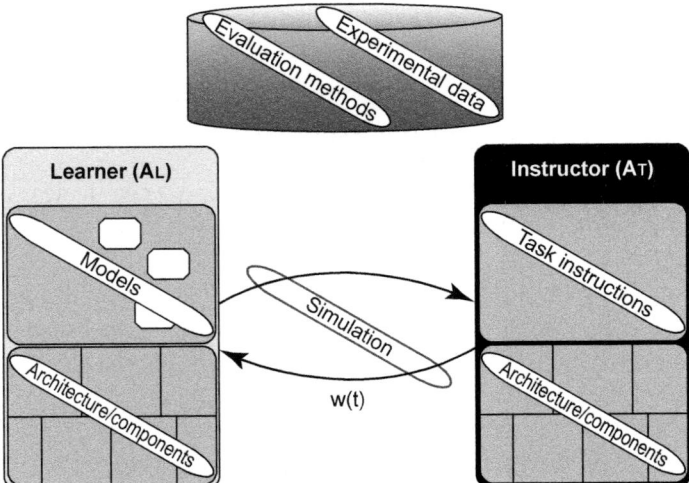

Figure 3.1   Conceptual framing of ITL: ellipses indicate computational artifacts and knowledge products that could be shared across the ITL research community.

Previously developed components may be especially useful for moving into ITL from other areas of research, such as in constructing a natural language understanding system. However, there are many existing components from which the capability for natural language understanding can be constructed. The specific components to choose (and the gaps that must be filled between them) depend on the level of sophistication and completeness required by the research.

Knowledge Components and Models

Knowledge components and models are elements expressed in some representation (typically, the representations of the underlying architectures or components). They are likely to be difficult to share directly unless different researchers are using the same architecture, but some functional descriptions could be shared. It may also be useful to share and exchange knowledge-level summaries of models and model components, even if the encoded representations are not directly sharable across researchers using different architectures and systems.

Simulations

Simulations are not required for ITL, but they may be convenient for speeding research and exploring options that are not cost-effective or safe to explore initially in physical settings. The sharing of simulations, such as Gazebo, has been useful in the robotics community for comparable reasons. Simulations could fall into any or all of the following categories:

- Those which simplify the real world to support or enable task learning in a simulated environment (e.g., a robot simulation or a virtual machine simulation of a mobile device).
- Those designed to mimic the role of a player or actor in the task learning interaction (e.g., a simulated instructor who interacts with a learning agent).
- Those that are the target domains or environment for task learning (e.g., a virtual human character).

Instruction, Instructional Strategies, and Simulated Instructors

Examples of instructions which summarize (and possibly codify) effective human instruction could be useful for bootstrapping and enabling systematic exploration of alternative instructional strategies for evaluating ITL. For example, Koedinger's SimStudent (Harpstead et al. 2015; MacLellan et al. 2014; Matsuda et al. 2007) captures and encodes human instructional steps. SimStudent's representation might provide an initial template for codifying and sharing instructional activity generally for ITL systems. Further, in the educational technology community, there are ongoing attempts to encode various human instructional strategies for use in intelligent tutoring and computer-based training (Wray and Woods 2013). An instructional strategy could be the pattern and choice of specific demonstration examples or the conditional introduction of special cases after a student demonstrates competence in the more prototypical examples. Tools and representations such as these could support the definition and reuse of instructional strategies for ITL. VanLehn (this volume) offers the perspective that ITL instructional strategies may not need to be terribly sophisticated, because actual human instruction is often not as sophisticated as it is hypothesized to be.

Research Data and Artifacts

There are many different classes of methods and data that could be shared within the ITL community, including evaluation data, tools and methods for data analysis, and repositories of task specifications. As we discuss further below, currently there is a gap in the community's ability to specify tasks in a way that meets all requirements. Task specification languages (Yost 1992) have been developed in the past, and recent efforts in general game playing and planning have included generalized representations for some kinds of tasks (Love et al. 2008). A more general task specification language is, however, needed for effective sharing of task specifications.

Challenge Problems for ITL

Above we listed a collection of tasks to describe the scope of ITL. Here, we recommend specific tasks that are plausibly feasible in the near term for the community to work on and share progress. In particular, a few well-specified "challenge problems" could help focus the community. Defining such problems is difficult: we want the challenge problems to be limited enough to be achievable, yet complex enough to push research in directions that will encourage progress toward large-scale tasks that we eventually want to be able to handle through ITL.

To keep these eventual goals of ITL in mind, we believe challenge problems can be identified in the five different domains identified in the goals of this Forum: assistive robotics, healthcare, education, training, and gaming. Developing specific problems in each of these complex domains should be possible to achieve in a two- to five-year research horizon, and would both stretch and focus the field. The complex real-world interaction problems and practical problems that are present in these domains would force the field to go beyond laboratory simplification. Focus would come through multiple research groups simultaneously investigating solutions to similar problems, enabling more immediate and impactful sharing of results.

However, there is an important practical problem to consider as well. The first four domains all involve real-world interaction. *Assistive robotics* and *healthcare* applications generally involve robotics, whereas *education* and *training* involve interactions with human participants. This creates a large barrier to entry for these four areas. Thus, a potential concern is that this additional complexity, inherent in the first four domains, will result in an overemphasis on *gaming* applications.

It is certainly the case that a gaming challenge task would be excellent for ITL. Indeed, large numbers of applicable games are already available in a convenient format (e.g., GitHub, FreeCiv), and games are already being widely used by deep reinforcement learning research. By using such games, it should be possible to show the advantages of ITL, especially if gaining expertise in such games through ITL occurs in a much smaller number of trials. Current noninteractive learning of these games often involves millions of trials, several orders of magnitude more than would be expected by ITL systems, or than is tolerable by people.

However, if gaming has a much lower barrier to entry than the other four domains, then the field runs the risk of spending too much effort in one area and ignoring the inherent challenges of the other areas. It may also fall into research traps, where some ITL research ends up only being applicable to the gaming domain and thus of little use to researchers in other domains.

For that reason, we believe it is vital to identify particular problems within the other four domains—assistive robotics, healthcare, education, and training—that also have a low barrier to entry. In particular, we are looking for problems where research progress can be made without research requirements that involve physical robotics or human subjects, due to the time needed to coordinate and to conduct studies with those participants. This may involve publicly available robotics simulators and publicly available data sets from human participants. We will also need researchers working in these domains with real robots and real interaction with human participants. By identifying helpful challenge tasks that do not have this large barrier to entry, researchers can contribute to the community without extensive overhead.

An example of a possible low-entry challenge domain is to build a trainable "pet" agent on a mobile phone or tablet. For instance, it could be taught new skills through the input/output capabilities of a phone or tablet. Input might consist of text, drawings, speech, or photos. The agent could be taught to answer simple questions or to play games. It could produce text and speech output or draw things itself. It would store knowledge gained through interaction and build on that in subsequent sessions. A potential useful application of this capability would be in primary education: children could teach their "pets" a particular skill and then all the agents in the class could then compete with each other to see whose agent has best learned the skill. For example, in a teachable agent project, students would "teach" their systems about a domain and then compete with other agents, either alone or in teams (Biswas et al. 2016).

Summary of Challenges and Opportunities

In this chapter we have reviewed and synthesized perspectives from multiple disciplines on the functional roles of knowledge and architecture to increase understanding of ITL systems and to assist in the development of synthetic ITL systems. Here we summarize the major themes that emerged from our analysis, focusing on the most important gaps that were identified in the existing state of knowledge as a basis for future, impactful research contributions.

What Are Suitable Formalisms for the Representation of Task Knowledge?

To understand how someone or something can learn a task, ITL needs a shared method of formalizing and describing tasks. In our discussion, we have attempted to characterize a "task" but acknowledge that this description is not sufficient, especially in terms of formalization. Formalization is important not only for practical purposes, such as for knowledge sharing and comparisons of task learning, but also to enable more productive communication and clarification within the field. Questions requiring further consideration include:

- Can suitable formalisms, ones that are comprehensive (express most tasks) and function-general (not assume specific architectural approaches or underlying computational representations), be constructed?

- How can tasks be formalized so that requirements for performing the task specify or indicate incremental modification of the representation of task behavior within the learning or performing system? A task specification may need to include task-level primitives, task concepts and terms, and task strategies as well as the specification of intermediate concepts that should be developed during task optimization. One example is the ability to recognize the board configurations, as in master chess players (Chase and Simon 1973).
- What methods can be developed to enable systematic characterization of tasks and the challenges for an agent learning that task? The community would benefit from being able to evaluate individual contributions focused on varying tasks if there was some way to understand the relative complexity of the tasks being learned. Relatedly, a researcher unfamiliar with a new task may wish to gauge the complexity of a task prior to attempting to get an agent to learn and to perform it.

Are Existing Computational Architectures Sufficient for ITL? What New Architectural Primitives Are Necessary for ITL?

Do existing cognitive and computational agent and learning architectures meet the functional requirements for ITL as summarized in this chapter? This is an empirical research question. As we noted, however, the highly dynamic and extremely fast dynamics of human speech, which is required for many kinds of ITL, may not be feasible in today's architectures. Some potential limitations may be attributable to a mismatch in the architectural primitives needed to support such interaction, rather than atheoretical constraints, such as available processing power. Additional questions that arise from this architectural perspective include:

- What constitutes a functionally sufficient or appropriate set of architectural primitives to support human conversational interaction at human timescales (as determined empirically, analytically, and from social psychological research)?
- Are novel architectural representations and mechanisms needed to support automatic prediction in computational agents? How can tractable automatic prediction, useful for agent learning and understanding, be supported at the many different timescales available in human prediction?
- How should architectures (better) support robust sharing of attention and reference during interaction? How can architectures more directly support the establishment and maintenance of local common ground during interaction?

- What constitute suitable representations and mechanisms to support behavior composition (and decomposition) in a scalable way? Are such functions and representations (necessarily) architectural?

What Novel Requirements for Learner Knowledge Representation Result from ITL?

Knowledge representations are shaped by the tasks and domains in which they will be applied, the ontological commitments of the representation to the situation that the representation will express (Davis et al. 1993). ITL forces a new consideration of the requirements for knowledge representations because specific task requirements cannot, by definition, be known in advance for a (general) ITL system. The psychology of learning and developmental psychology may be particularly useful in helping researchers understand how task knowledge is formed and shaped during all phases of task learning. Specific questions include:

- What are the requirements for a knowledge representation so that it will readily support generalization and transfer of task knowledge (task primitives, task terms and concepts, and task strategies) to new tasks? This question assumes that ITL requires that some distinct agent tasks share task knowledge.
- What are the necessary and sufficient knowledge primitives that will support efficient and general ITL across a wide range of domains?
- What is the ontology/taxonomy of task knowledge that should be acquired by a learner? How do the representations of different tasks interact with one another?
- How can an agent learn relatively complex procedures (such as counting) on the basis of instruction? In the human realm, it takes several years of experience for young children to learn to count (in a general way). How much time and effort is needed for an instructor to convey complex procedural information? How should such information be incorporated into existing agents (e.g., to what extent are production rules, often used to model procedural representation in humans, apt for procedural representation in ITL)?

What Are the Requirements for the Instructor and Learning Environment to Support ITL?

Unlike most artificial learning systems, ITL systems will, by definition, co-construct a learning environment with an instructor/teacher. In ITL, this will generally involve a triadic relationship, where the learner, the instructor, and the shared environment exert influences on one another to produce task learning (for more on this topic, see Shah et al., this volume). Such an arrangement is relatively novel for artificial learning systems. However, research from psychology and education regarding teacher–student interactions (see Van Lehn, this volume) may provide guidance in creating effective learning environments for artificial agents. Additional questions to consider include:

- How do human learners and teachers co-construct learning tasks during learning? How can ITL environments support such co-construction?
- How can ITL technologies take advantage of the troubleshooting and repair processes inherent in natural human dialogue to accelerate learning in ITL systems?
- What methods other than, or in addition to, natural language can be used to achieve common ground in instructional communication? Are there domain- and task-neutral technologies or methods that would allow a human instructor and agent learner to achieve effective common ground without using natural language?

How Might ITL Contribute to the Broader Goal of AI Systems?

The use of ITL for agent learning tends to assume that ITL will be a consumer of AI algorithms and tools, such as specific machine learning algorithms or functional components, like simultaneous localization and mapping or a natural language understanding component. We suggest that ITL may contribute both to the solution of outstanding challenges in AI as well as to the conceptualization of future AI systems (even if ITL is not used). Key questions to address include:

- How can an agent extract or produce symbol-like representations from continuous sensors and actions? Such symbol grounding—a long-standing goal of AI research—is still unmet, although significant progress was recently achieved in grounding for language learning, leveraging the recent availability of cheap and pervasive video (Perera and Allen 2014). ITL stresses symbol grounding, both in its requirements for conversational interactions at human scale as well as in the need to learn new task concepts, which may be expressed symbolically by an instructor but grounded in the perceptual and motor experience of the agent.
- What methods are fruitful for the integration and control of multiple learning systems? Most artificial ITL systems can be described as integrated, cognitive systems, a sub-area of research in AI. The focus on integration for a particular functional purpose (learning from instruction) may contribute to improved understanding in the composition of integrated cognitive systems generally.
- How might a human interact with, teach, and learn from nonhuman intelligence? Much of today's AI is dominated by methods that are functionally powerful, but they lack transparency and understandability. Increasingly, AI-inspired algorithms influence the day-to-day lives

of most living humans, sometimes with unintended and perverse consequences (O'Neil 2016). Thus, it is important that we understand their function and to direct them if necessary. ITL offers an approach that could serve as a foundation for more effective interaction, instruction, and understanding between human and nonhuman systems.

- What methods and techniques are useful for evaluating the generality of AI systems? Much of AI is focused on optimizing solutions to specific domain problems. Still, the goal of AI, as originally conceived, was to produce artificial general intelligence (McCorduck 2004). Because ITL is, by definition, not focused on single tasks or domains, it is likely to encourage generalization. If ITL is successful, it may also nudge the AI community away from its current single-task focus. The development of methods and tools to support the evaluation of ITL may offer a path to evaluate artificial general intelligence as well, by being more targeted and measurable.

4

What People Learn
from Instruction

Christian Lebiere

Abstract

Computational models offer a precise, quantitative way to represent the cognitive processes and representations involved when an agent interacts with another agent: from the receiving of instructions, to their interpretation, to the processes involved in learning to perform a task. This chapter discusses various forms of knowledge and skills involved in interactive task learning (ITL). It describes the components and processes in cognitive architectures relevant to ITL, organized around dichotomies of declarative knowledge and procedural skills, symbolic representations and subsymbolic statistics, as well as cognitive, perceptual, and motor processes. One specific cognitive architecture, ACT-R, serves to focus discussion. Using a model of interactive learning in decision making, it demonstrates how these components and processes interact. Representation, learning, and processing issues are discussed both in isolation as well as in the context of this integrated task learning model.

Introduction

A substantial part of human knowledge and skills comes from instruction, formal or otherwise. Children in modern societies spend at least a dozen years in school, sometimes significantly more well into adulthood. After school, people enter the workforce where a significant part of their knowledge and skill acquisition comes in the form of training sessions, followed by practical experience in the field. Outside of school or work, our environment in recent years has become increasingly characterized by the ubiquitous availability of information at the touch of a button (or touch screen). This information stems from an exponentially increasing array of facts about topics common and obscure, whether organized in formal repositories such as Wikipedia or marshaled in by search engines from the most remote corners of the Internet. It includes tutorials on how to perform any number of practical tasks, from cooking recipes to replacing a cracked screen, as well as entire classes on a range of topics from websites such as Khan Academy to university online courses. The idea that humans derive new knowledge and skills predominantly through painstaking reasoning and experimentation is becoming outdated, if it ever was accurate in the first place. Information is there for the taking, but the central question remains: How do people assimilate and integrate new facts with existing knowledge to understand, instantiate, and apply task instructions in both familiar and challenging settings?

Human-centered fields such as education and psychology contain a wealth of information on these questions, yet much of it is qualitative in nature, limited in scope, and thus cannot be easily integrated into actionable form. For instance, psychology experiments in the laboratory or naturalistic interventions in educational settings can measure the effectiveness of various forms of instructions but cannot, in general, explain how and why they work. What we need to guide interactive task learning (ITL) is a formalized, computational representation of the human learning process. It must be emphasized that ITL is fundamentally different from the typical forms of learning exemplified by machine learning techniques. Those approaches are typically based on algorithms that crunch through very large, uniform sets of context-free training instances. For humans, the task of learning from instructions and interaction with the world, by contrast, requires making sense of and organizing a quite limited and sparse set of information, applying it, and generalizing it on the fly. Requirements for a computational representation of human ITL must thus include a number of cognitive capabilities, including memory, action selection, decision making as well as perceptual and attentional processes and their interactions. The best match for these requirements is the concept of cognitive architectures (Newell 1973b), which originated in response to ad hoc domain-specific task models. A cognitive architecture is a computational instantiation of unified theories of cognition (Newell 1990) that integrates the various cognitive, perceptual, and motor mechanisms into one general, domain-independent framework. An increasing number of cognitive architectures have been proposed over the years. Here, discussion is grounded in one specific architecture, ACT-R (Anderson and Lebiere 1998; Anderson et al. 2004), although increasing evidence suggests that the cognitive architecture program is converging on an implementation-independent consensus (Laird et al. 2017b).

There is reasonable concern that cognitive architectures provide a too detailed, mechanistic account of human learning from instruction to be relevant in this setting. Newell (1990) formalized the idea of levels of description in his concept of bands of cognition, distinguishing between subsecond mechanisms in the neural and cognitive bands and phenomena in the rational and social bands, which can take place over days, weeks, months, or years. Anderson (2002), however, argued that while not every aspect of the fine-grained mechanisms will be relevant at the larger timescales, key characteristics will still be relevant and even determinative. A demonstration of such detailed relevance can be seen in the application of cognitive models to intelligent tutoring systems (e.g., Anderson and Gluck 2001). In that application context, cognitive models are formal representations of the student learning process; they are comparable to actual student behavior and can be used to infer their current state of learning and guide interventions (e.g., selection of problem set, hints, review of material) to improve the learning process. In particular, cognitive models have been shown to provide the right abstractions to enable proper understanding of performance curves as a function of the underlying knowledge and skills at the grain scale provided by cognitive architecture mechanisms and representations (e.g., Corbett and Anderson 1995). Thus, it is not much of a leap to suggest that cognitive architectures also provide the best tool to guide the ITL process, both in terms of design as well as run time.

Below, I describe components and processes in cognitive architectures that are relevant to ITL, organized around dichotomies of declarative knowledge and procedural skills, symbolic representations and subsymbolic statistics, as well as cognitive, perceptual, and motor processes. The interaction between these components and processes is exemplified in a model of interactive learning in decision making developed in my lab (Lebiere et al. 2013b). I conclude with a general discussion regarding potential avenues of research in developing cognitively based approaches capable of supporting ITL.

Learned Knowledge and Skills

Our discussion begins with a computational description of the knowledge and skills that are learned in the context of ITL (for further discussion, see also Wray III et al., this volume). Following the distinction made in many cognitive architectures, declarative knowledge is separated from procedural skill and symbolic structures from statistical parameters. For each category of knowledge, I describe the characteristics and instances of how it can be learned and supported through ITL:

	Symbolic	Subsymbolic
Declarative	Chunks: instructions, general knowledge, situation knowledge, beliefs	Activations: environmental statistics, quantitative uncertainty, distributional semantics
Procedural	Productions: heuristic strategies, general procedures, compiled skills	Utilities: learned rewards, probabilistic selection, adaptive generalization

The terminology used follows that of the ACT-R cognitive architecture, although similar constructs exist in most other popular cognitive architectures (Laird et al. 2017b). ITL systems have been developed using a number of different cognitive architectures, including ACT-R (e.g., Taatgen et al. 2006) and Soar (e.g., Kirk and Laird 2014; Mohan and Laird 2014; Mininger and Laird 2016).

Symbolic Declarative Knowledge

Declarative knowledge represents explicit, conscious knowledge as symbolic structures, equivalent to propositional representations. These structures, called "chunks," bind together a set of values (which can themselves be chunks) into specific roles. Chunks are created and stored automatically in long-term memory by recording the state of working memory at specific points in time. Chunks can represent any type of explicit knowledge related to instructional task learning: knowledge of specific situations, general factual knowledge, knowledge of procedures and instruction steps, as well as knowledge of other people's beliefs.

Cognitive architectures make strong mechanistic commitments as to how knowledge is learned, represented, and accessed but typically do not make ontological commitments as to what the knowledge types are and how they are organized. This largely reflects the historical split between mechanistic and ontological approaches to cognitive science and artificial intelligence. A number of attempts have been made to import knowledge ontologies into cognitive architectures and, more generally, to adopt stronger and more systematic knowledge commitments (e.g., Ball et al. 2004; Wray et al. 2004; Best et al. 2010; Oltramari and Lebiere 2012; Salvucci 2014). Recently, Lieto et al. (2018) examined issues systematically with knowledge representation in cognitive architectures and compared the approaches as well as potential solutions to those problems in various architectures.

Subsymbolic Declarative Knowledge

Symbolic representations provide structure but not all (or even most) of the knowledge. Much real-world knowledge is not conscious but rather implicit, resulting from automatic learning of statistical patterns. Anderson (1990) argued that human cognition adapts to the statistics of the environment to optimize its performance given limitations (e.g., working memory capacity, attentional limitations, and various other architectural bottlenecks). This adaptivity is fundamentally heuristic in nature and can lead to errors and cognitive biases (e.g., Lebiere et al. 2013b) when its assumptions do not match the actual nature of the environment. Although "soft" aspects of human cognition (e.g., adaptivity, generalization, and stochasticity) can be described at higher (e.g., Bayesian) or lower (i.e., neural) levels of abstraction, in cognitive architectures such as ACT-R and Soar they are represented in terms of subsymbolic mechanisms that control access to symbolic knowledge structures. Specifically, they are formalized in terms of activation processes that determine which chunk is selected from long-term declarative memory for a given retrieval request. Activation is represented by a base-level term that captures pervasive regularities, such as the power law of practice and the power law of forgetting. Strengths of associations from context elements to memory structures enable task-sensitive knowledge access. Generalization from similar situations is achieved through a partial matching mechanism which leverages semantic similarities between memory chunks. Stochasticity, useful in exploring solution spaces, results from a noise term that makes retrieval probabilistic.

While these soft characteristics of cognition seem to have little to do with learning to perform tasks (i.e., taking instructions and executing them), purely symbolic information processing is not enough in naturalistic settings. Intelligent behavior, by definition, is ill-defined; otherwise traditional algorithmic or optimization techniques could be applied, as has happened in many domains, from chess to program trading. Performing tasks effectively in the real world requires the very characteristics that human cognition has evolved in that environment. For instance, adversarial behavior requires sensitivity to the statistics of the opponent's actions (Lebiere et al. 2003). Moreover, learning complex patterns of events requires associative links to capture context sensitivity (Lebiere and Wallach 2001). Generalization to similar situations is essential for making decisions (Sanner et al. 2000) and controlling dynamical systems (Wallach and Lebiere 2003) in complex environments, where exhaustive or even extensive experience is impractical to achieve. Stochasticity is essential not only to avoid being exploited in adversarial situations (e.g., West and Lebiere 2001) but to explore complex spaces and plan solutions in combinatorial environments (e.g., Lerch et al. 1999). Even tasks that can be defined in a purely symbolic manner (e.g., learning arithmetic tables in school) leverage cognitive characteristics such as stochasticity, adaptivity, and generalization to overcome our cognitive limitations (Lebiere 1999). Indeed, ontologies can be most effectively integrated in cognitive architectures by representing not only the symbolic knowledge itself but by expressing the underlying statistics and regularities of the knowledge. Those regularities can be represented using the subsymbolic parameters that control retrieval and application of that knowledge (Oltramari and Lebiere 2013). Next I will discuss in more detail how these various characteristics can be integrated into ITL.

Symbolic Procedural Skill

While knowledge contributes crucially to performance in many tasks, skill is even more essential. Skill refers to the procedural control of behavior involving both external actions (e.g., manipulating objects in the environment) and internal operations, which include a broad set of internal cognitive capabilities. The ability to manage attention, given well-known limitations in scope and time, requires paying attention to the right things at the right time. The maintenance of appropriate context involves strategies for allocating limited working capacity to preserve the relevant parts of the task representation at all times. The control of knowledge requests from long-term memory involves constantly deciding what type of knowledge is relevant, how to access it most efficiently, and how to make use of it to perform required inferences, such as anticipating future developments. These three components, and related abilities, are often referred together under the term *situation awareness* (Endsley 1995). Those basic functionalities support key cognitive skills that implement complex task functions. Decision making involves selecting among competing courses of action, based on estimations of their outcomes. Planning results in a potentially complex sequencing of actions that alter the state of the environment, by oneself and potentially in concert with other agents. Managing communication with other agents requires skills that involve not only the generation and understanding of natural language, including resolving references to objects and events relevant to the task, but also social skills such as taking another's perspective (theory of mind), reasoning about their intent, or engaging in persuasive arguments.

Since the origin of cognitive architectures, the most popular form for representing procedural skills has been *production rules* (Newell 1973a). Formulated as condition-action pairs, production rules are not logical rules but rather tiny pieces of skills that have to be acquired in each step of the task learning process and then assembled at run time into coherent behavior. The flexibility and modularity of the framework is its major source of power in generating robust real-world behavior. Acquiring and assembling those skills can be a challenging and time-consuming process. Sometimes skills originate by random trial and error. Other times they originate through a more directed search process. The most direct process of skill acquisition is learning from step-by-step instructions that specify precisely how to accomplish goals (e.g., Taatgen et al. 2006). While other approaches to learning are possible, they tend to be complementary to the instructional process. For instance, if only the goal is specified in the instructions, a problem-solving process can be added to generate the instruction steps. Similarly, learning by demonstration can be conceptualized as requiring an additional process to generalize observations into instructional steps.

The path for this direct instruction process that is supported by the cognitive architecture is to learn the instructions declaratively (e.g., from interacting with a teacher), then retrieve each step from memory one by one and interpret it. The cognitive architecture, through a process called proceduralization (Anderson 1987), then speeds up the process of instruction interpretation and execution in at least two ways. The first, which applies primarily to the interpretation aspect, is to compile away the retrieval of each instruction step into a new unit of skill (a production rule) which directly applies that step without the need for explicit retrieval from memory. For instance, with enough practice, we can type in a password to one of our devices without having to recall it consciously: we just let our fingers "do the walking." The second step applies primarily to the execution aspect: consecutive steps are assembled into a single production rule, thereby making the process increasingly efficient (within the limit of cognitive resource conflict and external task structure, of course) by constructing a set of increasingly complex hierarchical task skills.

Subsymbolic Procedural Skill

The challenging part of skill acquisition, therefore, is not the specific process of creating the skill itself, since the proceduralization process is automatic, but rather the creation of the mental environment in which that process takes place. This involves either knowledge-rich methods (e.g., reasoning and inference) as well as weak methods (e.g., trial-and-error or means-ends analysis) or the much more efficient method of receiving and retrieving the proper task instructions through interaction with an expert teacher. Once new skills have been acquired, applying them properly is not a trivial matter. The new production rules have the same conditions of applicability as the previous, less efficient rules and will have to demonstrate their effectiveness. As for declarative knowledge, production rules are not selected purely symbolically but utilize subsymbolic processes similar to memory activation. The basis for selecting competing rules is that of utility, acquired for each rule according to reinforcement learning processes (e.g., Fu and Anderson 2006) that reflect the effectiveness of each rule in attaining positive rewards. As with memory activations, production utilities include a stochastic component which makes rule selection probabilistic and allows for an exploration-exploitation trade-off that allows for the most efficient skill to emerge gradually and be selected reliably.

A final challenge is that the context of the current problem might not necessarily perfectly match production rule conditions, for instance, because of minor variations in the task environment. As for memory retrieval, a partial matching process has been integrated in the production utility calculus to enable approximate matching of rule condition to the situational context. Analogous to how a matching penalty reduces chunk activation in the retrieval process, imperfect matching to a production condition reduces its utility by the degree of mismatch (Best and Lebiere 2006). What results is an adaptive process of skill application: production rules expand their range of generalization as long as they are successful; then they contract it once they reach situations for which they are poorly suited. Therefore, the process of task learning is not characterized by an all-or-none development of perfect skills; it is a gradual process in which independent skills are acquired and composed to find their range of applicability. Below I will describe two instances of how this approach can be applied to an experimental task.

Learning Complex Decision Making

Depending on the nature of both the task and the interaction, ITL can take on many distinct forms (see Wray III et al., this volume). Different tasks—from decision making in abstract spaces, such as games, to the manipulation of physical objects in 3D space—display fundamentally different characteristics, which will be reflected in the knowledge and skills that are acquired.

Similarly, modes of interaction—from passive instruction interpretation to bi-directional building of common ground through interaction in shared physical spaces—involve distinct processes and representations (Chao et al. 2011). To illustrate the concepts introduced above and sketch out the space of mechanisms involved, I will describe two ways in which these tasks involve substantially different domains and processes to demonstrate the generality of the approach.

The focus is on procedures for general decision making, independent of domain, and was initially defined in the context of a human–robot interaction scenario (Lebiere et al. 2013a). The task to be learned is framed as making arbitrarily complex decisions by decomposing them into a complex process that involves simpler decisions. The scenario is represented by a set of quantitative variables, each describing the value of a particular feature of a specific object or entity. The perceptual and attentional processes that lead to the encoding of each scenario are not explicitly represented. Instead, each situation variable for a given scenario is encoded as a separate chunk in memory, representing the outcome of the encoding process. Chunk activation values are set high enough to assume that access to the situation chunks is fast and reliable. This approach abstracts away from potential perceptual issues, such as grounding problems, that would require distinct solutions, such as perceptual learning.

The decision process is represented by a treelike chart that involves flowing through a set of basic decisions which lead to a number of actions. For each basic decision in the process decomposition, the situation variables entering into that decision are explicitly specified as part of the instructions. Again, the entire decision structure is represented as memory chunks, abstracting over an explicit interactive instruction process. While specific issues might arise as part of the interaction itself, there is nothing inherent in the approach that would present a limitation, as the sequential instruction retrieval process could easily be replaced by an interactive dialogue with an instructor.

Each instruction chunk in memory represents one particular step (e.g., encoding a situation variable, making the decision, or deciding on the next decision to make based on its outcome) of a basic decision process. Again, activation values of the instruction chunks are set sufficiently high as to ensure that they can be reliably retrieved.

The overall decision process (see Figure 4.1) follows the general pattern laid out in the previous section: each individual instruction step must be retrieved before a corresponding action can be taken. Each basic decision entails the following steps:

1. The value of each relevant situation variable is retrieved from memory.
2. A decision chunk is gradually composed.
3. Once that chunk is complete, a decision is made.
4. The outcome of that decision serves as the basis to move on to the next decision.

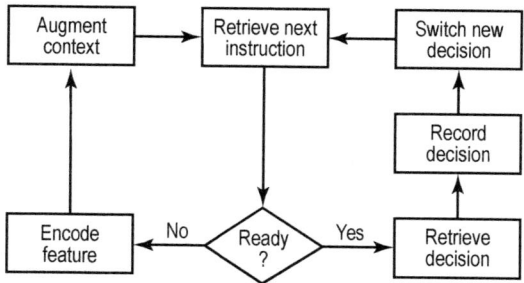

Figure 4.1 Decision-making process for instruction execution.

Each decision is made not according to a specific rule, but by leveraging memory retrieval processes to generalize from a set of previous decision instances held in memory. Those decisions could have resulted from demonstration by experts as part of the instruction process or experienced by the learner as part of a trial-and-error process. This process of instance-based learning (Gonzalez et al. 2003) leverages both the symbolic representation of previous experience in memory and their subsymbolic activation parameters. Thus, the task learning process is a combination of instructions and experience, reflecting a pervasive pattern in both formal education and field training. The result of that overall decision process generates a set of outcomes that closely reflects how humans gain expertise: the model follows the same decision procedure as experts, as specified by the instructions, and has also been trained from their decision instances.

Nothing about this instructional decision-learning process is specific to the domain or scenario. We applied the same instructional learning engine to an entirely different task of making probability judgments among hypotheses in a geospatial sense-making domain and obtained excellent results that match human performance, in particular, by exhibiting similar types and degrees of cognitive biases (Lebiere et al. 2013b; Thomson et al. 2015). The structure of the task consisted of a sequential presentation of spatial layers of information: each layer resulted in a probability judgment about a discrete set of hypotheses. Subjects were then instructed on how to combine judgments from the various layers to refine their probability distribution over the set of hypotheses. Those instructions were converted into the format described above. Only two minor generalizations to the approach used in the first domain were necessary. The first extended the test on the result of an intermediate decision to quantitative nonbinary outcomes. The second extended the set of values used in decisions to include the outcome of previous decisions in addition to the external environment variables. These changes were not domain specific in any way but rather increased the generalization of the instructional process. Interestingly, while the decision process in the original domain was specified in tree form, it could be generalized in this domain to a graphlike iterative procedure without requiring any change because of the nature of the representation of instructions as independent memory chunks. This illustrates the power of cognitive techniques to provide humanlike flexibility.

Conclusion

Computational models offer a precise, quantitative way to represent the cognitive processes and representations involved when an agent interacts with another agent: from the receiving of instructions to their interpretation and the processes involved in learning to perform the task. Computational models can be used in a number of ways to predict human performance in interactive learning tasks, to design ITL systems that work naturally with humans, to implement intelligent assistants that can learn as a human assistant would, as well as to implement intelligent agents and robots that behave as a human instructor.

5

An Ontological Perspective on Interactive Task Learning

Charles Rich

Abstract

Learning and teaching are best viewed as a collaborative interaction. As participants, both the teacher and learner share the goal of increasing the learner's abilities. Yet what does it mean to *know* how to do something? This chapter analyzes the abstract form, nature, and organization of task knowledge and illustrates these concepts using a shared task of tire rotation. It applies a hierarchical decomposition of knowledge for interactive task learning that involves three levels: domain knowledge, procedural knowledge, and metaknowledge. In addition, the traditional distinction between symbolic versus nonsymbolic task knowledge is noted. Representative examples are given, and open questions and unresolved problems are highlighted as suggested directions for future inquiry.

Introduction

What does it mean to know how to do something? This question concerns the ontology of task knowledge (i.e., the abstract form, nature, and organization of this knowledge) and answering it is a logical prerequisite to understanding how to learn or teach such knowledge. Although our ultimate goal is to identify the frontiers of the current state of the art, I have organized this discussion around the main ideas and trends in current and past work on task knowledge. I will not undertake a comprehensive literature review and will cite only representative examples in each topic area, highlighting open questions, unsolved problems, and controversies where appropriate.

Another underlying premise of this chapter is that learning and teaching are best viewed as kinds of collaboration: the teacher and learner are two participants in a collaborative interaction in which the shared goal is to increase the learner's abilities. Furthermore, learning and teaching are often naturally interleaved with other kinds of collaboration, such as delegation and supervision. Finally, because discussions of ontology have a tendency to become very

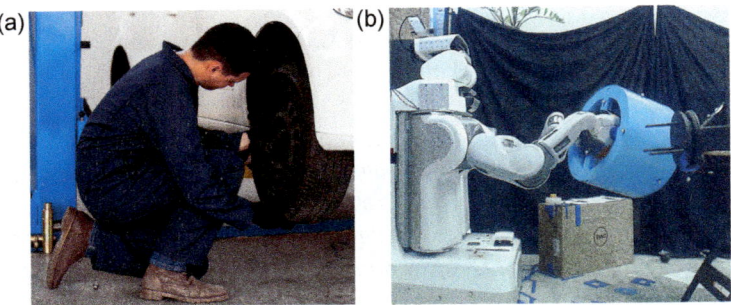

Figure 5.1 Example of tire rotation task performed by a human (a) and a state-of-the-art robot (b).

abstract, I will attempt, whenever possible, to illustrate ideas using the shared task example of tire rotation.

Rotating the tires on a car is a task commonly performed in an automotive service center (Figure 5.1a). It entails unscrewing the four or five lug nuts on each wheel, removing the wheels, and then remounting them on different hubs, according to a rotation pattern, such as back-to-front or crossover (e.g., left rear goes to right front). The overall task involves many individual steps, with repeated subsequences (such as screwing or unscrewing all the nuts on a particular hub), alternatives (the rotation patterns), and the use of a tool (lug wrench or power wrench). Recently, Mohseni-Kabir et al. (2015) taught a PR2 robot (Figure 5.1b) a greatly simplified laboratory version of this task using a combination of demonstration and instructions.

Levels and Types of Knowledge

To start, it is useful to think about the knowledge involved in interactive task learning (ITL) as stratified into three levels: domain knowledge, procedural knowledge, and metaknowledge. Discussion here is organized according to this breakdown, although there are many other ways to conceptualize and organize the knowledge necessary for task learning (see, e.g., Chai et al., Laird et al., and Wray III et al., this volume).

The most fundamental level, *domain knowledge*, provides general knowledge about the world (or parts of the world) and pervades all aspects of the tasks being learned. This knowledge can sometimes be assumed to be already shared between the teacher and student, thus providing "common ground" (see Mitchell et al., this volume). Alternatively, the concepts and relationships that are necessary to achieve common ground during task learning may need to be explicitly taught (see Shah et al., this volume). Physical manipulation tasks such as tire rotation, the example used in this chapter, require a commonsense understanding of everyday physics, such as the fact that an object will fall if you let go of it. This domain knowledge is used in both levels above it.

The middle level, *procedural knowledge*, comprises most of what is usually focused on when teaching tasks: the ordering of steps, preconditions, task hierarchy, and so on. For example, the first step in removing a wheel is to unscrew the lug nuts.

The upper level, *metaknowledge*, is knowledge about the procedural knowledge. For example, knowing how long it typically takes to rotate the tires on a car is metaknowledge about the tire rotation procedure.

In addition to these three knowledge levels, an independent categorization to keep in mind is *symbolic* versus *nonsymbolic* knowledge. (One might alternatively characterize the distinction as *discrete* versus *continuous*). The crossover tire rotation pattern is an example of symbolic procedural knowledge, whereas the motion trajectory for unscrewing a lug nut exemplifies nonsymbolic knowledge. An example of nonsymbolic domain knowledge would be the maximum weight carrying capacity of a robot. Symbolic and nonsymbolic knowledge can also be intermixed, such as in a parameterized motion trajectory with semantic constraints.

Domain Knowledge

Domain knowledge is the foundation of all task learning, other than the most limited form of mimicry. The type and extent of the required domain knowledge varies, of course, with the task (or more typically, collection of related tasks) being learned. For physical tasks in the everyday world (e.g., tire rotation or household tasks), domain knowledge begins with commonsense physics, which includes knowledge of

- gravity (how objects support other objects, fall, etc.),
- motion (how objects move in free space and slide, roll, or bounce in contact with other objects),
- materials (how solid objects bend or break in response to squeezing or stretching or other manipulations; how liquids flow, pool, and fill containers), and
- causality (how actions on one object result in changes to other objects via connections between them).

Can commonsense physics be built into robots and, if so, will that make it easier to teach them physical tasks?

As one moves into more specialized tasks, the domain knowledge becomes more specialized. For example, the domain knowledge underlying cooking includes specialized knowledge about state changes associated with specific temperatures (e.g., boiling, freezing, and burning). In addition to commonsense physics, tire rotation requires some specialized knowledge about the appropriate forces for screwing and unscrewing lug nuts. A lot of such domain knowledge is nonsymbolic.

Not all tasks are physical. Consider, for example, teaching arithmetic or teaching a virtual agent how to perform online tasks on your behalf. In these cases, the domain knowledge is more abstract and mathematical, such as the application programming interfaces (APIs) for online services.

Formal logic is a useful way to think about symbolic knowledge, even if it is not the form in which the knowledge will actually be used in practice. From this point of view, domain knowledge provides the primitive object types (and perhaps some designated object instances), predicates, and functions that are used for specifying the procedural and metaknowledge, as discussed below. Friedman et al. (1999) have taken this approach for learning probabilistic relational models.

What is the appropriate use of formal logic in research on ITL?

Procedural Knowledge

Procedural knowledge is the knowledge required to execute tasks; this does not include being able to teach, learn, or reason about the tasks. This knowledge has been studied most intensively in the artificial intelligence subfield of planning (Coles et al. 2017) and consists of the following key elements: primitive actions, ordering, conditionals, parameters, constraints, hierarchy, and motion trajectories.

Primitive Actions

The most basic element of procedural knowledge is the *primitive actions*. It is important to note that the notion of "primitive" is contextual: what is primitive in one situation may not be primitive in another. For instance, in teaching a robot how to rotate tires, unscrewing a lug nut and putting it down may be a primitive action, whereas for a different robot or teacher, there are two separate primitive actions: unscrewing and putting down.

In terms of computation, primitives are essentially symbols, although this assertion may be controversial from other points of view (see Wray III et al., this volume). However, "primitive" does not mean predefined or built in. A particular primitive may be directly associated with a precompiled motor (or cognitive) program. It may also be necessary to learn how to execute the primitive, in which case the primitive is ultimately a mixture of symbolic and nonsymbolic knowledge.

How can/does the notion of primitive actions apply to "continuous" activities such as ballroom dancing or dribbling a basketball?

Ordering

The next most basic element of procedural knowledge is *ordering*. The most minimal form of a plan is thus a sequence of primitives. Ordering knowledge also commonly takes the form of a partial order. For instance, in tire rotation, the order in which you unscrew the lug nuts does not matter, but all the un-screw actions must precede unhanging the wheel.

What are the natural kinds of parallel and overlapping execution models appropriate for interactive task learning?

Conditionals

To achieve a formally complete computational system, all that needs to be added to primitives and ordering is *conditionals*. Conditionals can take many forms and be used in different ways. Basically, however, they are all about changing which actions are executed depending on the state of the world.

The most common form of conditional is a simple if-then-else, which speci-fies two alternative primitives (or sequences of primitives), depending on the value of some Boolean condition. In tire rotation, you can use either a lug wrench or power tool to tighten the nuts, depending on whether the power tool is available.

Also commonly used are *preconditions*, which, when false, block the ex-ecution of a specified primitive or sequence. This is particularly useful when actions are partially ordered, because there may be something else that can be done while waiting for the precondition to become true. For example, after squirting oil on a rusty nut, the unscrew action on that nut is blocked until five minutes have passed; meanwhile, other nuts can be unscrewed.

A less commonly used form of conditional is a maintenance condition, which is a condition that is expected to hold throughout the execution of a long-running action. If the condition ever becomes false, the action is supposed to be terminated. In the tire rotation example, the pneumatic power supply is expected to be maintained to the power lug nut tool throughout execution of the screw or unscrew action.

Conditions need not make only binary choices. For example, imagine a sorting task in which the color of the object is supposed to match the color of the sorting bin, where there are more than two colors.

Conditionals are a key place where domain knowledge is used in procedural knowledge, because the domain knowledge specifies the relevant and perceiv-able properties of the world that are tested in the conditions. In tire rotation, it is domain knowledge that specifies that the position of the wheels on the hubs is relevant to the tire rotation task, and not the color of the tires. To frame this in formal logic terms, the domain knowledge specifies the predicates that are used in the conditions.

What are the natural kinds of uncertain and probabilistic conditional models appropriate for interactive task learning?

Parameters

Using only what has been discussed above yields a very limited form of procedural knowledge, because the actions can only be applied to specific objects in specific settings—think of programs without inputs. To achieve generality and reusability, most procedural knowledge also includes *parameters*. For example, the procedural knowledge of how to rotate the tires on a car can be applied to any car. Thus the car (and implicitly all of its parts, such as the hubs, wheels, etc.) is an input parameter of the task. Input parameters also often have restrictions on what values can be provided. These are very much like preconditions. For example, the input to the tire rotation task must be a four-wheeled vehicle.

In addition to input parameters, it is also common for procedural knowledge to include output parameters. Generally speaking, the outputs of a task are considered to be the objects whose properties are changed or that are created during the execution of the task. For example, the output of the unscrew action is the nut, because its location has changed.

How can procedural knowledge be parameterized automatically?

Constraints

Knowledge of parameters is important because it enables the specification of *constraints* between the input and output values of different actions. The simplest form of constraint is equality. For instance, if you are painting a matching set of chairs, the color input parameters of all the paint actions would be constrained to be the same.

An equality constraint between an output parameter of an action and an input parameter of a subsequent action[1] is called *data flow*. Figure 5.2 demonstrates the data flow in our tire rotation example: the arrow specifies that the output of the unscrew action (the nut) becomes the input to the putdown action. Data flow is an aspect of the causal structure of a procedure. Causal structure is important knowledge for, among other things, recovering from errors.

How is causal structure used in interactive task learning?

Constraints other than equality can also be part of procedural knowledge. For instance, in a robotics procedure, for stability, the weight of an object placed on top of another object may be constrained to be less than the weight of the first object.

Hierarchy

Hierarchy is important because humans approach complex tasks by breaking them down into subtasks. Figure 5.3 illustrates one way to break down tire rotation into five levels of hierarchy. The tree notation in this figure, called a

---

[1] It is logically possible, but unusual, to constrain two output parameters to be the same.

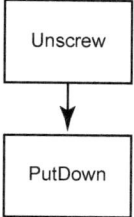

Figure 5.2   Schematic data flow constraint for the tire rotation example.

hierarchical task network, has been widely used in artificial intelligence—a version has even been formalized as an ANSI standard (see Rich 2009). The fringe of the tree specifies the sequence[2] of 64 primitive actions required to rotate the tires on a car (some repeated structure is elided in the figure to save space). Notice that there are multiple instances of the same type of primitive (e.g., Unscrew) with different parameters (i.e., different nuts).

The nonprimitive nodes in the tree are called *abstract* tasks. The top-level abstract task in this example is "Rotate tires." Other abstract tasks at intermediate levels of the hierarchy include "UnscrewHubs" (unscrewing all the lug nuts on one wheel) and "UnhangHubs" (removing the wheel from the car and putting it down). Although knowing these abstract tasks is not fundamentally necessary to execute tire rotation, it is important for two reasons: reuse and communication.

Regarding reuse, not only are there multiple occurrences of the same type of primitive in this task hierarchy, there are also multiple instances of the same type of abstract task. For example, UnscrewHub, UnhangHub, HangTire, and ScrewHub each appear four times (once for each wheel on the car). Furthermore, it is common for the same abstract tasks to be reused in other top-level tasks in the same domain. For instance, these four abstract tasks are also steps in changing a flat tire.

Abstract tasks also provide a richer communication vocabulary than just the primitives for collaborative interactions, including teaching and delegation. For example, if there is a breakdown, rather than saying "pickup nut step 49 failed," one can say "pickup nut failed while unscrewing the first wheel." Similarly, one can delegate an entire abstract task, rather than specifying the sequence of primitives (e.g., "now, please unhang all the hubs").

Conditionality in hierarchical task networks is usually conceptualized in terms of decomposition choices, called *recipes*. Notice that Figure 5.3 is a tree with two kinds of nodes. The primitive and abstract tasks discussed above are rendered as rectangles. Each abstract task is decomposed into subtasks, each of which may be abstract or primitive. The oval recipe nodes in the tree

---

[2]   Technically, in this case, it is not a sequence but rather a partial order, since some actions (e.g., unscrewing lug nuts on a single wheel) are unordered.

70

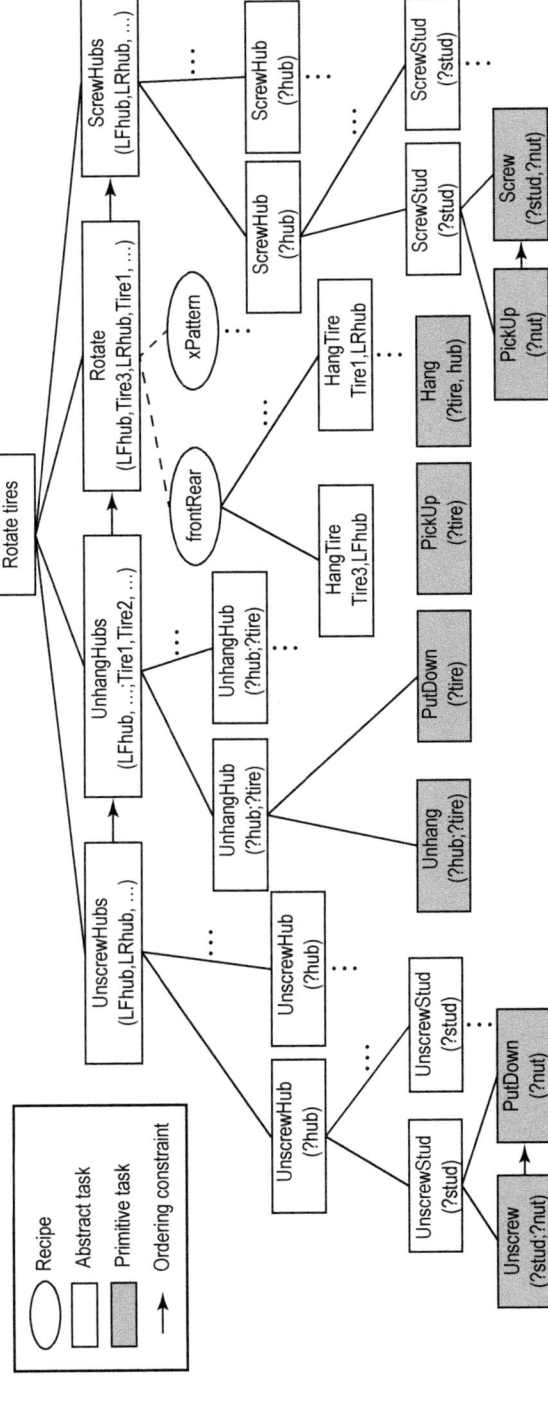

Figure 5.3 Hierarchical procedural knowledge required for tire rotation.

indicate alternative decompositions.[3] For example, in Figure 5.3 there are two alternative recipes for the rotate task, frontRear and xPattern, corresponding to the two standard patterns for rotating tires on a car: front wheels to rear wheels on the same side, or left-front to right-rear, etc. (The figure does not show the constraints that specify the details of these two patterns.)

The condition that determines whether a particular recipe can be chosen is called an *applicability condition*. Like the other conditions discussed above, applicability conditions test the state of the world and are built on domain knowledge. At execution time, if the applicability condition of more than one recipe for a given abstract task is true, then the procedural knowledge allows any of the applicable recipes to be chosen.

What is the best way of teaching recipe applicability conditions? Like abstract tasks, recipes also serve an important communication function. For example, it is efficient to simply say, "let's use the x pattern."

Finally, a cautionary note: the hierarchy in Figure 5.3 is not the only way to break down tire rotation into subtasks. Some people might insert an extra abstract task (say, RemoveHubs) and group together UnscrewHubs and UnhangHubs. Someone else might not group Unscrew and PutDown into UnscrewStud. Even the choice of primitives cannot be assumed to be universal. For some people, Pickup and Screw may be a single primitive, rather than a sequence of two primitives. This adds a lot of challenges to communication and collaboration, beyond even the problem of different people using different words for the same task concept (discussed further below).

The Policy View

In reinforcement learning, which is increasingly being applied to robotics, procedural knowledge is viewed somewhat differently than above. Instead of embedding the predicates that check the state of the world in conditional structures on sequences of actions, a single *policy* function, π, is used that maps from (all possible) states of the world, S, to the possible actions, A:

$$\pi : S \rightarrow A.$$

Torrey and Taylor (2013) have explored combining this approach with interactive teaching.

The policy view and the conditional plans view are mathematically equivalent in the sense that each can be mechanically translated into the other. However, they are very different in terms of the affordances they provide for human interaction. The policy view offers only a single monolithic function, π, to discuss, teach, correct, and so on, whereas the conditional plans view

---

[3] Such trees are also called *and-or* trees, where the abstract tasks are *and* nodes, because all of their children are executed, and the recipes are *or* nodes, because only one of their children is chosen.

exposes much more structure for interaction. There has also been work on learning parameterized policies and hierarchical policies.

Can reinforcement learning algorithms be applied directly to conditional plans?

Motion Trajectories

All of the kinds of procedural knowledge discussed above are essentially symbolic. However, there is clearly also a kinesthetic dimension to "knowing how to do" a task. For example, as Figure 5.4 suggests, there are only certain motion paths through space that work to unhang a tire. Unlike the action sequences and hierarchies discussed above, this knowledge is in the form of continuous functions. At its simplest, such knowledge is a single motion path through three-dimensional space (e.g., implementing a primitive action such as Unhang). More complicated versions of such knowledge may be in the form of regions in three-dimensional space that constrain the motion, or trajectories/ regions in the higher-order joint configuration space of the manipulator. Sometimes it is also important to know the appropriate forces to be exerted at various points in the trajectory (e.g., how much to tighten the lug nuts).

What is the relationship between symbolic and continuous procedural knowledge, and how can they be learned together? Mohseni-Kabir et al. (2018) have recently begun to explore this question.

Metaknowledge

Metaknowledge is knowledge *about* other knowledge. Here I review some key categories of knowledge about procedural knowledge that are generally

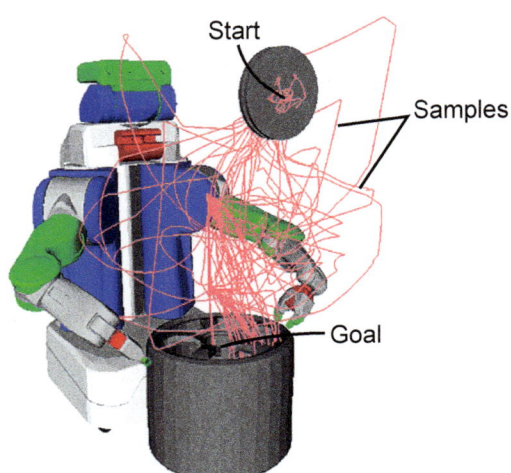

Figure 5.4 Motion trajectories for unhanging a tire.

important for learning/teaching and collaboration: capabilities/tools required, postconditions, failure recovery, duration, difficulty, uncertainty, and pedagogical knowledge.

The preceding discussion of procedural knowledge assumes that the agent performing the task is *capable* of executing all of the primitives. This, however, is not always true. In fact, a common motivation for collaboration is that one particular agent alone does not have all necessary capabilities. Knowledge of which primitives a particular agent is capable of performing (or restrictions on the parameters on the primitives) is thus an important category of metaknowledge. For instance, if a human and a robot are collaborating on tire rotation, the robot may be capable of performing all of the screw/unscrew primitives but any action that involves physically removing the tire must be done by the human, because the tire is too heavy for the robot to lift. The availability of a particular tool, such as a lug nut or power wrench for tire rotation, is a closely related kind of capability metaknowledge.

Should metaknowledge about capabilities include preferences (for different types of tasks)?

Actions do not always succeed. Knowing how to decide whether an action has succeeded or failed is a type of metaknowledge about the action: the *postcondition* of the action.[4] Both primitives and (in the case of hierarchical tasks) abstract tasks can have postconditions. Like the other conditions discussed above, postconditions test the state of the world and are built on domain knowledge. For example, the postcondition of the Pickup primitive is that the input object is in the agent's hand or manipulator. This condition could fail, for example, if the agent has not properly grasped the object before it moves its arm.

In which kinds of procedures is it most important to know postconditions?

Action failure introduces the issue of *failure recovery*. One of the differences between a novice and an expert in a particular task is that the expert knows more about how to recover from specific failures. For example, suppose Unscrew fails because the lug nut is frozen. An expert knows to squirt some penetrating oil on the nut, let it sit for five minutes, and then try again. Failure recovery knowledge is obviously of great practical importance and, in some cases, can be much larger than the basic knowledge of how to do the task when everything goes well.

What is the role of generalized failure recovery knowledge?

When planning for tasks in the future, especially in the context of teaching or delegation, it would be very useful to have estimates of the *duration* and *difficulty* of the tasks. For hierarchical tasks, one might have independent estimates for abstract tasks or derive them by combining estimates for the primitives.

---

[4] Unlike preconditions, which are needed to control execution (as discussed in the section on Conditionals), postconditions are not strictly needed to execute a procedure if everything goes well. This is why postconditions are being categorized as metaknowledge here.

What is a natural metric with which to estimate task difficulty?

All of the types of metaknowledge discussed above are subject to *uncertainty*. As a simple example, estimated task duration may be plus or minus 10%. More significantly, even the success of a task may be uncertain, for instance, due to sensing difficulties. To return to the Pickup example, what if the robot's hand does not incorporate a sensor to detect whether it is holding something? Here, in order to proceed intelligently with the task, it would be helpful to know the *a priori* probability of Pickup succeeding. What if the sensor is unreliable? It would then be helpful to know the probability it actually succeeded, based on the sensed information.

Is probability the best approach to expressing uncertainty for interactive task learning?

Knowing how to do something is not the same knowing how to teach it. A good teacher has extra *pedagogical* knowledge about a task, such as appropriate fading and scaffolding strategies, the mistakes students typically make, and so on (for more on pedagogy in the context of ITL, see Shah et al. and VanLehn, this volume). Like error recovery knowledge, this pedagogical knowledge can be larger than the basic knowledge of how to do the task.

How do teachers learn pedagogical knowledge?

Conclusion

In this chapter, I have reviewed the ontological aspects of task knowledge and have excluded other important issues in ITL, most notably the use of natural language. I have delineated the levels and types of knowledge needed for ITL and provided examples of each using a tire rotation example.

In hierarchical tasks, although the recipes and abstract tasks (as well as the primitives) provide an important communication "vocabulary," the reference here is to concepts, not the specific words or phrases being used. We would not expect, for instance, anyone to spontaneously say "UnhangHubs" as the second top-level step of tire rotation. Understanding what people actually say is a major research challenge. Similar natural language challenges apply for all the types of knowledge discussed here.

Finally, I have, as much as possible, skirted the issue of knowledge representation in the discussion above. You have to know what the knowledge is before you think about how to best represent it for particular computational purposes.

6

The Representation of Task Knowledge at Multiple Levels of Abstraction

Niels A. Taatgen

Abstract

Interactive task learning requires knowledge and skills that are highly flexible and composable, and a cognitive architecture to support this. Cognitive architectures aim to bridge the gap between the brain and intelligence, providing a formal level of description for rigorous theories of behavior. Architectures typically operate on a single level of abstraction, but this may be too limited for interactive task learning. Instead, architectures with multiple levels of abstraction should be considered, each with their own formalisms and learning mechanisms. Each level should be able to explain the abstraction level above it, thus creating a reductionist hierarchy of theories to model human intelligence, not with a single formalism, but with several.

Introduction

Suppose a teacher were to give the following instruction: "Using this box of paperclips, create an outline of a rabbit." This instruction would not, in general, pose an immense challenge to the learner, even though the task itself may be novel, because the concepts involved are understood by the learner.

Now consider a second example of instruction: "Please use these paperclips to woffle a dingdong." Here, the directive would be far more problematic, because the concepts "to woffle" and "dingdong" are unknown. The instruction would need, therefore, to include a verbal definition of the concepts (e.g., "to woffle" means to create an outline). Words alone, however, may not suffice: the learner may require a demonstration. In addition, if an unknown concept (e.g., woffling) entails complex actions, several demonstrations and learning attempts may be required before the learner is able to replicate it successfully.

These two examples demonstrate different types of learning: In the first, the learner is able to draw on existing skills. In the second, a particular sequence and combination of mental operations is required, but if these mental operations cannot be expressed in words, other instructional methods may also be needed.

Interactive task learning (ITL) can also require a lower level of abstraction, as in the following: "Please use the paperclips to 信息 the 面写." Assuming that the learner has no experience with Chinese characters, s/he would first need to learn the meaning of the words involved as well as be able to recognize the particular characters. In ITL, the goal of the teacher is to supply the learner with the knowledge needed to perform the task. As illustrated in this example, the knowledge that the learner lacks may stem from different levels and require a different type of teaching: a verbal explanation consisting of a few sentences, an extensive demonstration, or a large set of training materials.

In this chapter, I investigate the hierarchy of abstraction of task knowledge through the lens of cognitive architectures. The goal of cognitive architectures is to provide a platform to model all aspects of human intelligence, including human ITL. Although architectures have been very successful in modeling wide ranges of phenomena, no current architecture is close to having the capability of learning arbitrary new tasks from instructions. A possible reason is that most architectures limit themselves to one or a few levels of abstraction. As a consequence, many architectures are very good in explaining certain phenomena, but may struggle to explain others (Newell 1990).

Take a prototypical cognitive architecture: ACT-R (Anderson 2007). The majority of research into ACT-R involves modeling human data from behavioral experiments. Models created with ACT-R, therefore, usually address phenomena that concern a single constrained task that can be completed within a minute, is repeated (possibly with variations) over a modest span of time (e.g., 1–2 hours), and requires as little background knowledge as possible. ACT-R assumes that our brains are capable of representing symbolic knowledge in the form of declarative chunks and production rules, without being overly concerned (but not indifferent) to how our brain implements these representations (see, e.g., Stocco et al. 2010). More abstractly, ACT-R is less suitable in covering the long-term aspects of knowledge and learning and how knowledge of different tasks is interconnected. ACT-R does not, however, provide representations that directly support compositionality; that is, the human ability to understand and carry out novel tasks quickly as long as the components of that task have already been mastered.

Consider the two tasks in Figure 6.1: although both may be completely novel to most readers, each of us should be able to perform them effortlessly, because we can easily combine skills such as counting, selecting items with a particular attribute, and determining whether something is more than something else. Moreover, after completing these tasks, we have not acquired a new "count the spoons on placemats" or "more red fish" skill. In terms of ITL, as adults, all we needed was the instruction written above the picture. Young

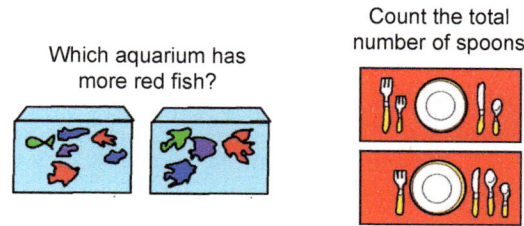

Figure 6.1   Two novel tasks that require no training for a typical adult.

children, by contrast, could have difficulty solving these tasks, if they have not mastered certain skills (e.g., counting, reading). Thus, for children to solve these tasks, additional instruction of a different type (e.g., demonstration) may be required.

This is not a critique of ACT-R: it performs excellently at its chosen level of abstraction. However, if we want to make progress on architectural approaches to ITL, we have to consider using architectures with multiple levels of abstraction, each with their own representation and learning mechanisms. Take, for example, symbolic-level architectures, which give a modeler complete freedom in how knowledge can be specified. A model of subitizing (i.e., counting dots on a screen), therefore, would not necessarily strengthen a counting skill that can be reused for other tasks. This is not a problem if our focus is on modeling subitizing phenomena, but it becomes one when counting is viewed as a skill that is itself a unit of representation. Critically, different levels may be strongly linked together; that is, a higher level of abstraction can be produced at a lower level. Let us, then, explore the interconnection of levels through composition, where a single unit of abstraction at a certain level can be decomposed into several units at a lower level.

The idea of multiple levels of abstraction is, of course, not new. In 1990, Newell described 12 levels of abstraction, subdivided into four bands (neural, cognitive, rational, and social) that operate at different timescales: from microseconds (neural) to extended periods of time (social). Newell, however, encouraged researchers to "carve nature at its joints"; that is, to settle on a level of abstraction sufficient to model intelligent behavior, below which processes are integral only to implementation. Given our current understanding of neural-level representations, we should not ignore any level of abstraction but rather look for representation, learning, and transfer at different levels of abstraction.

There are, of course, many ways to define a multilevel architecture. Here I wish to initiate the discussion with a proposal that builds on several existing architectures and modeling approaches. Since I will be using ideas from many different sources, I do not wish to "brand" the idea and will refer to it as a "multilevel architecture" (MLA), considered here in the context of ITL.

A Multilevel Architecture for ITL

Global Workspace Assumption

The starting assumption for my proposal is that the human brain is to some extent modular and can therefore be subdivided into distinct areas (e.g., perception, motor areas), different types of memory, areas related to control, and possibly (but not necessarily) specialized cognitive functions (e.g., language and numerical cognition). This assumption is visible across several (although not all) levels of abstraction, thus making it necessary to specify how information is made available to different modules, and to have a way of controlling the flow of that information.

A useful starting point is the global neuronal workspace (Baars 1997; Dehaene et al. 1998), a common area in which information is exchanged between modules. But how do modules "know" when to use information from one another? For this, (procedural) knowledge decides, based on the current contents of the workspace, which information has to be moved to a particular area within the workspace. Figure 6.2 illustrates the general concept of the workspace: different modules occupy their own subspace and procedural

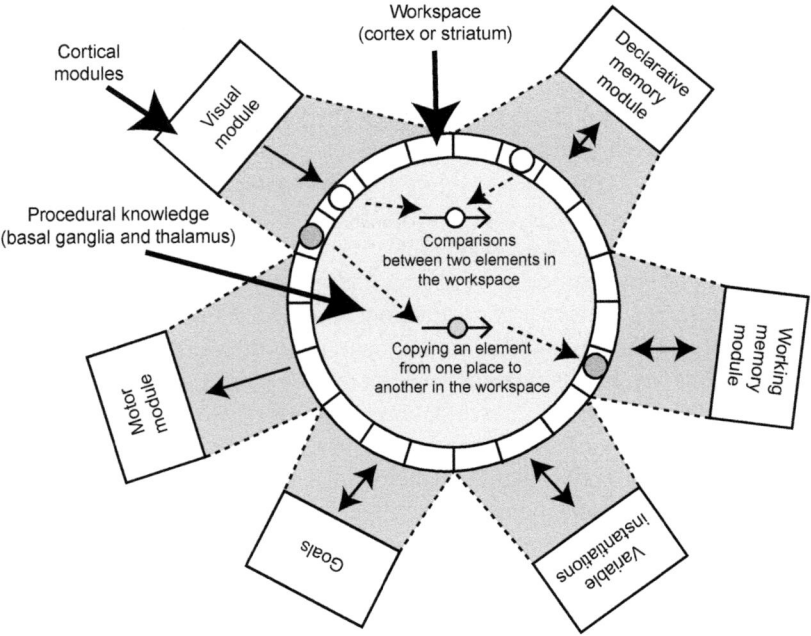

Figure 6.2 The global workspace model (PRIMs version), reprinted with permission from Taatgen (2013).

knowledge (in the middle) transfers information between modules. The subspace for each module consists of a number of "slots" which can be used to represent separate information items.

The concept of a global workspace has been applied differently in cognitive science and artificial intelligence, for example, in blackboard architectures (Hayes-Roth 1985) and in buffers in ACT-R (Anderson 2007). I use it here as the central organizing element in the MLA.

Five Levels of Abstraction

As illustrated in Figure 6.3, the MLA consists of five levels of abstraction. Building up from the bottom, these levels are neuronal clusters, primitive operations, operators, goals, and tasks leading to higher-level representations (e.g., language). The general idea is that the units at a particular level serve as the building blocks for a level higher. Between levels, learning mechanisms

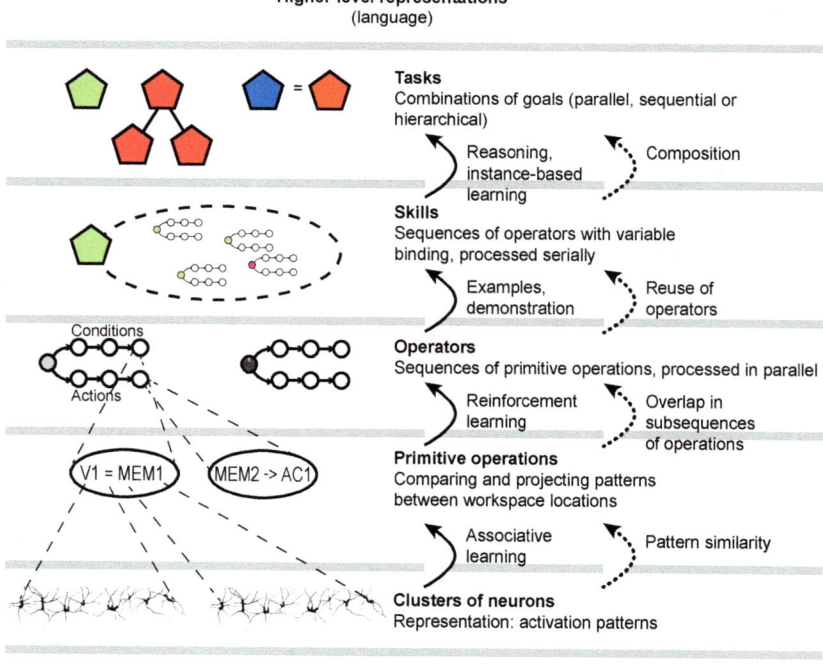

Figure 6.3   Proposal for an MLA with five levels of abstraction. Each level provides the building blocks for one level higher and learning mechanisms operate between levels. The figure indicates some tentative learning mechanisms on the arrows. Also indicated (dotted arrows) is how knowledge can be reused to produce transfer of learning. The activation patterns of the neuron clusters at the bottom of the figure represent the contents of global workspace buffers: V (visual), MEM (memory), and AC (action).

operate to build new knowledge representations. Figure 6.3 also illustrates how knowledge can be used for different purposes, providing knowledge transfer between tasks.

Below, I describe how the different levels represent, use, and learn knowledge using two examples: an aural-vocal task and the simple tasks depicted in Figure 6.1. In the aural-vocal task, the model is presented a tone of low, medium, or high pitch and must respond by saying "one," "two," or "three," respectively. I use this example because it requires few procedural steps and has been successfully modeled in a neural architecture (Stocco et al. 2010).

Clusters of Neurons

A generally held assumption in neural networks is that meaning is represented by activation patterns in clusters of neurons (e.g., Stocco et al. 2010; Eliasmith et al. 2012). At this level, each slot in the global workspace consists of a cluster of neurons, and different meanings can be attached to different firing patterns. In the aural-vocal task, aural processing of the tone produces a particular firing pattern in the aural module. For this pattern to take on meaning, it must be able to activate certain memories linked to that tone. For this to happen, the tone has to be used as a cue for memory recall. For this, the system has to be able to consider the firing pattern as a single unit that can be used at a higher level of abstraction. Yet the neural representation of, say, a low-pitched tone in memory may differ from that produced by the aural module. Thus, the aural pattern of a low-pitched tone has to be transformed into a memory pattern of a low-pitched tone, which is something the network has to learn through some form of associative learning, or by using an algorithm (Eliasmith et al. 2012). A single symbol at the higher level may therefore correspond to many patterns at the neural level. Learning at this level is typically very slow, requiring many practice trials. If knowledge at this level is missing in an ITL setting, it has to be painstakingly trained before higher levels can integrate and use it (e.g., processing Chinese characters).

The model by Stocco et al. (2010) assumes three modules: a vocal module, a memory module (prefrontal), and an aural module. Coordination between these modules is performed by a loop through the basal ganglia and the thalamus. Figure 6.4 illustrates how this model performs on the aural-vocal task. The gray boxes represent three sub-areas (vocal, prefrontal, aural) in the global workspace, each with 100 simulated neurons. In the top box (Stage A), the aural module has perceived the tone and produces an activation pattern that represents that tone. This representation has to be transformed into a memory query to determine which word corresponds to the pitch of the tone. The basal ganglia maps the input from the aural input onto the prefrontal area that is linked to memory, producing the activation pattern in Stage B. The prefrontal activation pattern corresponds to a memory query that asks: "To which vocal output does this tone pitch correspond?" In Stage C, memory has produced

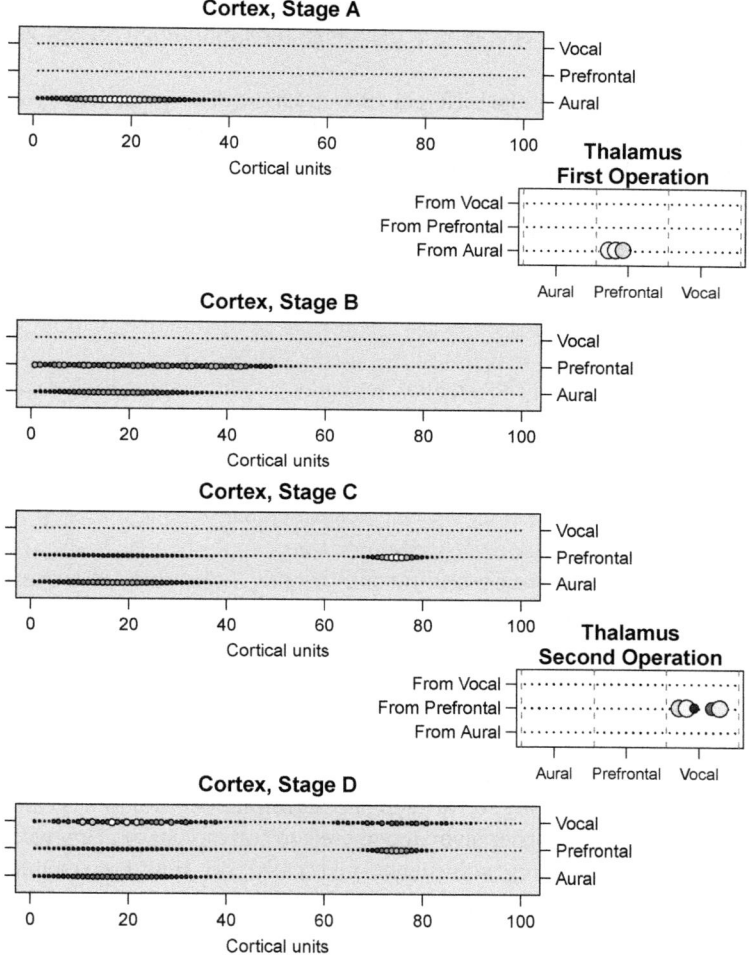

Figure 6.4   Activation in Stocco et al.'s model of the aural-vocal task (Stocco et al. 2010), reprinted with permission.

the relevant memory trace, and a new pattern is created that represents the word that must be spoken. It is now the basal ganglia's turn to transform that memory pattern into an actual speech representation that can be put into the vocal part of the global workspace (Stage D).

Primitive Operations

At this level, neural firings are replaced by symbols that represent specific firing patterns. Instead of specifying the particular pattern of neurons that represents a low pitch, we will just specify "low pitch" in a slot of the aural part of the workspace. By doing so, we discard information (e.g., the ability to determine whether a low pitch is more similar to a medium pitch than it is to a high pitch). This information is still available, assuming that we simulate the lower levels.

In addition, the same representation for the low pitch in the aural part of the workspace is used in the memory part, even though both parts were represented by different patterns in the neural model, and these patterns had to be learned through associative learning.

Activity in the basal ganglia can now be simplified into primitive operations (Taatgen 2013): First, given an input in the first slot of the aural (AU) workspace, copy that representation to the first slot of the memory (MEM) system. Second, given a representation in the second slot of the memory system, copy that representation to the first slot in the vocal (VOC) system:

$$\text{AU1} \rightarrow \text{MEM1}$$

$$\text{MEM2} \rightarrow \text{VOC1}$$

In addition, we need to specify that memory contains three items: (low-pitch, one), (medium-pitch, two), and (high-pitch, three).

For a very primitive single-task model, this may be enough. However, actions are always tied to a particular context and may also be augmented by particular conditions. Thus, an aural-vocal task is performed based on skills (to respond to an aural stimulus with a prememorized response), supplied with the particulars (tones with certain pitches, certain words) and guided by the goal to actually perform this task. To do all this, more levels of organization are needed as well as an additional primitive operation; namely, the ability to compare elements in the global workspace. For example, in the aural-vocal task, we only want to respond to specific tones—not to just any sound. Suppose that the aural system places a type of sound in AU1 and, if it is a tone, the pitch of tone in AU2. In that case, we want to make a comparison first, using a comparison primitive operator:

$$\text{AU1} = \text{tone}$$

The problem with this comparison is that it is not a primitive operation, because it contains a particular value ("tone"). At this level of abstraction, we do not want particular values because we do not want to define primitive operators for every possible value. The solution is to create an additional subspace in the workspace ("C") to represent currently relevant particular values. If we do that, the comparison becomes:

$$\text{AU1} = \text{C1}$$

The nice property of primitive operations is that, given a particular size of the workspace, their number is fixed. This fixed-sized set provides the building blocks for the next level: operators.

Learning at the level of primitive operations involves finding the right combinations to perform something meaningful, probably through trial-and-error combined with reinforcement learning. Such a combination provides the unit for the next level of abstraction: operators.

Operators

Operators are perhaps the most common form of representation found in cognitive architectures. Soar (Laird 2012) uses operators (although they work differently), but in ACT-R and most other architectures, they correspond to production rules. Operators combine several primitive operations, all of which are typically carried out in parallel (or in quick automated succession). Initially, operators carry out a number of primitive operations that test conditions; if these are satisfied, they carry out operations that perform actions.

For the aural-vocal task, we need two operators that assume that C1 contains "Tone," C2 contains "Associate," and C3 contains "Say." The tone is represented in AU1 ("Tone") and AU2 (e.g., "Low"). Memory contains chunks like ["Associate," "Low," "One"], which will end up in MEM1, MEM2, and MEM3, respectively. The vocal module can take commands such as ["say," "one"], represented in VOC1 and VOC2.

Operator 1:	Operator 2:
AU1 = C1	MEM1 = C2
C2 → MEM1	C3 → VOC1
AU2 → MEM2	MEM3 → VOC2

Although operators are like production rules, they are not directly linked to a particular task or skill. This makes it possible to reuse operators in many different contexts. To promote reuse of operators, all the particulars are represented in the "C" part of the workspace. Another property of operators, contrary to productions, is that they do not bind variables; instead, they assume relevant variables are supplied in the "C" subspace. Thus, the variable binding problem does not need to be solved at this level of abstraction. The next level of abstraction, the skill, is responsible for supplying values to that space.

Skills

To perform all but the most elementary tasks, several processing steps have to be taken. In other words, several operators are needed that are carried out in sequence. Given that operators only transfer information in the global workspace, operator activity is interleaved with module activity (as in the neural example in Figure 6.4). To get from operators to tasks, we need an intermediate level—the skill level—because even moderately complicated tasks need many operators, so connecting operators straight to tasks is too large a jump.

Skills organize which operators are needed and are responsible for instantiating particular values (i.e., binding is carried out at the level of skills). In the aural-vocal example, we specify that the skill is associated with two operators, and that it can be instantiated with an aural type ("Tone" in the example), an index to the type of associate fact that needs to be retrieved ("Associate"), and the action that needs to be performed ("Say"). However, we can also change these bindings to modify the skill's behavior, providing a means to use the same skill for different purposes.

For the elementary aural-vocal task, one skill is enough. However, to perform the simple tasks depicted in Figure 6.1, multiple skills are needed in combination to explain why we are able to solve these tasks without prior learning.

Skill-level learning is important in ITL if the learner has not yet mastered all skills required to perform a task. Referring to the paperclip example given in the introduction, a learner might respond: "But I don't know how to make an outline." To remedy this, the teacher could demonstrate how to make an outline of a house using a pencil or verbally offer detailed step-by-step instructions. An effective, but not unique way to teach a skill is to show by example.

Tasks

At the uppermost level of MLA, multiple skills are linked, often in unique combinations, to perform tasks successfully. The underlying idea derives from the concept of compositionality in language, where words and grammar allow us to produce and understand sentences that we have never before heard. This level offers the same versatility: skills and instantiations of skills are used to accomplish tasks never before encountered.

Within a task, skills can be organized in several ways: they can follow each other sequentially, be active in parallel, or organized hierarchically. Consider the simple task in Figure 6.1: "Which aquarium has more red fish?" To respond, we need a *compare* skill, a *count* skill, and a *has-property* skill, organized hierarchically from top to bottom, each instantiated with the appropriate bindings.

To illustrate the compositionality property, consider the other task in Figure 6.1: "Count the number of spoons." This task can be carried out with some of the same skills: the top skill is now an *add-all* skill, but the sub-skills are the same as in the aquarium task, albeit with different bindings.

Figure 6.5 shows the two tasks across several levels of abstraction. Not only do the two tasks share two out of three skills, the two skills which they do not share still overlap in terms of operators as well as in the sequences of primitive operations.

In an ideal ITL situation, the instruction is enough to mobilize the appropriate skills and instantiate those skills to create a task structure. Accordingly,

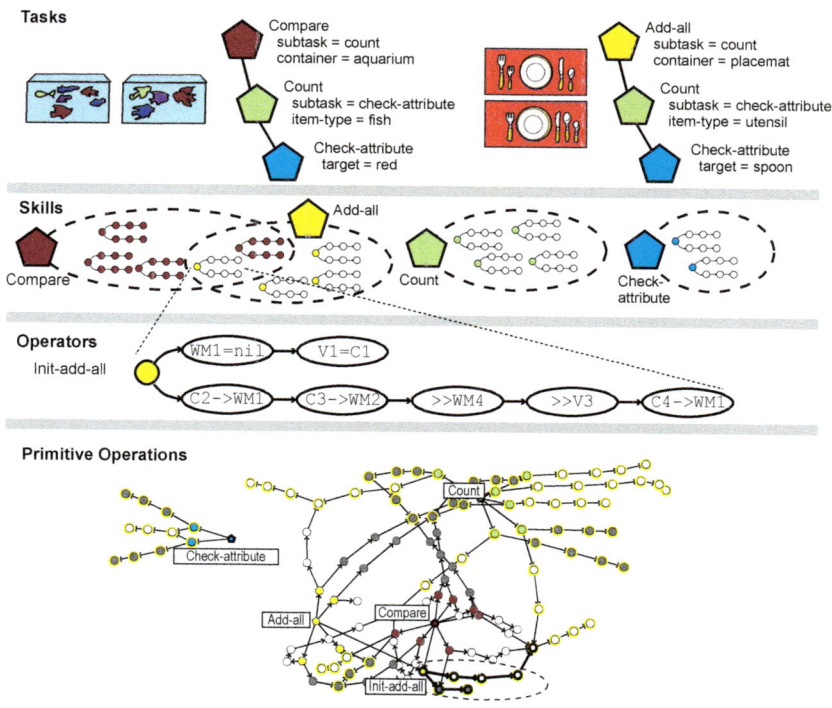

Figure 6.5   Illustration of the representation of the two tasks from Figure 6.1 across levels of abstraction. At the task level each task has a hierarchy of three skills with particular bindings. At the skill level, each skill is associated with a set of operators, which can potentially have overlap. At the operator level, several primitive operations are bound together, but particular substrings of primitive operations can be reused (note that, for brevity, we have not explained the >> primitive operation). The bottom section, "Primitive Operations," is output from the PRIMs application (https://github.com/ntaatgen/PRIMs-Tutorial) in which both tasks are modeled. It shows all the operators with their strings of primitive operations, and where these overlap. For example, the Init-add-all operator shown at the Operators level has two conditions and five actions. This operator is also shown at the Primitive Operations level, where it is outlined with bold lines. Some of the primitive operations at that level are shared with other operators; for example, the two conditions are also used by one of the compare operators.

learning a new task is straightforward, similar to the situation in which a subject in an experiment receives and carries out instructions. A misunderstanding of instructions may be due to a problem of connecting the language of the instruction to the appropriate skill, or because the subject lacks a skill. In that case, the learning process has to drill down to the appropriate level of abstraction in order to identify and fill in the gaps, and the mode of instruction has to be adjusted appropriately.

Multilevel Processes for ITL

An interesting property of the MLA is that at the bottom levels, the system
is quite mechanical and syntactic in nature (i.e., it lacks "meaning" of what
it is doing) whereas the highest levels closely resemble semantics in natural
language. The learning mechanisms linked to the different levels should reflect
this. So why don't we just model the highest levels of abstraction and consider
the lower levels as implementation details? As described above, learning and
transfer happen at all levels of abstraction, and thus leaving out levels results
in an incomplete system.

Learning

In machine learning, three types of learning are typically distinguished: unsupervised learning, reinforcement learning (i.e., supervision by reward only),
and supervised learning (correct answer is given as feedback). I would like
to add a fourth type to this: learning by instruction and explicit reasoning.
Learning involves two aspects: how to assemble elements from one level to
create a unit one level higher in the MLA, and how to evaluate whether existing elements of knowledge have to be retained or discarded.

At the lowest level of the MLA, unsupervised learning is the most likely
candidate for learning, because processing at that level is too fine grained to
distinguish correct from incorrect. Instead, co-occurring patterns are associated with each other, creating links which the next level above it (primitive
operations) needs to function. Moving up in levels, reinforcement learning
combined with genetic algorithms provides a possible candidate to construct
operators out of primitive operations. At the highest level of abstraction, tasks
can be assembled from instructions (by parsing natural language into task
representations), from examples (by deriving what operators are needed to
achieve a certain thinking step, or by imitation), past experiences, or through
reflection (by mentally simulating what the outcome of a certain skill will be).
For high-level learning, there is much to learn from Soar (Laird et al. 1986),
a cognitive architecture that has focused on higher-level reasoning processes.

Transfer

One of the conclusions we can draw from the concept of an MLA is that tasks
are not learned in isolation; knowledge is interconnected at all levels of abstraction. Thus, at every level, there are opportunities for knowledge transfer.

At the neuronal level, similarity between activation patterns signals that
these patterns are processed in similar or even identical ways. This similarity
originates in perceptual and motor systems, which are connected to the real
world. It can, however, also result from associative learning processes that
preserve some of the pattern similarities in higher-level representations.

Primitive operations often end up in the same clusters. Through compilation processes, these clusters can be executed more efficiently and may therefore be preferred in the case of choice. This type of low-level transfer can be quite powerful, as has been demonstrated (Taatgen 2013); it is capable of modeling transfer among different text editors as well as in different tasks that require cognitive control. Similarly, skills can reuse operators that have been learned for a different purpose, thus sidestepping a lengthy bottom-up learning process.

The largest potential for transfer is in the reuse of skills. By instantiating and connecting skills that it has already acquired, the system is able to address novel tasks.

Conclusion

Because learning and transfer happen at all levels of the proposed MLA, it is important to remember that all levels of abstraction are necessary. There is no natural intermediate level under which one can say that everything is simply implementation detail. The MLA framework has the potential to unify many approaches to cognitive modeling, whether neural, symbolic, hybrid, or otherwise. By creating a strong reductionist framework, the pitfall of exploding complexity can be avoided. Some of these ideas have already been implemented in the PRIMs architecture (Taatgen 2013), but many open questions remain to be explored with respect to representation, learning, and transfer.

Interaction

7

Interaction for Task Instruction and Learning

Andrea L. Thomaz, Elena Lieven, Maya Cakmak,
Joyce Y. Chai, Simon Garrod, Wayne D. Gray,
Stephen C. Levinson, Ana Paiva, and Nele Russwinkel

Abstract

This chapter considers the qualities of human interaction and learning that will be most effective and natural to incorporate into any interactive task learning agent, and focuses specifically on the interactions involved in learning from explicit instruction. At the center of this interaction is a process that brings the common ground between a teacher agent and a learner agent into alignment. Errors or misalignments to this common ground drive the interactive learning process. The importance of timing is highlighted as is the dynamics of an interaction, as a communication channel itself, in this alignment process.

Introduction

What are the most effective and natural methods for humans, robots, and AI agents to interact in support of instruction and learning? To address this central question, we begin by establishing a context to frame the scope of our discussion, defining the landscape of tasks and learning interactions that we are considering for interactive task learning (ITL). We then introduce a model for ITL and discuss both the support for this model in natural human interaction and its implications for learning agents.

Group photos (top left to bottom right) Elena Lieven, Andrea Thomaz, Stephen Levinson, Joyce Chai, Simon Garrod, Ana Paiva, Nele Russwinkel, Stephen Levinson, Wayne Gray, Andrea Thomaz, Joyce Chai, Nele Russwinkel, Wayne Gray, Maya Cakmak, Elena Lieven, Simon Garrod, Ana Paiva, Maya Cakmak, Joyce Chai, Andrea Thomaz, Elena Lieven

A question of contention raised throughout is: For ITL to be successful, is full human-level capability required? Our goal is to lay out the key features of *human–human* interaction and discuss ways in which principles of these interactions should be replicated in *human–agent* interactions. Importantly, these principles of interaction are implemented in a variety of ways with a variety of communication modes in human interaction. Thus, we expect that artificial agents may use any of a range of modes of communication to implement naturalistic communication principles.

Types of Tasks

In thinking about interactive tasks, it is useful to consider a "task space" that expresses gradients of difficulty on different scales. For instance, tasks can vary from a "simple" interaction with a physical domain (hammering in a nail), through an antagonistic interaction with an animate agent (zero-sum games), to a cooperative interaction where complex issues of joint control, synchronization, perspicuity of contributed actions, and plan reconstruction become especially crucial. Another dimension might be a hierarchy of modes of transmission of task knowledge between agents: from reverse engineering of a product, to emulation of an observed agent, through minimal (e.g., gestural) instruction, to full verbal and multimodal demonstration and instruction. A third dimension might be computational complexity, which is partly dependent on the other two hierarchies; for instance, cooperative interaction tends to be more complex than interaction with a physical domain because it also involves modeling another agent. Of course, physical tasks vary in their own complexity, too: picking up an egg and separating the yolk from the white, for instance, will require extreme delicacy of physical manipulation compared to picking up a ball and throwing it away. This three-dimensional space provides a way to think about the landscape of interactive tasks.

Consider the following different tasks, representing some of the variety encapsulated in this task landscape. The first is an example of a coordinative joint task: a healthcare robot that should help an elderly person who is not able to use his or her legs and needs help transferring in or out of a wheelchair. The actions to be learned (or the interaction with the person) should be flexible; for example, the action "lift the person up" would be quite different depending on whether the person is to be lifted up from a chair, from a bed, or from the floor. Here we need learning as a first step to train the robot to do a safe action and also to adjust the action according to the feedback of the person (e.g., "slower," "careful"). This interaction could take place via language (assuming the robot has some knowledge about what these adverbs mean) or by demonstration. The robot must not necessarily be able to talk, but it requires feedback mechanisms to give the person being cared for a secure feeling that the task will be carried out as expected.

The second is an example of an agent system that could be an assistant system in a car that has to learn when to offer help or information to the driver. The system should also learn or anticipate when the driver is engaged in a task and should not be disturbed. The system should offer information when it is needed, at the appropriate time (e.g., "construction work on the road," "traffic jam on this route"). Here the agent needs to learn to model the driver and the tasks in which the driver is engaged, as well as the individual preferences of the driver. This system may need language processing to interpret the driver's language input and generate language feedback to the driver. This is not a cooperative task in the sense of equal partners; the agent system is providing support and should learn about the needs and goals of the driver through experience and feedback.

Although the agent systems from both examples share some aspects, teaching these two systems would be very different.

Another consideration is that tasks to be learned are often compositional and can be represented by, for example, and-or graphs (Liu et al. 2016) or hierarchical task networks (Mohan and Laird 2014; Mohseni-Kabir et al. 2015). An overall task can be broken down into subtasks (which can possibly be further reduced into other subtasks) with temporal and spatial constraints. A subtask can also be decomposed or implemented by primitive actions. Thus, learning a task will involve learning how to perform actions at different levels of abstraction, and this may require different forms of teaching. For example, a primitive action can perhaps be best taught through physical guidance, whereas a high-level task with partial order of subtasks may benefit most from language instructions (Chai et al., this volume).

Types of Interactive Learning

Let us now consider the types of learning found in humans and more precisely define which are most relevant to the ITL problem. Forms of interactive teaching and learning differ in two ways: (a) whether or not the modeler/teacher has an intention to teach and (b) how it is that the learner learns.

The action of a modeler/teacher may not consciously intend to teach but can still afford learning in the learner. One example in natural learning is chimpanzees learning to crack nuts: chimps crack nuts and young chimps stay close to their mothers, who tolerate "nut stealing," as follows:

Stage 1: Initially, the young chimp makes no efforts to crack the nuts but only to eat them once the mother has cracked them. This keeps the chimps near the mother and may lead to observational learning: the young chimps learn that *nuts can be cracked.* This form of learning is often called "stimulus enhancement" (Tennie et al. 2009).

Stage 2: The young chimp starts to bring nuts to the mother, demonstrating an understanding of the "goal" of the task (Boesch 2003).

Stage 3: At a later stage the young chimp starts to try to crack nuts. Note that they do not coordinate the type of hammer, the type of anvil, or the nut type, and it takes many years of "trial-and-error learning" to achieve success. However, many conclude that "emulation" may be involved since the young chimp may have learned something about the kinds of actions needed as well as connecting this to the goal of these actions—the cracked nut (Tennie et al. 2009). They may, for instance, use the hammer/anvil that the mother has left. In this example, the issue is whether chimpanzee mothers are deliberately teaching. Boesch (2003) claims they are, whereas other researchers (e.g., Tennie et al. 2009) contest this interpretation. In terms of how learning from another agent takes place, Tomasello (1990) and others make the following distinctions:[1]

- Mimicry: copy the action; the goal is the action in itself.
- Emulation: copy the result of the action using other actions.
- Rational imitation: copy the modeler with an understanding of the intention behind the actions. For example, in Gergely et al. (2002), children are asked to imitate turning on a light when the modeler turns it on with her head, either (a) when she cannot use her arms because they are covered or (b) when her arms are free. Children turn the light on with their hands in the arms-covered condition ("she can't use her arms but I can"). They also turn on the light with their head when the modeler's arms are free ("she could use her arms but doesn't, so I should probably do the same").

Finally, natural pedagogy (Csibra and Gergely 2009) suggests a human-specific type of social learning through communication that speeds up learning (and avoids trial-and-error learning and statistical observational learning). This is a form of imitation in which the modeler presents a demonstration accompanied by cues that focus attention on specific elements, thus signaling that the cues are "intended" for the learner: slowing down, pointing, eye gaze, exaggerated actions.

Of course, not all learning takes place through interactions between agents, and each type of learning outlined below will have implications for how the learning is structured:

- *Entrenchment:* simply being immersed in the environment, the learner is exposed to experience from which to learn. For instance, children learn language through being entrenched in positive examples, not through explicit instruction.
- *Self-exploration:* the learner is on his/her own in the environment, learning through self-discovery and interactions with the world.

---

[1] Note: the precise definition and use of these terms is debated in the literature.

- *Structured discovery:* something more like the self-learning that happens in a preschool, whereby the environment has been arranged to support particular kinds of self-learning.
- *Apprenticeship:* learning happens through emulation and imitation, by the learner doing the tasks that are modeled by an expert, but the expert does not necessarily have explicit instruction interactions with the learner.
- *Explicit instruction:* the teacher and learner enter into an explicit communication about the learner, with the joint goal of the learner obtaining some new task model. We are limiting ourselves to dyadic interactions, but this type of interaction can also happen in groups.

The learning interactions we consider most relevant for ITL are those with *explicit instruction.* When a human partner is interested in transferring task knowledge to an artificial agent, we expect this to be the most common form for that interaction to take. There may be contexts in which apprenticeship is appropriate, where a person wants to have their artificial agent learn by example without explicit teaching. For the purpose of this chapter, however, we focus on the scenario where explicit teaching is taking place.

Making the Task Learning Problem Interactive

Baseline Model: Two Agents plus World

Consider the following first approximation of ITL, shown in Figure 7.1 (see also Figure 2.1, Mitchell et al., this volume). For two agents, the learner (A_L) and the teacher (A_T), learning by A_L can be considered as the improvement of the performance metric (P) in the completion of a task through experience. Natural interaction between A_L and A_T shapes the experience, thus increasing the potential for interactive learning by A_L to occur. Note that both A_L and A_T are agents that have the ability to act in the world, W, and can thus change it

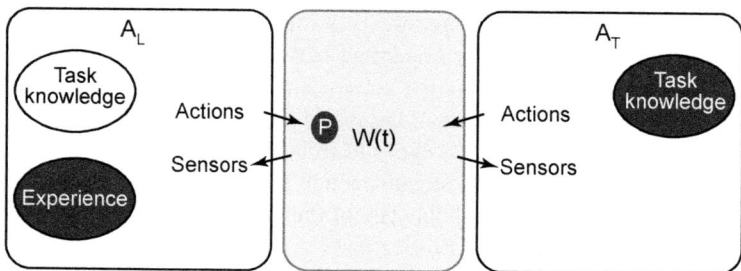

Figure 7.1 Baseline model for learning between two agents. A_L (the learner) and A_T (the teacher) interact in a shared world over time, W(t). Learning is considered to occur when A_L improves its performance (P) of a task based on experience.

as well as each other. The learner perceives the world (and the communicative actions of A_T) as changes which occur in the world over time, $W(t)$. A_T has the possibility to change the experience observed by A_L in a way that makes task learning more effective and efficient. Shaping the experience through the A_T's actions in this ITL framework aims to permit A_L to improve task performance. For that, communicative actions also function to keep the communication channel open and make the learner more engaged in the task learning.

Update to the Baseline Model

Many human joint activities involve what Bratman (1992) calls *shared cooperative activity* (SCA). He argues that SCAs depend on both agents recognizing that they are part of the SCA, and that they are committed to the goals of the SCA as well as to carrying out their part in the SCA. SCAs tend to have a clear lifetime with a beginning, middle, and an end. The cooperating agents may have to negotiate the beginning and the end of the SCA and thereby define the period over which their shared commitments apply.

Further developing the two-agents-in-world model of interactive learning, we propose the following changes to capture this commitment to the joint activity that is required for natural style interactions with a human teacher:

- Enrich the world state by including, for instance, more than just $W(t)$.
- Enrich the representations within each agent.

In relation to the first change, enriching the world state, the knowledge and information that is shared between A_L and A_T should be explicitly represented; we refer to this as *common ground*. In Figure 7.2, we model common ground as a bulletin board that would minimally contain all the task-relevant information. Such a bulletin board would include the results of A_L and A_T's actions as well as any relevant props for their joint actions (e.g., furniture being constructed, bed and chair when lifting grandpa). In turn, we can distinguish between W_{AL}, W_{AT} (corresponding to A_L's and A_T's own representations of the world) and $W_{AL1'}$, $W_{A2'}$ (corresponding to the common ground of A_L and A_T's knowledge of each other's world state). In general, joint activities are likely to be more successful when there is greater alignment between W and W' (Sebanz et al. 2006). Thus, each agent has its own actions and perceptions of the world as well as a shared model of the state of the world that is built up between the two agents. There are several different types of information contained in this common ground representation. The agents need to maintain agreement/alignment on the task-relevant world state, the state of the learner's task knowledge, and the state of the interaction itself. Notice that ITL *requires* common ground, but also that the learning process itself *results* in new common ground through additional task knowledge.

In relation to the second change, enriching the individual agent's models, we need to differentiate actions into communicative and noncommunicative

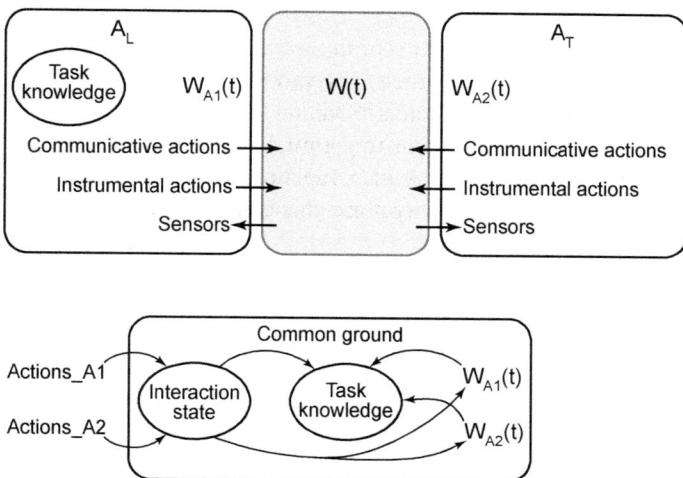

Figure 7.2   Enhanced model with explicit shared knowledge (common ground) between two agents, A_L (the learner) and A_T (the teacher), in a shared world over time, W(t).

actions (and perhaps communicative and noncommunicative sensing as well). More importantly, each agent could have mental representations of the information associated with planning the joint action. For instance, this could include information related to A_L and A_T's joint action planning (e.g., A_L needs to hold the pieces of furniture together for A_T to drill a hole for the screws). This could also include information about how A_L and A_T will implement their joint action (e.g., A_L holding the pieces of furniture at the same time and location as A_T drilling the hole). In general, joint activities are likely to be most effective when these representations are the same or aligned between A_L and A_T (Garrod and Pickering 2004).

Given this new formulation of the ITL agent model (Figure 7.2), we can now consider several different ways that the teacher agent can possibly shape the experience (E) of the learner:

- *Directing attention/referring*: this can be done by (a) manipulating the world to make some features, objects, or processes, which the teacher wants the learner to focus on, more salient; (b) using nonverbal communication (directing eyes, pointing) to the salient elements in the world (and in communication); and (c) by communicating verbally to focus attention. No matter what means of communication is used, referring has the result of bringing the two agents' common ground world state into closer alignment.

- *Manipulating the timing by which some actions occur* (communicative and environment changing actions): if a backchannel is communicated with some delay, for example, different meanings can be considered. These backchannels can provide signals like "keep going," "continue," "give me more information," or "stop going on about this." This represents the dynamic nature of the interaction state, whereby time can implicitly move the interaction state forward. Nonaction must be considered an intentional action itself.

- *Providing a task description through verbal communication using natural language or some description language* (which at some extreme can be a programming language): this would have the goal of changing the state of task knowledge represented in common ground.
- *Demonstrating a task to the learner*: this requires synchronization between perceptions/actions/responses of the two agents and should involve alignment along all parts of common ground. The interaction state should have agreement on whose turn it is to act, the agents' world states should have agreement on what aspects of the perceptual space are relevant to the demonstration, and the demonstration itself should make a modification to the shared understanding of the task knowledge state.

We argue that this update to the learning agent model is what makes the learning process interactive, particularly when the interaction is with human partners. In the remainder of this chapter we describe ways in which human interaction is driven by the need to maintain these different aspects of common ground between agents. We first lay out more specifically what common ground means for an interaction, followed by how common ground is repaired when errors or misalignments arise. Thereafter we discuss how many important elements and characteristics of human interaction stem from the rather strict timing dynamics involved in maintaining common ground over an interaction. Finally, we close with a discussion of the multiple modalities humans use for such learning interactions and the extent to which agents need to have humanlike interaction abilities.

Common Ground

The "classical" version of common ground comes out of philosophy as a way of handling some of the ways in which languages distinguish material that is new to the interaction from older information that is already established (e.g., an oversimplified way of mapping onto the distinction between "a" and "the"). Common ground is normally considered to have different components:

- Basic facts taken for granted (e.g., we are in Germany, it's summer, gravity reigns on Earth, the capital of the U.S.A. is Washington, D.C.)
- The things in front of us that are mutually manifest (e.g., the books on the table in front of us both)
- The activity (or task) in which we are engaged
- How far we have progressed on that task

- Tacit pacts about referring expressions (e.g., once we have referred to object X as "the O-ring" we should continue to use that expression)

Think about this as a blackboard: each time agents introduce entities and propositions, we add them to the list, so common ground is quintessentially incremental.

This classical version has problems. It assumes, for example, that everything interactant A believes, B should believe as well; there is no possibility of a quarrel or disagreement in such a world. Or, more relevantly for robotics, there is no individuation of what A and B know from what a new agent C knows. Thus, in such a learning environment, each agent must have its own "commitment slate": C's slate can be updated by interaction with A or B. For informational and skills learning tasks, the whole point of interaction is this updating, but the updating will happen in all cases: if we are moving a table together, we will serially update the location and orientation of the table.

In human–human interactions there are some categories of things that are assumed to be in common ground, such as naïve physics of the world/objects, referents which by default have space/time recency, actions that are usually rational (efficient with respect to goals), and other forms of common sense for humans. Agent designers should keep these aspects of human common ground in mind when considering what must be included in the innate abilities of an effective interactive agent. It would be great to have agents with these aspects of prior knowledge, but at the very least, agents should be able to communicate their limitations in background knowledge.

Developmental Trajectory of Grounding, Informing, and Referring

Babies start to take part in synchronized, dyadic exchanges early in life; there are arguments as to how early this happens, but definitely by two months of age (Feldman 2007). By around six months, dyadic games are well established (e.g., peekaboo, vocalization exchanges) (Rochat et al. 1999). At around nine months, infants show the beginnings of "intention reading/social cognition"; that is, the understanding that both they and others have communicative intentions (Tomasello 2008). This is signaled by episodes of joint attention in which the child and caregiver are sharing attention to the same object (Cameron-Faulkner et al. 2015) and making this manifest by gaze checking, mutual laughing, and pointing (Callaghan et al. 2011). Pointing has been much studied, and by about 12 months, babies point to share information, to inform, and to correct. Evidence for this comes from changes in the frequency/intensity of points dependent on caregiver response (Liszkowski et al. 2012). Tomasello (2003) stresses that the "gavagai" problem (Quine 1960) of referential ambiguity can only be solved through this system of intention reading, which allows the child to map the form of another's utterance to an inference about their intention in uttering it. Further, he argues that the rapid increase in word learning that occurs in the second year of life is dependent on intention reading.

In the early stages of language development, caregivers are usually the ones to provide common ground. Knowing what a toddler means can be very problematic without shared background knowledge and common ground. For instance, toddlers will use referring expressions like "it" without having previously used an identifying noun. But children start using determiners and pronouns in discourse fairly correctly (e.g., the use of "a" to introduce a referent and "the" for further mention) before the age of three. However, if usage is probed in experimental situations, it becomes clear that the subtleties of use (e.g., sensitivity to whether the interlocutor can or cannot see what is being referred to) take quite some time to develop (Matthews et al. 2006). If children are put into "standard" referential communication tasks where they have to identify a minimally different referent to a partner, their performance is surprisingly poor, both in terms of the identifying descriptions that they give and in terms of asking for clarification (Lloyd et al. 1995).

Grounding and Egocentricity

When a speaker refers to something as "the cup" (as opposed to "the big cup" or "the middle-sized cup"), this is supposed to indicate that there is only one cup in the common ground. However, when a speaker says "the middle-sized cup," this is supposed to indicate that s/he is referring to only one out of many possible cups in the common ground (i.e., the second largest of three cups). In other words, speaker and listener are assumed to take into account both the speaker's and the listener's perspectives on what they are looking at when making such references. However, there is much evidence to suggest that even skilled adult speakers and listeners do not always take common ground fully into account when referring to objects in a scene. Keysar et al. (2000) have carried out various experiments in which they arrange tableaux of objects, such that the speaker and listener have different views of those tableaux, and it is quite apparent to both that they have different views. For example, the speaker may see three cups (a big, middle-sized, and small one) whereas the listener can only see two (a middle-sized and a small cup). When describing the cup that is "middle-sized" from the speaker's perspective, which is actually the largest cup for the listener, the speaker will often refer to it as "the middle cup." In other words, the speaker will often refer from an egocentric perspective, not taking into account what is in common ground and available to both speaker and listener. Conversely, if the arrangement is the other way around—the speaker sees only two cups while the listener sees three—when the listener hears "the big cup," s/he will often look at what is the largest cup from his/her perspective, but not from the speaker's. In other words, the listener also tends to follow an egocentric perspective on what should be in common ground when s/he interprets the speaker's references. The effects are particularly pronounced when the speaker and listener are under time pressure, suggesting that common ground is not always fully taken into account even by skilled adult speakers and listeners.

This process is even more complex for robots and AI agents in that being physically present in the same space does not mean that humans and robots have the same perceptual access to the shared environment. A robot has much different perceptual, motor, and reasoning capabilities than a human. The robot's representation of the shared world is significantly misaligned from the human's. The lack of a joint representation makes grounding between humans and robots extremely challenging, yet essential for the success of an interaction and a baseline from which learning can take place. As shown by Chai et al. (2016), humans and robots will need to make extra effort to bridge the gap and strive for a common ground of shared representations.

In the context of ITL, it is reasonable to assume that humans may be better than robots at detecting and remedying missing common ground. Thus, it is important for the robot to take extra effort to provide sufficient cues to assist the human in detecting and repairing missing common ground in a timely fashion. One potential device is to make the robot's internal representations transparent to the human through, for example, language description or visual display (Alexandrova et al. 2014; Hayes and Shah 2017). Another device is through confirmation, commonly employed in dialogue to help establish common ground. There are two types of confirmation: (a) explicit confirmation, where an agent always explicitly asks for a confirmation about its understanding (e.g., "you are talking about this cup, correct?"), and (b) implicit confirmation, where the agent provides an implicit confirmation of understanding, as through the relevant actions on their next turn (Litman and Pan 2002). There are trade-offs between explicit and implicit confirmations. Explicit confirmations make it easier for humans to correct mistakes and can lead to better task success rate, but they are cumbersome and can result in lengthy interactions. Implicit confirmations are much more natural and quick, but there is risk in the delay of detecting and repairing mistakes. In human–robot referential communication, studies have shown that the robot's belief of the reference (referred to by the human) may often not be the same as the reference intended by the human (Chai et al. 2016). In this case, if the robot replies a generic "got it," this may fool the human into believing that common ground has been established when in fact it has not. This simple acceptance from the robot is more detrimental to common ground than a simple rejection ("I don't get it"). When the robot provides information about its internal representation of the believed reference (e.g., through language descriptions), common ground can be significantly improved. Thus, it is important for the agent to adapt different types of confirmation under different situations.

In addition to handling explicit communication about common ground, agents need to account for the fact that much of the common ground between humans is inferred implicitly rather than explicitly discussed, as in the goal-directed action inference example mentioned previously. When children see a woman hitting the light with her head when her arms are free, thus indicating that the use of the head is intended since her hands could have been used, they interpret this as an instrumental action (hitting the light) *and* a non-instrumental action (not using the most efficient means available), which implicitly communicates that the goal of the task *includes* the action of hitting with the head, not just turning the light on (Gergely et al. 2002).

Bringing the perceptual and action capabilities of robots or AI systems to human levels will certainly help with most of the common ground issues and could enable interactions at the level of human–human interactions. However, realizing this is extremely challenging. Alternatively, if robots had all the capabilities necessary to perform the range of tasks they might need to learn and there was broad acceptance and usage of best practices for ITL implementations across robots and AI agents, humans might get used to the robot's limitations and still interact with them smoothly. This is similar to how humans interact with pets, where the expectations are lower but accurate.

Humanlike transparency mechanisms (e.g., gaze, pointing, head gesturing) can exploit people's ability to interpret mechanisms without any instruction or training. Implementing those mechanisms with precision in robots, however, is difficult. Also, robots performing these actions might bring an additional expectation that the robot can also process such information from the human, which is even more challenging. Instead, visual transparency channels on a robot, such as a screen or projection from the robot onto the environment (which are not afforded in human–human interactions), can afford high bandwidth visual information transfer and might be effective in human–machine interactions.

Repairing Misalignments in Common Ground

Next let us now turn our attention to the concept of "errors" in common ground. In a sense, this is the entire reason for a learning interaction at all: to repair the misalignment in common ground between a teacher and a learner. Errors trigger the requirement for an ITL process. Without errors there would be no need for ITL. Continually throughout an interaction, misalignments in all aspects of common ground are being detected, diagnosed, and repaired.

Detection happens through self-monitoring (broken expectations in perceptions or actions), implicit feedback (back channel), and explicit feedback ("stop, that was wrong"). There are many different sources of error to be detected and diagnosed:

- *Perception errors* occur during perception (e.g., the agent fails to perceive an object in the world).
- *Task execution errors* are caused by faulty execution of an action in the world (e.g., the robot does not reach an object in the world).

- *Representation errors* encompass misconceptions as well as incomplete or incorrect task representations. This is related to the Brown and VanLehn (1980) generative theory of human misconceptions, or bugs, which can be seen as resulting from procedural skills acting on incomplete or incorrect procedures (i.e., tasks).
- *Communication errors* result from a lack of common ground or inadequate mechanisms (e.g., feedback) to sustain the flow of communication with the teacher agent.
- *Anticipation errors* occur in the prediction mechanisms of the agent and lead to expectations (future state of the world) that are not possible to meet, given the available actions and resources.

Once an error is detected, it can be self-diagnosed and repaired or the diagnosis and repair can happen through subsequent interaction. In human conversation, self-corrected repairs can be nearly immediate (within ca. 700 msec, at a rate of about once every 80 sec; Dingemanse et al. 2015). But situations that Norman (1981) refers to include longer time-frame errors, as do many that Reason (1990) discusses.

Human Interaction Is Built to Minimize Errors

When repairing errors or misunderstandings in dialogue, effort is distributed between the two interlocutors. In the following telephone conversation (from Drew 1997), for instance, the listener helps the speaker detect the source of confusion (the word "gorillas") by interrupting with "forty-nine what?"

Hal: .an' Leslie 't was marv'lous (.) D'you know he had (.) forty nine [?]
g'rillas. .hh th-there. (b) (.) br[eeding in ( ) [?]
Lesley: [pf- f- Forty nine what? [?]
Hal: G'rillas. [?]

Dingemanse et al. (2015) argue that the effort of identifying the source of the problem to be repaired and repairing it is nicely distributed between the two interlocutors to minimize the time spent repairing the dialogue.

In thinking about how errors in robot–human interaction could be handled, it is useful to refer to the human–human system and to try to extract general principles that might be useful. Let us take the language user and consider the production or execution system. The person starts out with an intention, recodes this into a semantic specification, which in turns gets recoded into a syntactic specification, which is then fleshed out in an abstract sound system (phonology), which in turn is recoded into articulatory (muscular) instructions. At every representational level, the human system probably does error checking (e.g., legal expression checking, checking the derivation form for each of the prior levels). What we know with certainty is that there are two self-monitoring levels: a so-called "inner loop," which is prearticulatory, and an "outer loop," where as one says something, one checks that it corresponds to the intended sequence. In many cases, the inner loop can catch errors before there is any overt sign at all. At other times, there may be some mild perturbation (e.g., a pause or vocalization, "um") in the overt signal. Where the error is detected by the external loop, the speaker interrupts herself with an audible glottal closure, signaling "oops," and then recycles the earlier delivered chunk back to the point of the error: "You go left at the corn... you go RIGHT at the corner." At this point the speaker may think she has completed her turn, but the interlocutor may now miss the beat (ca. 200 msec after the turn ends), at which point a response is expected (perhaps also leaning forward or frowning), indicating some possible hitch in comprehension. This provides a space (ca. 300 msec) for the error speaker to self-repair or augment, "You go left at the corn... you go right at the corner, (500 msec) at the intersection." If the speaker misses this opportunity, the addressee in difficulty can still trigger (initiate) self-repair by the original speaker:

A: "You go left at the corn... you go right at the corner, (500 msec) at the intersection."
B: "The intersection of Bryant and East Street?"
A: "Yes, by the 7Eleven."

Sometimes, the misunderstanding only becomes apparent later:

A: Do you know who's going to the meeting?
B: No, who?
A: I don't know...

Here the response to B's turn displays a misunderstanding that A's turn was a preliminary to a telling, when in fact, as turn three makes clear, it was a simple question. This illustrates the utility of a communication system that alternates short turns across speakers; the responses indicate whether the prior turns were understood as intended (Sacks et al. 1974).

By looking at the whole system of incremental possibilities of repair, one sees that the whole interaction system is designed to give multiple successive opportunities to catch misunderstandings and errors (Schegloff et al. 1977). The system is optimized for efficiency, first within the speaker's self-monitoring loops, then through overt self-repair (where the speaker foresees errors or upcoming misunderstandings), then through a pause inviting self-repair, and finally (and reluctantly) through other-initiation of repair. The latter involves an inserted, potentially disruptive sequence. The disruption is minimized by an ordered preference of types of other initiation: the listener that does not understand gives a precise localization of the problem ("the intersection of Bryant and East Street?" is more efficient than "Where?"). Interestingly, the sum of the length of initiator and repair tends to be no longer than the original troublesome utterance (Dingemanse et al. 2015). One reason for this efficiency is that other-initiation of repair occurs every 80 sec in natural conversation.

Learning from Mistakes

What does it mean to be wrong? What do errors tell you? One explanation is that what you did was wrong and requires follow-up; however, there are several learning paradigms in which errors are integral parts of the learning process (Lorenzet et al. 2005). There is utility in exploration. Importantly, though, there must be bounds beyond which the learner knows it cannot be allowed to go, based on safety or cost (or other criteria). The role of errors may be different for humans and agents. For example, children may need to make an error to learn effectively from it. However, a robot may have a learning mechanism that allows it to learn based on communication that substitutes for the experience of actually making the error.

Children sometimes make errors on purpose, to gain a better understanding about the consequences of some action and better predict the future. If an action and its consequence are not known by the child, the child will be curious to explore it. What happens if the glass falls? Does it break? Does it not break? What happens after a glass has broken? Exploring this gives the child a more complete picture of the world.

For robots it is also important to explore the environment and to explore what works and what does not (within a certain range, of course). If a glass is grabbed too hard, it will break; if the door handle is turned too gently, it will not move. Exploring the range of different outcomes from an action gives a better and richer representation to operate more flexibly.

For these kinds of "errors," which function more as exploration actions, the robot/child will probably need to realize on their own when an error occurs. For some types of errors, this might not be obvious to the robot, thus necessitating the support of the teacher. Support is also needed if the reason for an error is unknown to an agent. In the end, it is crucial to know what action would *not* cause the error.

Controlling and guiding this exploration process is an important role that the human teacher can play for the learner; that is, helping the learner collect the most informative "near miss" examples that will lead to generalization.

Learning systems can end up in a sort of good enough, less than optimal state (e.g., Klein and Perdue 1997). Consider a landscape of solutions where there are many, locally optimal, sort of OK solutions, but only one or two optimal solutions. For a machine to escape from its less than optimal cul-de-sac, it will need to be nudged out of its local optimum and forced to explore the larger landscape of possibilities. This can be achieved, as in Bayesian modeling of fitness landscapes, by perturbing the current state: the system is forced to go downhill for a bit and then starts to climb another incline that may turn out to be the global maximum (e.g., Markov chain, Monte Carlo, Metropolis coupling in Bayesian phylogenetics, as in Reesink et al. 2009). Gray and Lindstedt (2017) talk about "plateaus, dips, and leaps" (see also Gray et al., this volume). "Plateaus" contrast with "asymptotes" (Gray 2017). Asymptotes reflect performance at a theoretical limit whereas plateaus are periods of stable but suboptimal performance. Better methods can yield better performance but the agent may not have knowledge of the better method or possess the skills required to master the method, or simply not care to get better.

Different teaching styles result in different ways that a misalignment on task knowledge is handled. Consider the following task instruction scenarios:

- Marie is a preschool teacher who wants to teach her students how to mix colors to obtain other colors. Instead of just telling them the different combinations, she uses a technique called "provocation." She presents the students with an uncolored picture of a frog and two tubes of color: blue and yellow. She says "this is all we have, what should we do?" Some students will give up and say that it is not possible; others will paint the frog blue or yellow (which is not considered a mistake); a few will try something new and mix the colors to obtain green. Children who go through this discovery process are much more likely to remember the outcome than those who are just told about it or observe it (Craft 2001).

- Kenan is a carpenter certified to teach woodworking. He regularly works with apprentices at his workshop. When apprentices start working at the shop, Kenan first assesses skill levels by having them perform basic actions, such as cutting with a saw or sanding. He gives them a task like "cut all this wood into 16 cm pieces," commensurate with skill level. When teaching a new task, Kenan uses a supervised discovery process. He tells the apprentice to attempt the task and interrupts them when they go wrong. For example, he may tell an apprentice, who has never previously done the task, to glue two pieces of wood together. If the apprentice goes down a wrong path, such as starting to apply glue before making certain that the surfaces to be glued have been sanded completely flat, he then interrupts the apprentice's work to bring him/her back to the correct path. When teaching how to use more dangerous cutting and milling machines, Kenan first demonstrates the use of the machine and then tightly supervises apprentices as they try to use the machine.

One key difference between these scenarios is the extent to which mistakes by the learner are allowed, or even encouraged, as a function of situational characteristics. The same provocation-based discovery learning process that is entirely appropriate in the context of interactions meant to improve understanding of color may be inadvisable in the context of interactions meant to improve woodworking knowledge and skill. Kenan cannot afford to let his apprentices explore possible paths that might lead to waste of expensive resources and, perhaps even more importantly, an increase in risk of harm. The woodworker's blade is less forgiving than the artist's brush. The difference between these two scenarios may greatly impact the role of exploration needed for ITL in robots and AI agents. It also emphasizes the important role of situation understanding and contextual reasoning in selecting modes of interaction.

Interaction Timing and Synchronization

So far we have focused primarily on the substantive context and content of interaction, leaving implicit in the discussion an inherent characteristic of all interaction: it progresses over time. Timing of an activity is separate from content. In this section we detail the importance of timing as a first principle of the interaction.

Correct timing and synchronization are crucial in many aspects of human interaction. The turn-taking system in conversation operates with ca. 200 msec turnaround (Stivers et al. 2009), the normal human minimum response time for the simplest preplanned response. This is far too short for planning spoken utterances, which generally require at least 1 sec (600 msec for a single word, 1500 msec for a simple clause) before output can begin (see Levinson, this volume). This implies that speakers are predicting the ends of the incoming turn and planning their own so that they are ready to respond on time. The speed may have origins in the phylogeny of our communication system, before the complexity of linguistic signals developed (Levinson 2016), but it is also maintained by the semiotics of delay. For example, neuroimaging has shown that as a gap after a prior turn lengthens, expectations change: we usually formulate questions to favor a *yes* answer, which is expected at the normal ca. 200 msec response time; if the answer *no* comes in at the normal response time, it evokes an N400 or surprisal reaction, but this evaporates over time as *no* becomes more probable (Bögels et al. 2015a).

One issue for timing of interactions, especially when a machine is teaching a human, is that human tutees often interpret a slight pause before positive or neutral feedback as signaling negative feedback (Fox 1991, 1993). For instance, if a student answers a tutor's question and then gets a slight pause before hearing "yes" or "umm," the student will often infer that the answer is incorrect.

In general a delay in response after a response-requiring turn signals that an unwelcome response is likely. Withholding response after any turn can signal that its import was not clear. In general, then, timing is part of the signal in human interactions. This may be very problematic for human–machine interaction, but it is worth noting that there are classes of humans, most notably children, who may be much slower than the human norm, and adults are pretty good at adjusting expectations to childhood norms. It may be better for ITL systems to signal their timing limitations (perhaps by junior stature) than to attempt to meet full normal human speed of response.

Synchronization is a low-level, probably largely unconscious, process whereby two agents come to coordinate. The simplest case can be modeled as coupled oscillators, as when the seventeenth-century Dutch scientist Christiaan Huygens noted that two pendulums mounted to the same structure come to synchronize over time. In biological systems, quite complex behaviors, like the synchronized firing of a swarm of fireflies, arise in a similar way (here, by resetting a biological capacitor when the neighboring firefly turns on). Humans playing music together, for instance, tend to harmonize brain oscillations, thus providing a shared internal metronome. Finely timed coordination may well depend on this, but for various reasons it will rarely be sufficient: fine timing may also depend on predicting the other's action culmination and even preplanning one's own productions so they are ready to go. Coordination is prediction plus generation in a joint activity, requiring mental simulation of the other.

Human speech communication has the distinctive property of alternating communication bursts between speakers. During an incoming utterance, an addressee may be signaled out by gaze and normally gives feedback signals at major chunks of incoming material. This becomes especially obvious if speaker A is delivering a story which, given the turn-taking structure, is often implicitly negotiated at the beginning:

- A: "Did you hear what happened to Joan?"
- B: "NO, what?"
- A: [chunk1], [chunk 2], [chunk 3].

In overlap with the end of chunk 1, B is likely to say "mm"/"uhhuh" or the like, or nod, thereby recognizing (a) that chunk 1 has been received, (b) there is nothing in chunk 1 that is causing comprehension difficulty, and (c) that B has nothing compelling to say at that point. The opportunity to do this recycles at the end of chunk 2, 3, and so on (Schegloff 1982). The class of relatively content-free back channel responses is fairly limited per language, with upgraded versions also available (e.g., surprise markers "wow," empathy markers "oh dear"). These signals do not count as turns, which is why they typically occur in overlap. Conversely, longer phrases (less limited of course) are likely to count as turn initiators, leading to expectations of possible speaker switch. Back channels of this kind, as the name suggests, thus essentially signal "channel open, message received," while implying by virtue of the activities that were not done instead (e.g., initiation of repair or a new major response) that full understanding has occurred. B recognizes that A is producing a longer stretch of speech that has not finished.

Through a back channel, a listener demonstrates to a speaker his/her continued interest in communication. As back channels play an important role in coordinating human–human conversation, it becomes important for agents to have a capability to generate back channels during human–agent communication. This involves making a decision on when to generate back channels (i.e., the timing of back channels). In human–human communication, back channels occur very fast and seem to be elicited by the speaker based on a variety of prosodic, verbal, and nonverbal cues (Schroder et al. 2012). Previous works in conversational virtual agents have developed predictive models (e.g., sequential probabilistic models) to predict the timing of where a back channel should be generated (Morency et al. 2008). As there is evidence that human listeners generate back channels even without attending to the content of communication, these previous predictive models often only consider surface features, such as prosody, pause, gaze, and direction from the speaker. Their empirical results have demonstrated that generating back channels that are synchronized in time with speaker contributions is an extremely challenging task.

The nature of communication in ITL as well as in conversational virtual agents (e.g., in the context of social communication for negotiation, consultation, and therapy) are quite different. It is not clear whether previous work on virtual agents can be directly applied to ITL. The prediction may no longer depend on acoustic but rather visual features (e.g., observed from human demonstrations). In ITL, the chance that an agent might misunderstand task instructions given by a human (whether verbally or through demonstrations) is high. Since humans may perceive back channels as an explicit confirmation of understanding, generating back channels will need to be tightly linked to content processing (rather than surface cues). This could delay the appropriate timing for generation. Without connecting to content processing, a back channel may have a danger of leading the speaker to believe a task instruction has been successfully understood and then later discover otherwise, causing a high cost repair for the downstream communication. Thus, when to generate appropriate back channels in ITL remains a challenging research question.

Chao and Thomaz (2013) have shown the importance of timing control in human–robot interaction and its social impact. Their work on the CADENCE system shows that manipulating these turn-taking timing parameters (e.g., space between acts, likelihood of interrupting the partner) results in robot behavior that people perceive as being significantly different. Moreover, people attribute different personalities to the robot; changing the robot's personality by manipulating these timing parameters results in different behavior from the human partner, thus manipulating the social dynamics of the dyad.

Conclusions

In this chapter we have considered the qualities of human interaction and learning that will be most effective and natural to incorporate into any ITL agent, specifically focusing on the interactions around learning from explicit instruction. We argue that this type of learning is centered around bringing the common ground between these two agents (teacher and learner) into alignment. Thus, errors drive the ITL process by triggering interaction and learning. Without misalignment of common ground, either through errors or missing knowledge, there would be no need for ITL. Finally, we highlight the importance of considering timing and the dynamics of an interaction as a communication channel itself, as well as emphasize the importance of analyzing the extent of shared capabilities between the two agents.

What all of this argues for in ITL with robotic and AI agents, is that information from multiple short turns of interaction between the teacher and learner will have the best opportunity for minimizing errors in communication that will arise naturally. Short bursts of information between the two interacting partners is likely to be the most successful way to transfer task knowledge between the two, incrementally updating errors in common ground until the teacher and learner come into alignment.

8

Natural Forms of Purposeful Interaction among Humans

What Makes Interaction Effective?

Stephen C. Levinson

Abstract

The design of an interactive robot should make crucial reference to the observed properties of human interaction. Obviously, human communicative interaction varies across languages and cultures, but remarkably uniform is the basic organization of interactive language use: participants take short turns at talking while avoiding overlap; they utilize a basic inventory of action–response pairs (e.g., question–answer), which can be recursively employed; they have systematic backup systems for communicative difficulties and deploy multimodal signals (speech, gesture, facial expression, gaze) to disambiguate or reinforce intended content. This chapter spells out these design properties and makes the point that human comprehension is fundamentally predictive, and has to be so to achieve the typically rapid response times despite the large latencies involved in generating speech. These properties may pose a substantial, even insuperable, hurdle for a fully humanoid interactive robot, but fortunately humans are excellent at adapting to interactants with restricted capabilities, such as children, foreigners, or aphasics.

Introduction

Humans appear to involve themselves effortlessly in social interaction with their peers and are capable of rapidly integrating novel tasks and routines into these interactions. Clearly, efforts to build machines that might assist humans in varied tasks have much to learn from an analytic grasp of what makes humans so seamlessly able to conduct cooperative task management, even when it is novel. Right at the start, a caveat is in order: the domain of the study of human communicative interaction is still in its infancy; it is a field that has been dominated by a few maverick pioneers, and has only recently acquired the extensive public databases and measurements typical of cumulative science. Although our understanding of human–human interaction is still quite limited, what we know already makes the prospects for seamless human–machine interaction quite remote. I will at the end, however, suggest that emulating humans may not in fact be the most productive use of new technologies, if indeed it is possible at all. The body of this chapter tries to delineate what we know about human interactive skills (see also Thomaz et al., this volume).

The proverbial Martian arriving on Earth would quickly notice that humans have the propensity to huddle in a face-to-face arrangement and engage in a curious exchange of alternating short bursts of communicative activity. S/he would also note that in this regard there are plenty of parallels with other species (e.g., the vocal duetting of many types of songbirds and many species of primates). Those other species tend, however, to have a small, relatively fixed repertoire of signals, whereas it would rapidly become self-evident that such is not the case for humans. Human repertoires are not only immense, they also vary significantly across ethnic groups or cultures. Moreover, it would be obvious that human exchanges happen in myriad different contexts, apparently aiding numerous types of endeavor. Beyond that, the system might be inscrutable. Let us try here to analytically unpack this a bit.

Basic Ethological Properties of Human Communicative Interaction

The fundamental niche for human communication is social interaction in a face-to-face context: this is the context in which language is learned, the bulk of usage occurs, and almost certainly the context in which it has evolved. It is characterized by the rapid exchange of alternating short bursts of communication (averaging ca. 2 sec) as well as by multimodality: the face, the hands, the deployment of the trunk as well as the vocal organs are typically all in play at once. One can look at the system from the point of view of comprehension, in which case it is clear that the incoming multimodal signal is parsed in parallel and integrated extremely fast, also combining with numerous aspects of the context. Gestures, for example, can be shown to be unified with the linguistic message just as fast as they happen (Özyürek et al. 2007). From a production perspective, the language system may be more serial (Indefrey 2011): a message is composed and serially encoded, although the processing of each successive chunk can proceed in parallel as it passes through the many stages of encoding from message to linguistic form to articulation. However, once multimodal production is considered, it is clear that facial expressions, manual gestures, and other bodily components must be produced in parallel but temporally integrated, much like a chamber orchestra would perform a score.

Although the strictly linguistic aspects of comprehension have been extensively studied experimentally, the multimodal aspects remain fairly obscure. Linguistic "strings" are often treated as linear—phonemes follow phonemes, words follow words—but clearly multimodal signals are delivered simultaneously and, in this regard, are like prosody and voice quality, offering different kinds of units at different parallel levels, but somehow integrated semantically and temporally. In addition, comprehension and production must work closely in consort, as the following considerations show, and this area has only recently started to be explored.

As mentioned, interactive communication involves the rapid alternation of speaking roles. What is interesting about this is the cognitive load that is involved. Across languages and across conversational corpora, the modal gap between turns is only 100–200 msec, quite literally in the blink of an eye. It takes at least 600 msec to crank up the language production machinery; that is, the time it takes from knowing what word you want to say until the time anything comes out of your mouth. For a simple clause, the latency is more like 1500 msec (for references, see Levinson 2016). The implication is clear: to respond so rapidly, the speaker must predict the content of the incoming turn and start early preparation of a response, as illustrated schematically in Figure 8.1. We have shown that the point of an utterance or the speech act is often predicted from the very first words of an utterance (Gisladottir et al. 2015). We have also shown, by using EEG, that the production system starts as soon as the point of the incoming turn becomes clear, as indicated in Figure 8.1 (Bögels et al. 2015b) and then proceeds all the way through the various encoding stages. Thus, predictive comprehension and language production have to work in overlap and in consort. Humans are not generally good at multitasking,[1] due in part to the working memory bottleneck, so turn-taking must impose a heavy cognitive load.

The structure of interaction involves sequences of speech acts (i.e., actions packaged up in linguistic and multimodal format). Humans clearly map linguistic utterances into something action-like. Consider, for instance, "Can you reach the wine?" and its nonverbal response action. The pragmatic thrust, the point of an utterance, is only very indirectly related to its form. Thus English yes–no questions typically come in declarative format with falling prosody; the giveaway is often the manner in which the declarative is a statement about something that is more within the addressee's epistemic domain (e.g., "you are feeling better"). The many-to-many mapping between linguistic form and speech act has been explored in corpora (e.g., Couper-Kuhlen 2014; Levinson 2013a, 2017), and although inference can make use of varying probabilities, it is abductive in character involving many contextual parameters. For example,

---

[1] It is generally agreed that multitasking slows performance and increases errors, but the idea that true multitasking is impossible has required recent revision; for a review, see Fischer and Plessow (2015).

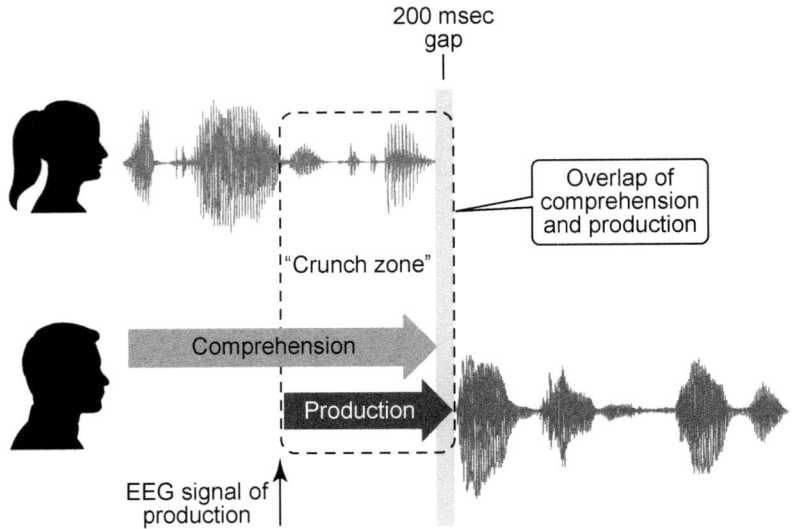

Figure 8.1   Overlapping comprehension and production processes during turn-taking (after Levinson 2016).

an utterance like "Can we lift this?" could be a request for permission to remove something, an offer to help you with your suitcase, or an enquiry about the weight of something; apart from prosody, only contextual factors are likely to disambiguate. Outside restricted domains (e.g., travel agents, bank enquiries) these inferences will be very challenging to model in human–machine interaction.[2]

As noted, speech actions come in sequences and often occur as two paired actions (a so-called adjacency pair): following a question, an answer is due; following an offer, an acceptance or refusal is expected; following a request, an action or excuse will be forthcoming (see Exchange 1 and 2 below). Actions are thus often contingent on prior actions: "yes" only makes sense in relation to the prior query. The simple device of paired actions can, however, be recursively applied according to the template shown in Figure 8.2 (Kendrick et al., in prep.): FPP marks the first part of a paired sequence of actions and SPP marks the second part, the response; each of the expansion types can also consist of pairs of actions (Schegloff 2007).

Consider the following exchange involving a question–answer pair embedded within a question–answer pair:

---

[2]   A reviewer made the interesting point that the complexity of inference is narrowed in humans by the matching evolved design of sensors and effectors, and no doubt the cognition connecting them: I know what is visually salient to you without complex calculation, but this may not be available to machines with their different perceptual systems. This may be a more serious barrier than inference to constructing satisfactory interacting machines.

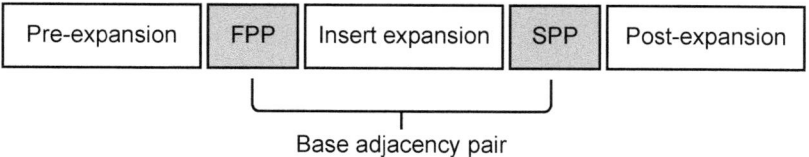

| Pre-expansion | FPP | Insert expansion | SPP | Post-expansion |

Base adjacency pair

Figure 8.2 Basic template for sequence organization. FPP marks the first part of a paired sequence of actions, and SPP the second part, the paired response. Each expansion type can also consist of pairs of actions. After Schegloff (2007).

Exchange 1
    A: "May I have a bottle of Mich?"
    B: "Are you twenty one?"
    A: "No"
    B: "No"

Interestingly this kind of center embedding can go on indefinitely—naturally occurring cases of at least six degrees of center embedding have been found in conversation, far more complex than anything found in natural language syntax where such embeddings are capped at level 3, and then only in written language (see Levinson 2013b). A lot of ink has been spilt on why the limitation is found in syntax, and it is usually attributed to short-term memory problems. What is interesting, then, is that these limitations do not hold in a joint task, even though each party must hold in mind the same pushdown stack model. This is a first indication that there is something rather special about the human capacity for joint undertakings, something that seems to be largely absent from other species: it is a capacity to "distribute cognition" over individuals (Hollan et al. 2000; Hutchins 1995), thus allowing interactants to form a joint computational device, which seems able to overcome the stack-maintaining limitations of individuals.

Center embedding is also exploited for a fundamentally important function; namely, interactive repair. This typically involves a repair initiator (e.g., "huh"? or "what"?) from the recipient followed by a repeat or clarification of what has just been said by the first speaker. Repair becomes much more difficult if displaced from the slot immediately after the troublesome turn. Thus there is pressure to solve understanding or hearing problems as soon as possible. Without this basic repair mechanism, our understandings would rapidly diverge; repair, therefore, plays a crucial function in maintaining intersubjectivity or mutual understanding, occurring roughly every 80 sec in any language (see Dingemanse et al. 2015). This kind of repair uses the "insert expansion" slot in Figure 8.2 and can be recursive, so this can get quite complex, as in the following center-embedded insert expansions (Merritt 1976; see also Levinson

2013b), where the pushdown stack character is indicated by the indentation and line numbering:[3]

Exchange 2

> S: Next ← *Request to order*
> 0 C: Roast beef on rye ← *Order*
>
>   1 S: Mustard or mayonnaise? ← *Q1*
>   2 C: Excuse me? ←*Repair–Initiator (R1)*
>   3 S: What? ← *Repair–Initator (R2) on R1*
>
>   3 C: Excuse me? ←*Repair of R2*
>   2   I didn't hear what you said
>   1 S: Do you want mustard or mayonnaise? ← Repair of Q1,
>   requested in R1
>   C: Mustard please. ← *A1 to Q1*
>
>   0 S: ((provides)) ← *Compliance with order in line 0*

Other kinds of sequence exist. For example, one possibility is to exploit the preexpansion slot in the schema above (Figure 8.2), as in Exchange 3 (below), where a prior adjacency pair checks the preconditions for the main action (in this case, an offer). Here, the pre-FPP ("Say, whadja doing?") is plausibly a pre-offer, but it could also be a pre-request (a preliminary to, e.g., "Want to come and help me clean up?"): B's response is not a straight answer (compared to "doing homework" or the like), but is couched to encourage the following action, which has clearly been foreseen. Sequences thus involve a kind of look ahead, with responses geared to inferences about what is likely to be coming up next. From the perspective of artificial intelligence (AI), this can be thought of as plan reconstruction: B infers that A has a plan to offer something nice, and so encourages it; likewise, A projects from B's response that B intends to do nothing to impede the offer and is therefore likely to accept it. This meshing of mutually inferable plans is a crucial property of human interaction, both verbal and nonverbal (e.g., constructing something together) (Bangerter and Clark 2003).

Exchange 3

> 1 A: "Say, whadja doing?"
> 2 B: "Not much"
> 3 A: "Y'wanna drink?"
> 4 B: "Yeah"

---

[3] One reviewer queried the recursive structure here, suggesting a possible list representation; see Levinson (2013b) for many examples, which I think have to be understood as interactively checking preconditions on preconditions on preconditions… (up to 6 deep) to actions. A list structure cannot capture the way in which each "push" has to have its own paired "pop," as in the structure $[A_1 [B_1 [C_{1}...C_2] B_2] A_2]$; see Levinson (2013b).

How exactly these plans are made mutually manifest, and how far downstream interlocutors predict the likely courses of action, is a puzzle (see Levinson 2013a, b, 2017). Clearly, as in the example above, sequences may have a recurrent pattern, and just as observers of primate behavior have noted under the rubric of *ontogenetic ritualization*, the initial action can come to project the entire sequence. In human interaction, however, such projection is often much less clear, and as the potential ambiguity of the pre-offer/pre-request above indicates, the stakes for misconstruals can be relatively high.

In work we have conducted across languages and cultures, all of these patterns (turn-taking, repair, sequence organization) are strongly universal, following the same principles in many detailed ways (e.g., Dingemanse et al. 2015; Kendrick et al., in prep.; Stivers et al. 2009). This contrasts markedly with the diversity of languages, which differ at every level, from the sound system, to the combinatorics involved in phonology, morphology and syntax, to the meanings that are conveyed. We believe these interactional principles form a strong infrastructure for language, which in turn makes the learning of languages possible and influences their structure in subtle ways. In addition, all spoken languages exhibit deep similarities in the use of multimodal resources despite the occasional cultural taboos to be found here: gaze, gesture, and facial expressions play an important role in framing and supplementing the linguistic content.

The properties of human communicative interaction reviewed here can be abstracted into a set of design features (see Table 8.1). These are some of the basic desiderata that any interactive computational agent will either have to mimic or be endowed with ways to achieve equivalent functionality. Table 8.1 sketches the functions of each design feature, which we briefly review.

Multimodality offers redundancy or complementarity in, for instance, gesture, facial expression, and speech. Crucially, the taking of short turns will make clear, in your immediate response, whether you understood me correctly ("legibility"). Action sequences, such as question–answer (Q–A), structure interaction by setting up expectations for responses; note that in Exchange 1, the expectation can be postponed but still persists. Although spoken turns minimize overlap, nonvocal signals (like laughing, smiling, shaking hands) may get their significance precisely through simultaneity. Communication about the state of communication (metacommunication, as in repair initiators) plays a crucial cybernetic role in guiding utterance interpretation and signaling communicative success. The fact that conversation has an expected "clock speed" allows participants to sense an interpretative problem simply from a delayed response; the expected simultaneity of, for instance, laughter is a further check on temporal meshing. Feedback signals like "mm-hmm" or nodding typically occur in overlap at the end of an utterance constituent, and their timing allows the speaker to proceed rapidly. None of this would work without a presumption of engagement and engagability, which makes it possible to enter an interaction with a stranger on the street. This presumptive mutual regard and helpfulness

Table 8.1 Key design features of human interaction: a checklist for constructing intelligent interactive agents.

1. Media	
• Function: communication, redundancy	Multimodal signals, language
2. Action sequences	
a. Alternating turns	Speech act mapped to language
• Function: includes legibility, opportunities for repair	
b. Action sequences	Adjacency pairs (e.g., Q–A)
• Function: includes structuring exchange	Complex sequences, e.g., insert pairs (e.g., Q–Q–A–A)
c. Simultaneous	Complex sequences, e.g., shaking hands, laughing
• Function: coordination, ritual	
3. Metacommunication	
• Function: check communication	Repair
• Function: confirm message receipt	Feedback tokens, e.g., *uhuh*
4. Timing	
• Function: indicates state of processing is "on time"	Turn-taking timing
• Function: "clock speed" check	Synchronicity, e.g., shaking hands
• Function: "message received now"	Timing of feedback
5. Motivation	Shared goals
• Function: enter and maintain engagement	Affect/attachment, politeness, specific rewards
6. Legibility of	
• Function in attention: indicates current focus of processing	E.g., gaze readability
• Function in intention: aid predictive processing	E.g., gesture signal vs. instrumental action

seems unique to humans, but of course is cemented when there are proximate shared goals or rewards. In general, for an interaction to work, each agent's actions must be performed in such a way to make them legible (i.e., interpretable) as an action intended to be perspicuous for its purpose. This is obvious for communicative signals, but holds for other kinds of joint actions, so that when, for example, you and I carry a table together, the direction of my gaze can indicate the intended direction of motion. It follows that instrumental actions which are not part of the joint endeavor (e.g., scratching one's head or coughing) should also be clearly legible as irrelevant for the joint purpose in hand.

Affect, Empathy, and the Human Person

A robotic assistant can clearly be of great utility without having any deep understanding of humans; after all, humans will readily adapt to its limitations. But any machine that wishes to "pass" as capable of humanlike interactive task learning will have to know a great deal, not only about human communication but also about human nature. Human interaction is, in fact, replete with "ritual" aspects which no successful interactant can ignore. A century ago, the sociologist Durkheim (1912) suggested that religious beliefs personify society, the collective consciousness of consciousness, as he put it, so that individuals come to have as social persons a kind of sacred quality. Goffman (1959) built on this in his analysis of "interaction ritual," noting how we often treat persons with elaborate care: we pretend not to notice others' slips, belittle ourselves (e.g., walking into a lecture late, stooped over), and worry about our perceived social competence to ensure that mutual dignity or "face" is maintained both by self and other on each other's behalf. Brown and Levinson (1987) elaborated this account in a theory of "politeness" in which individuals' ritual or "face" requirements could be maintained in two rather different ways: by claiming empathy and fellowship (Durkheim's "positive rites") or by giving the other maximum *Lebensraum* (Durkheim's negative rites or avoidance rituals). Which kind of ritual is deployed depends on social closeness versus social distance (vertical or horizontal), together with some measure of the weightiness of the action or imposition. This translates directly into the choice of linguistic expressions: if I want to borrow the pen of my neighbor in a plane to fill out a landing form, I might say "Hey, I need your pen for a moment" if he's a friend, but "Excuse me, could I possibly borrow your pen just to fill in this form?" if he's a stranger. Although the whole business is wrapped up in culture-specific conventions (e.g., the bowing and honorifics of Japanese), there does seem to be a universal basis to these mini-rituals of the person. Further corpus work suggests that in choosing linguistic expressions we make quite elaborate computations of rights and duties, epistemic territories or domains of expertise, and estimations of effort or contingency (Drew and Couper-Kuhlen 2014). We do this because the failure to recognize the other's right to self-esteem is cause for offence; thus, one works with what Goffman (1959) called "the virtual offence" (the worst construal of what one is doing) and tries to stop it from happening. When it is necessary to invade another's domain, as in medical examinations, elaborate circumspection is required.

An aspect of recognizing the other as a "sacred being" is the recognition of the need for empathy among close associates. Failure to greet a person or omitting to extend condolences or congratulations is also cause for offence, as is the failure to laugh at people's jokes, appreciate their stories, or empathize with their travails. Interacting successfully with a child may involve entering, for a while, into its momentary make-believe world. In general, any interactional success here will be achieved by keeping tabs on the life courses of all significant others. Interestingly, it is also crucial to keep track of common ground—the things we have told each other—requiring a record of informational exchange (complete with the reference forms used) for each such person. Failure to do so will have you classified as the party bore or the senile kinsman who repeats information already imparted. All this is self-evident to us, but will likely be opaque to an interactional machine: it will not be feasible to build in all the sensitivities and particularities of the social world that parameterize underlying social principles. Perhaps some future machines will be able to learn some of these mores, although that, in itself, poses formidable inferential problems of the kind explored in studies of child development. Still, modeling the stiff politeness of anonymous service staff may be within our grasp and a likely prerequisite for a successful interactive robot.

Learning Interactively

Understanding how new tasks can be learned in and through interaction is the focus of this volume. From experiments on joint action, it seems that cooperative interaction relies on each participant modeling the other's unfolding action plans, transposing themselves into the other's footsteps as it were, and so co-representing the joint action. This seems to be so even if my half of the joint task has an independent timing and function (Sebanz et al. 2003, 2006; Vesper et al. 2016).

There have been interesting reports from cross-cultural studies of societies where children, for example, acquire most adult tasks simply through observation, not instruction or demonstration (Gaskins 1999; Rogoff et al. 2003). So how important communication, or indeed interaction, is in learning new tasks is perhaps not clear. In an interesting preliminary study, Laland and associates tested the learning of flint-tool knapping under different conditions: reverse engineering (by inspection of the tool), emulation (by watching production), restricted gestural communication, and vocal communication (Morgan et al. 2015). They found that the task was easily learned only when there was full communication, because certain critical tricks are not easily extracted through direct observation (e.g., in this flint-knapping case, preparing a striking platform under 90 degrees). More studies of this kind are needed, with different types of tasks, but generally this work suggests that directed communication may often be essential to the learning of skills.

Here I would like to make the point that any kind of sustained cooperative interaction presupposes communication, even if it is subliminal. For instance, maintaining synchrony in a chamber orchestra involves visual cues, as in the exaggerated, but precise, lifting of a bow to indicate an impending entrance ("now is when we begin, at this tempo"); carrying a table together involves some low-level signaling of direction ("go left now"), evidenced by an exaggerated tilt of the head to the left. These signals work by having a shape that is not purely instrumental: the exaggeration of bow movement beyond what is needed for sound production or the tipping of the table greater than is needed to get around the corner. The detection of the not-just-instrumental quality of a basically instrumental action can be very subtle but allows our species to collaborate in a coordinated fashion (see Bangerter and Clark 2003; Grice 1975 on the Maxim of Manner). More generally, if there is a way to indicate "this is the way to do it" by the manner of doing, it may play a crucial role in human cultural transmission. This kind of signaling has been turned into a theory of "natural pedagogy" (Csibra and Gergely 2009), and the distinction between instrumental and noninstrumental action has been held to be central to the learning of culture, where both causally transparent and causally opaque actions need to be learned; unlike instrumental activities, table manners, for instance, should be followed, not innovated or improved upon (Clegg and Legare 2016).

The Cognition behind the Ethology: The Gulf between AI and Human Interaction

It is not easy to reconstruct the underlying cognition that makes human interaction possible. Apparently, effortless coordination requires complex inference: if we are putting together an IKEA bookshelf, and I put the screwdriver down with the handle toward you, I may be signalling what you need next (Bangerter and Clark 2003). Such wordless communication is challenging to model. The philosophical reconstruction by Grice (1975) remains the best general attack we have: the signaler intends to cause an effect in the mind of the recipient just by getting the recipient to recognize that intention. But how does this happen? The general answer seems to be: by an inflection of manner that suggests the action was not purely instrumental. Why take the extra effort to turn around the screwdriver? Alternatively, take the case of a student who arrives late to class with a cappuccino moustache: a fellow student might vigorously wipe her own lip while gazing across at the new arrival, such that the latecomer wonders, "Why is she doing that?" This may lead to the realization that there is something anomalous about the latecomer's own lip. This seems to work by an implicit comparison to the simplest instrumental version of the action: any excessive elaboration of manner suggests communicative intent. This kind of incidental communication lies behind nearly all our coordination: I wait to walk across the road at a junction until I have caught the driver's eye, so that he and I both know we are aware of each other, and it's safe to cross the road (building on the assumption of our common minds and bodies, in a way that is inherently problematic for a robot).

The Gricean analysis involves reflexive ratiocination: I plan my signal with its manner inflection thinking that you will reconstruct the communicative intention behind it, realizing it is a signal and not (or not only) an instrumental action (thanks to the manner inflection). Many psychologists have been doubtful of any such thing, pointing instead to our tendency for lazy egocentricity. To explore this, we designed the following task (de Ruiter et al. 2010): two players took turns to signal where the other should place a gaming piece on a checkerboard. They were denied all means of direct communication and could only indicate by the manner of moving their own piece where the other player should place hers. Indeed, one of them was in an fMRI scanner. Participants could solve the task, even though each trial had different properties that denied them the possibility of forming implicit conventions: they did so by choosing a route for their own piece that went over the square where the other should place her piece, and wiggling or rotating or otherwise by manner inflection indicating the solution. What we found is that the mentalizing brain areas activated by the signaler were reflected in the areas activated in the recipient, prima facie evidence that we do indeed attempt to mirror the other person's reasoning (de Ruiter et al. 2010; Noordzij et al. 2009). It would seem to work by an abductive leap from the mannered signal to the sender's likely intention, but an adequate computational model eludes us. Blokpoel (2015) proves formally that the inferences required are intractable (NP-hard) unless the set of goals and signals are highly constrained (see also Van Rooij et al. 2011). Attempts to model the interpretation of nonce or "one-off" signals by analogy also prove problematic (Blokpoel 2015).

These results hold just as much for linguistic communication as they do for nonce, nonconventional communication. That is because the *point* or *speech act* of an utterance is rarely explicit: the relations between linguistic form and speech acts are many-to-many (Levinson 2017). Consider "Say, whadja doing?" in Exchange 3 (above), which has the superficial form of a question, but its purpose is to pre-adumbrate an invitation or suggestion, a purpose perspicuous to the answerer: "Nothing much." The reconstruction of its intended point requires a projection of an upcoming sequence (offer and acceptance); that is, it involves plan reconstruction and plan-meshing, as illustrated in Figure 8.3. Here, from the utterance "Whadja doing?" Clara detects a possible plan to invite her out (the pre-invitation reading of the utterance), so she answers in such a way as to make clear that she has no impediments and is likely to accept (the go-ahead reading of her response, "Not much"). This kind of plan reconstruction was explored in early AI but has proved computationally tractable only when the alternatives were highly constrained (Allen and Perrault 1980), as Blokpoel's (2015) theoretical work predicts.

At present I believe we simply do not have adequate computational tools to model the mysteries of human communication, which fall in the domain of "the inference to the best explanation" or inspired abduction under an umbrella of reflexive reasoning. This is the kind of reasoning successfully employed in Schelling games of pure coordination, where we both get a prize if we can think of the same number without communicating (Schelling 1960).

We come now to the computational tractability of the ethological properties, like multimodality and turn-taking discussed above. Turn-taking with its

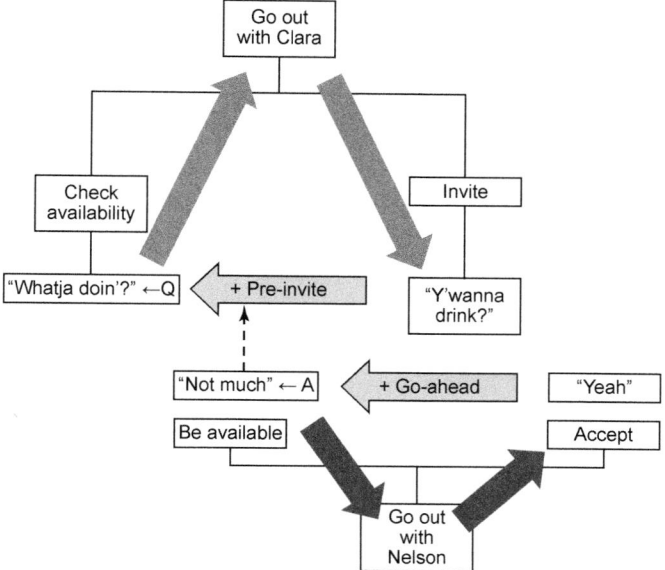

Figure 8.3   Sequence organization as plan reconstruction (from Levinson 2017).

split-second turnaround poses problems of sheer processing power. To respond, I argue, you must compute (most likely predict) the intended point or speech act of the incoming utterance early enough to begin your response preparation well before the end of the incoming turn. This problem is fraught with the many-to-many mapping between utterance and action mentioned above. Again, we can show the inference is often made very early using brain imaging (Gisladottir et al. 2015). In the case of responses to prior utterances, decisions about the speech act may be made by the first word. Comprehension is a massively parallel undertaking, which helps to explain its speed. In addition, the pace is pressured by the fact that delay has its own semiotics: a delay over 500 msec after a request, for example, will be interpreted as reluctance to comply; using brain imaging, one can track the hearer's moment-by-moment change of expectations (Bögels et al. 2015a).

In the end, linguistic production has to be serial due to the articulation bottleneck (Indefrey 2011), as reflected by the relatively huge latencies involved in linguistic encoding. The decision about how to respond, which words and sentence frames to employ, would counterfactually predict very slow responses (consider the psychological generalization known as Hick's Law, which holds that response times increase logarithmically with the number of possible responses from which to choose, and the 30,000 odd items in the average vocabulary). So, even though production latencies are large (up to 1500+ msec to code a simple clause), given the vast decision trees possible, they are still remarkable.

Multimodality compounds these problems, because one is dealing with the orchestration of a veritable ensemble of channels. For instance, recent work shows that

- long blinks toward the end of an utterance signal "go on" (Hömke et al. 2017),
- gesture holds signal "turn unfinished" (Torreira and Valtersson 2015),
- gaze aversion by responder signals "this is not the response you were hoping for" (Kendrick and Holler 2017), and
- frowns on a recipient or her head thrust and freeze of body position signal "repair request coming" (Floyd et al. 2016; Kendrick 2015).

These signals often overlap with the verbal signal, and the main clue to their scope is probably timing. As in the McGurk effect,[4] it is likely that such timing can be delayed by up to 200 msec without disrupting a sense of synchronization. While massive parallel computation may again solve the comprehension problem, the orchestrated production of these signals by an early utterance plan is not understood at all (for a discussion on gesture planning, see de Ruiter et al. 2012).

All this takes place within the multitasking environment of turn-taking, where halfway through an incoming turn a person is already planning a response. Figure 8.4 models this and is based on the processing of the "crunch zone," the latter part of an incoming turn (see Figure 8.1). Here, Bob is listening to Anne's turn; during the incoming turn, he first concentrates on comprehending what she is saying. As soon as he grasps the essential point or speech act, Bob calls on his production machinery and begins to formulate a response. This goes all the way down the chute to be clothed in the phonology and articulatory programs. Meanwhile, he is still listening to Anne and parsing the incoming turn, looking for points of possible syntactic completion, the end of a possible turn. As soon as one of these is detected, he checks for prosodic (or gestural) cues to turn closure. When such prosodic cues are detected, he launches his response. Given the natural limit to human response times, the response will emerge around 200 msec after the end of Anne's turn (as in the mode of the typical response times, shown in the inset histogram).

The upshot of this is that we may be deeply skeptical whether a machine can model all these processes in real time, in a way that could match a human interactant. "Deep learning" on vast databases of interactive discourse may help to capture interactive routines, but there is no prospect of a solution to the creative abduction of "the inference to the best explanation" under reflexive reasoning, the Gricean inference to intent, in the split-second manner typical of human interaction. Above all, we still have only the poorest descriptive

---

[4]   This is the effect where visual lip reading interacts with the acoustic signal to produce a blended perception, allowing fruitful experimentation of multimodal processing. Another unsolved problem is what counts as synchronization with nested structures; for example, an indexical point with "Is that big red truck over there John's?"

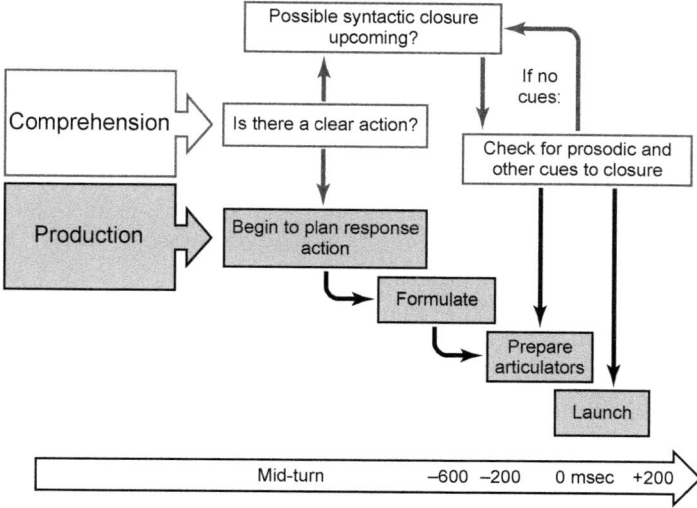

Figure 8.4   How comprehension calls production midway through an incoming turn (from Levinson and Torreira 2015).

coverage, let alone understanding, of the many features that generate human communicative interaction.

Fortunately, this does not spell doom for interactional robotics. There are two saving graces. First, humans are good at adapting their interaction to interactants with lesser skills: to infants, children, foreign language learners, and even dogs. A very useful exercise would be to try to work out exactly which properties of human interaction are indispensable, and through which outputs these would best be embodied in an interacting machine (for suggestions, see Thomaz et al., this volume). Second, humans have a natural instinct to attribute deep interactive cognition to simple machines, lower animals, and even causal events without any cognition involved. It is this tendency that makes humans susceptible to superstition, religion, witchcraft, and animism (Levinson 1995, 2006). The very properties that make human interaction possible are freely read into the world around us. This accounts for some of the signal achievements in robotic interaction, as in the finding that autistic children are brought out of their interactive shells by playing with robots (for a review, see Diehl et al. 2012). It also accounts for our wonder, like Ovid's Pygmalion falling in love with his sculptural bride, at our abilities to make even shallow simulacra: the medieval fascination with automata (Truit 2015) or "Erica" in Ishiguro's Symbiotic Human–Robot Interaction Project (Glas et al. 2016). In restricted domains (e.g., travel agencies, directory enquiries), where specific human goals can be presumed, robotic performance can pass muster because we imagine greater competence than is actually present. Such bridgeheads will help to increase the utility of our machine companions, but we should not delude ourselves that we can really simulate the complex behavior that we instinctively produce in our own actions, of which we have only the slightest analytical grasp.

Conclusion

Human social interaction is our elite capacity. Other species have spectacular navigation abilities, the ability to sleep on the wing, or swim faster than a frigate, but our highly evolved trick is communicative interaction. This has made cultural transmission possible and has propelled us into the position of the dominant species on the planet. The ingredients of this elite capacity are scarcely understood, but they include capacities to model each other's plans and actions, to foresee them, and to plan ahead accordingly. Human communication is based on this: given the asymmetry in processing latencies in comprehension and production, the only way we can maintain the observed pace of communicative interaction is by predictive comprehension and preemptive production. The modeling of another's actions requires deep knowledge of the individual (in the case of known social others) and extensive cultural expectations, including interpersonal rituals in all cases. The speed and depth of computation is not likely to be matched by any machine in our lifetimes. Perhaps in some future world, it may be partially modeled by "breeding" machines using an analogue to unnatural selection, Darwin's selective breeding, to recapitulate the evolution of our elite capacity. In the meantime, human–machine interaction can trade on our charitable overattribution of interactional intelligence to anything that moves or squawks.

Acknowledgments

I wish to thank Elena Lieven for early comments on a draft and to the members of my working group (Thomaz et al., this volume) for spirited discussion, to the reviewers who forced clarifications, and to Edith Sjoerdsma for her help with the manuscript. Finally, many thanks to Julia Lupp, Aimée Ducey-Gessner, Marina Turner, and Catherine Stephen for making participation a delight.

9

Teaching Robots New Tasks through Natural Interaction

Joyce Y. Chai, Maya Cakmak, and Candace L. Sidner

Abstract

This chapter focuses on the main challenges and research opportunities in enabling *natural interaction* to support interactive task learning. Interaction is an exchange of communicative actions between a teacher and a learner. Natural interaction is viewed as an interaction between a human and an agent that leverages ways in which humans naturally communicate and does not require prior expertise. The goal of communication is to achieve common ground and allow the learner to acquire new task knowledge. This chapter outlines the different types of knowledge that can be transferred between agents and discusses the perception, action, and coordination capabilities that enable teaching–learning interactions.

Introduction

Extending the framework introduced by Mitchell et al. (this volume), our focus in this chapter is on *natural interactions* between a human and an agent that enable interactive task learning (ITL). Reflecting most prior work on this topic, we focus on ITL scenarios where the teacher is a human and the learner is a physically embodied agent (e.g., robot) as opposed to a software agent.

Imagine an elderly couple, Katie and John Smith, who purchased a robot "Mia" as their personal assistant. Mia comes equipped with general knowledge of household chores and perceptual capabilities to recognize common household objects, such as those sold in grocery and hardware stores. Mia also has basic manipulation skills like grasping common objects or opening different types of containers. Despite these preexisting capabilities, Mia is unable to perform many tasks at Katie and John's house right out of the box. Not only does Mia need to be taught the unique tasks that the Smiths desire, it must also acquire new knowledge and capabilities to enable those tasks. The process of learning these tasks as well as task-relevant knowledge and capabilities happens through various forms of interaction with people, as in the following scenarios:

On the day of delivery, David, an employee from the company that manufactured Mia, arrives at the Smiths' with the new robot. David has an associate degree in robotic technology and has completed training on how to teach robots. The process starts with teaching Mia a map of the Smiths' house. David manually drives Mia to different rooms to construct the map and verbally provides information about each room as well as different points and regions in the room, such as where the main entrance is and the locations of appliances, trash bins, tools, and supplies. Next, David programs a set of basic skills tailored for the Smiths' house: how to open or close their cabinets, drawers, and appliances, for example, as well as how to operate various tools and appliances. He teaches Mia these skills by moving the robot's arm to demonstrate them. Then, under various scenarios, David tests the learned skills to ensure they are robust.

Once Mia is settled in the new house, the Smiths continue to teach Mia new knowledge and tasks. For example, they show where to put groceries or kitchen tools through pointing and verbally describing their locations with natural language: "The waffle maker goes in the bottom cabinet next to the stove." Katie also teaches Mia how to make their favorite dish from a family recipe. Using natural language and deictic gestures, Katie shows Mia different ingredients and demonstrates how and in which order to mix the ingredients. Mia sometimes has difficulty understanding Katie's instruction. For example, when Katie asks Mia to "grind the onion," Mia does not understand what "grind" means and subsequently asks for further instructions. Katie then provides detailed step-by-step instructions to show Mia how to perform the action "grind": "cut the onion in half, put the pieces into the blender, and press the top button." By following Katie's instruction and observing the change of the onion, Mia learns the meaning of the verb "grind" with respect to how the corresponding action changes the physical world. Mia can now transfer this understanding and perform related actions, such as "grind the carrot," assuming that Mia understands what a carrot is. Through this type of interaction, Mia continuously optimizes its task performance based on feedback from Katie, such as: "That looks slightly overcooked. Try reducing the baking time next time around."

For outdoor chores (e.g., a simple car maintenance task), John instructs Mia similarly to how he taught his son: he demonstrates how to (a) open the hood of the car, (b) check the engine oil, (c) check the radiator coolant and fill if needed, (d) check the windshield wiper fluid and fill if needed, and (e) replace the air filter if it is dirty. John and Mia both use language and deictic gestures to establish shared attention during the teaching–learning process. Once John explains and demonstrates how to fill radiator coolant, Mia can apply the learned skill to fill windshield wiper fluid. To teach the task, John uses conditional statements (e.g., "if the oil is below this line, then add coolant") and purposive descriptions (e.g., "you hold it because the funnel is too big," "put it so that the screw comes through the narrow part," or "place it right where the middle center opens into the screw so that the screw goes through the middle hole where it's open"). Mia extracts causal effect relations and converts them into schemas to support action planning and execution. The process also involves learning background knowledge mentioned in conditional statements, such as a too large funnel, the air filter being dirty, the time needed to hold an object in place, or the colors of objects through demonstrations or examples.

To understand Mia's capabilities and limitations, the Smiths can ask Mia different questions about its knowledge and its representation of the shared environment and tasks. These questions not only include "what" questions, but also "why" and "how" to assess Mia's reasoning and decision-making capabilities. Mia also proactively communicates with the Smiths about its internal representations of the world and the tasks, as well as the underlying reasoning that might take place to reach certain conclusions or decisions. Mia can even teach the Smiths' grandson how to cook their favorite dish and how to do car maintenance.

These scenarios illustrate different types of natural interaction that humans can use to teach robots new tasks or task-relevant knowledge and capabilities: by performing the task themselves, by verbally or kinesthetically guiding the robot, or through situated language instructions and gestures. This natural interaction between humans and agents instantiates the general framework of ITL, as shown in Figure 9.1. The human teacher has some *target task knowledge* in mind and intends to transfer this knowledge to the robot through various forms of interaction. Let S represent the set of states of the physical world relevant to the task and S_c represent the set of states of communication, such as the verbal utterances or focus of attention of the teacher at each step of the interaction. The robot learner perceives a *task-related world state* $s \in S$ through its sensors and constructs a *communicative state* based on its perception of the teacher's communicative actions. Let A represent the set of task-related actions (e.g., pick up an object) and A_c the set of communicative actions (e.g., asking for confirmation for its interpretation of a world state) available to the robot through its effectors. At each step of the interaction, the robot needs to decide what *task-related actions* $a \in A$ and/or *communicative actions* $a_c \in A_c$ it should take, given its current state and learning goals. The sequence of states and actions that a robot goes through during ITL constitutes its *interaction experience*. The robot needs to then extract *learning experience* from its interaction experience to obtain examples, specifications, and feedback to acquire new *task knowledge*.

Enabling ITL in robots through natural interactions requires a wide range of capabilities for perception, action, reasoning, learning, decision making, and communication. Here, we discuss the challenges and open questions associated with these capabilities. Specifically, we explore

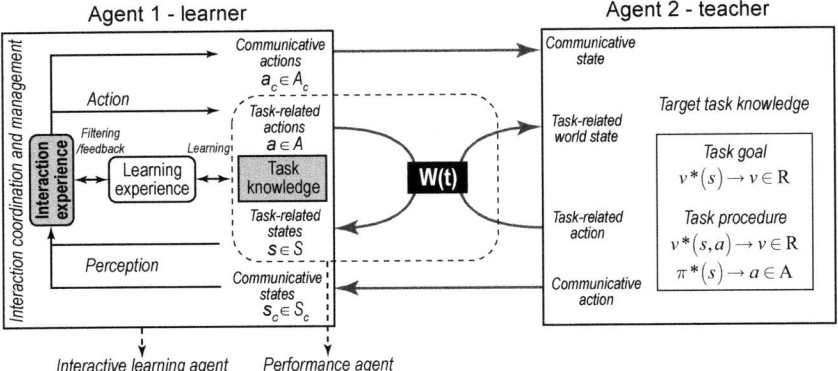

Figure 9.1 Extended two agent and world model separating task-related states and actions from communicative state and actions. Task-related states (S) and actions (A) are the minimal set of states and actions that an agent needs to perform the target task successfully. Communicative states (S_c) and actions (A_c) are what an agent needs to communicate to extract useful data and provide feedback for learning.

- forms of human teaching and the different kinds of knowledge that can be taught or learned through interaction,
- capabilities to perceive and infer task-related state and communicative state through sensors, including visual scene understanding, language understanding, and grounding language to visual perception (e.g., the environment, perception of human gestures, and perception of human actions),
- capabilities to act in the environment through effectors, including acting to manipulate the environment and communicating to the human during interaction, and
- capabilities to manage and coordinate interaction and establish common ground.

Types of Task Knowledge and Forms of Interaction

Humans can learn new tasks from other humans through various means: watching each other perform the task, doing the task themselves accompanied by instructions and guidance, or conversing and imagining the task without performing any actions (e.g., acquiring a new recipe). Similarly, as illustrated in our example scenario, robots can learn from humans in analogous ways.

As shown in Figure 9.2, in ITL the robot needs to extract learning experience from interaction experience through interaction. The learning experience can involve examples of goal states, examples of action sequences that lead to a goal, or evaluations of action sequences generated by the robot. These

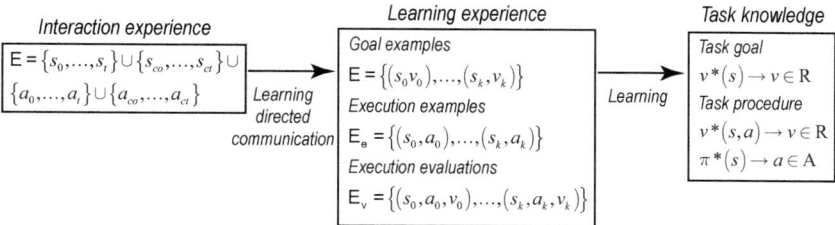

Figure 9.2   Interactive task learning is the process of converting the learning agent's experience into task knowledge. Different types of knowledge are learned from different types of example data.

different learning experiences can be expressed in terms of the physical world state (s_i), task-related actions (a_i), and values assigned to them (v_i). The goal of task learning is to extract different types of task knowledge such as task goal (e.g., $v(s) \rightarrow v$) and task procedure (e.g., a policy to perform the task $\pi(s) \rightarrow a$) from these experiences. Different learning algorithms require specific types of experience data (e.g., direct policy learning requires sequences of state-action pairs). The role of the communicative actions is to extract this data from the unstructured stream of data that the agent experiences. For instance, communicative actions by the teacher might indicate the start and end of a demonstration to help the learning process, even though the communicative states and actions are excluded from the learning data. As we discuss next, the way in which task knowledge is transferred and the role of communicative actions in that process largely depends on the type of task knowledge.

Task Knowledge Types

The main goal of task learning is to acquire *task knowledge*, which defines what a task is and provides sufficient information to permit the robot to perform the task on its own. There are different types of task-related knowledge and capabilities (described below) that can be acquired during interaction. As discussed by Laird et al. (this volume), task-related knowledge often includes goals, actions, and constraints which define the problem space as well as procedural/policy knowledge needed to perform the task. Here, we focus on two types of task knowledge and their representations: task procedures and task outcomes.

Task procedure information captures what the agent needs to do to complete the task, as shown in Figure 9.2. Most existing agent frameworks represent procedural information as a policy, which is a function that maps the perceived state to an action: $\pi(s) \rightarrow a$. Such functions can be represented by many different types of classifiers or regressors and be learned from examples. Process information can also be captured in more explicit forms such as plans, programs, finite-state machines, or hierarchical task networks. Although these different representations do not necesarily provide a full mapping of states to actions, they still capture procedural knowledge by specifying a sequence, a partial ordering, a schedule, or a hierarchical organization of actions in the context of a task. For example, Pardowitz et al. (2007) introduced task precedence graphs that capture ordering constraints between actions involved in a task. Similarly, Ekvall and Kragic (2008) represent tasks with a set of ordering constraints between pairs of actions. Alexandrova et al. (2015) use a flow diagram to represent tasks with actions that have pre- and postconditions that can cause branching in the program. Huang and Cakmak (2017) use the general-purpose visual programming language, Blockly, to represent various branching and looping tasks.

Task outcome information relates to the goals or desired outcomes of a task, independent of the process followed to achieve them. This is different from the actual outcomes when performing a task (which can be expected or unintended). Task goals are often captured by the reward or value functions associated with states and actions, assuming the agent is maximizing reward or value. In practice, task goals might be easier to express in terms of world states in which the task is considered complete; for instance, a conjunction of state variables that need to be true or other arbitrary functions that evaluate a given state in terms of whether the goal is achieved. A value can then be associated with each state based on how close they are to a goal state. The task "tidy up the living room," for example, could be specified with the list of items in the room and their desired locations, without any information on how to get them there. Such a representation was used by Chao et al. (2011) to represent simple object reconfiguration tasks. The ability to carry out tasks based solely on specified goals often requires the robot to have planning capabilities.

Some task representations involve combinations of process and outcome information. For example, a recipe for a particular dish specifies not only a sequence of actions but also mentions what to expect at the end of the process or when a task is considered complete.

Forms of Interaction in Transferring Task Knowledge

There are many forms of interaction that enable transfer of task knowledge. Our focus here is on two key types of information transferred in those interactions: *demonstrations* of the task and direct *specifications* of task constraints or properties.

In learning task processes from *task demonstrations*, multiple demonstrations provide alternative ways of achieving the same task (e.g., Argall et al. 2009). Different task representations capture this information in different ways. For example, a partially ordered plan captures alternative orderings of low-level actions. Hence different demonstrations of the task might involve a different ordering of actions. Similarly, a program with conditionals and loops captures alternative ways of performing a task, depending on a perceivable "condition" or different numbers of repetitions contingent on user-specified or environmental parameters. Different demonstrations of such tasks will involve alternative traces of the program. Task outcomes can also be taught by demonstration. Multiple examples provide variations of the states in which the task is considered successfully completed. One of the key computational challenges is to identify parts of the state that are relevant/irrelevant for the task. Thus it is important for demonstrations provided by the teacher to involve such variations.

Tasks can be demonstrated through different forms of interaction (e.g., by the teacher performing the task, or provided directly to the robot, with guidance from the human teacher). In tasks performed by humans, one of the most intuitive ways to demonstrate a task is for a human to perform it herself. For the robot to learn from this type of demonstration, the robot must be able to perceive the human's actions and/or the effects of human action on the environment. Perception of human actions can be facilitated through external sensors or wearable sensors on the human. Once a robot perceives human actions, they need to be mapped to corresponding robot actions. This is referred to as the retargeting problem. In some cases, perception of actual actions is not necessary, as long as the robot can detect the state changes that result from the task demonstration and learn the task based on that information (Baisero et al. 2015; Mollard et al. 2015).

For tasks performed by a robot with guidance from a human, the human teacher demonstrates a task to the robot by guiding it through the task. This mode of teaching bypasses the retargeting problem but requires the teacher to have a good understanding of the robot's action capabilities. The guidance to the robot can be provided in various ways, from kinesthetic movements to verbal instructions:

- *Kinesthetic guidance* involves physically holding the robot and moving its manipulators to perform the task (e.g., Akgun et al. 2012; Phillips et al. 2016).
- *Natural language guidance* involves instructing the robot on what to do to perform the task. Mohan et al. (2012) and She et al. (2014), for instance, use step-by-step language instructions to teach new tasks to a robot.
- *Multimodal language guidance* uses multimodal instructions (speech and gestures) to guide a robot through the task.
- *Gestures* often serve to reference parts of the environment.
- *Joystick-based guidance* involves driving the robot and triggering prespecified actions with the help of a special device to perform the task.
- *Guidance based on graphical user interfaces* (GUIs) employs a graphical interface to control the robot and trigger prespecified actions to perform the task.

In *task specifications,* alternative ways of achieving a task are directly specified by the teacher in a format compatible with the robot's task representation. For example, for a partially ordered plan representation, the teacher might verbally state:

> First *bring all of the ingredients and tools* to the kitchen counter (in any order).
> Second, pour *all of the dry ingredients* into the mixing bowl (in any order).

While teaching by demonstration inevitably involves a particular ordering of the actions and hence requires multiple demonstrations to capture order invariances, direct specification provides an efficient way to provide the same information. Similarly, if the representation is a program, the user can directly specify loops or conditionals by literally writing a program or verbally specifying those with instructions:

> Insert a toothpick into the center of the cake. If it comes out clean, take out the cake; otherwise continue to bake. Alternatively, for each cup on the muffin pan, pour until three-quarters full.

Similarly, direct specification of task goals involves the teacher directly indicating parts of the world state that are relevant or irrelevant to the robot's task, rather than trying to exemplify variations of positive and negative goal states. For example, the teacher may verbally describe the desired goal state when teaching a robot to set a table:

> The red bowl should be on top of the green plate, and the napkin should be placed to the right of the plate.

Task specifications can be provided through natural language or GUIs:

- *Natural language specifications* involve the use of language to specify directly certain properties or constraints about the task representation. For example, Cantrell et al. (2012) use natural language to specify precondition and effects of action schemas for task planning.
- *GUIs* can be used to specify properties or constraints about a task being taught to a robot.

Humans often combine these two means of communicating task knowledge (demonstrations and specifications). For example, a teacher might demonstrate the physical act of adding different ingredients to the mix in a particular order as part of teaching a recipe, while verbally specifying partial ordering constraints by saying: "Add all dry ingredients in any order." Similarly, a person might set up the table themselves to show an example of how they want the table to be set, but then specify invariance constraints by saying: "The salt and pepper can be anywhere in the center area of the table."

Regardless of the specific form of interaction, during learning, symbolic representations of human inputs (e.g., GUI, natural language) need to be tightly grounded to the robot's internal representations of perception and action.

Task-Relevant Background Knowledge and Capabilities

When we speak about a robot learning new tasks, we often assume that the robot has the necessary background knowledge and capabilities. The ability to perform new tasks, however, might equally be due to the acquisition of other knowledge or capabilities, not solely due to newly acquired task knowledge. Hence, the ability to acquire these different kinds of background knowledge and capabilities through interactions is also highly relevant for task learning. For instance, the capabilities of a robot that already knows the task of sorting objects, based on different properties, can be expanded through the acquisition of new perceptual capabilities (e.g., the ability to detect new object properties) or new action capabilities (e.g., the ability to manipulate new types of objects). Below we list four types of knowledge and capabilities relevant for task learning:

1. *Perception capabilities* refer to the ability to perceive the task-relevant environment and interpret human language, including
   - state and actions of humans,
   - state, properties, and affordances of objects,
   - scene composition (surfaces, objects, humans, and their relationships),
   - changes of the state that occurred to the environment, and
   - state of communication such as communicative intent and focus of attention.
2. *Action capabilities* refer to lower-level policies that control a robot's actuators to carry out tasks and/or communicate with humans. These include capabilities that allow robots to
   - navigate the environment,
   - manipulate objects in the environment, and
   - communicate with humans in the environment.
3. *Linguistic knowledge* concerns the meanings of words and phrases. For physical robots, which need to sense from and act upon the physical world, as opposed to the symbolic world, this knowledge cannot be purely symbolic as in a dictionary or thesaurus. Word semantics need to be grounded to the robot's sensorimotor skills.
4. *World knowledge* captures any other task-relevant knowledge about the world and how the world works:
   - *Facts* about the world and the robot's task environment: "my owner's name is Katie Smith" or "I was built in 2017."
   - *Commonsense knowledge* that allows a robot to interpret human language and reason about its perception and action goals (Al-Moadhen et al. 2013; Tenorth and Beetz 2009): to "boil the water," water must first be placed in the boiling pot.

- *Action knowledge* that captures the existing knowledge about subtasks and subgoals previously acquired or learned. Formal action models capture preconditions and effects of actions (Fox and Long 2003). Preconditions specify world states in which the action is applicable; effects specify the expected changes to the world state.
- *Domain knowledge* that corresponds to information specific to a particular task environment or user that a robot needs to perform its task. For example, a robot that performs object deliveries to hotel rooms needs to have a map specific to the hotel within which it is deployed, with room numbers annotated on the map.

Some of these types of knowledge and capabilities can be programmed into a robot. They can also be acquired through interactions with humans, although the means of acquisition is less clear than that for task knowledge.

Forms of Interaction for Learning Task-Relevant Knowledge or Capabilities

The types of interactions that support acquiring task-relevant knowledge and capabilities are similar to those involved in learning the task itself. As shown in Table 9.1, the forms of interaction often depend on the kind of knowledge or capabilities to be learned. For example, to help train the robot's visual perception capabilites, the teacher may use language descriptions and also show target objects from different angles. To acquire the navigation map, teleoperation (e.g., through joystick guidance) can be employed as well as language

Table 9.1 Example forms of interaction for different types of knowledge

Knowledge	Example Forms of Interaction
Perception capabilities	• Natural language and deictic gestures to teach labels of objects and indicate their relations • Natural language to specify object affordances
Action capabilities	• Kinesthetic demonstration to teach low-level control policies to generate arm trajectories or navigation strategies
Linguistic knowledge	• Natural language combined with deictic gestures to teach nouns and adjectives • Natural language combined with action demonstration to teach action verbs
World knowledge	• Natural language to specify order constraints among sub-actions • Natural language to specify causality (i.e., precondition and effect) of an action • Demonstrations performed by the human to show how basic actions/verbs change the state of the world • Joystick guidance to build a map of the robot's environment for navigation

descriptions. Acquisition of low-level action knowledge (e.g., lower-level policies to generate trajectories) may benefit from kinesthetic demonstration whereas higher-level task knowledge (e.g., partial orderings) may best benefit from language instructions. Linguistic knowledge certainly involves the use of language, which is often combined with deictic gestures or action demonstrations because the semantics of words need to be grounded to visual perception and the change of state in the physical world.

Open Questions in Enabling Effective Task Learning Interaction

Teaching Presupposed Task-Relevant Knowledge

While previous work has investigated the acquisition of many types of task-related knowledge and capabilities, the acquisition of commonsense world knowledge in task learning has largely gone unexplored. In human–human interactions, knowledge about the world and the domain is often presupposed. The speaker and the listener believe they share the same kind of world knowledge, so it does not need to be explicitly stated. However, in human–robot interactions, huge discrepancies in world knowledge can exist between humans and robots. Often, the robot does not have sufficient background knowledge to learn a new task. Thus human teachers need to be able to assess what kind of background knowledge the robot has and how to teach the robot background knowledge pertinent to the task at hand. For example, the result states of basic action verbs are not usually specified, and humans naturally take them for granted. Existing lexical resources (such as Verbnet, FrameNet) and preexisting knowledge bases (e.g., Google's Knowledge Graph, Freebase) do not offer the level of detail required for the robot to understand the very basic principles about the conditions for their actions (e.g., "put A on B" requires A generally smaller and lighter than B) and how their actions may change the world (e.g., slicing a cucumber may lead to the change of the shape, size, and pieces of the cucumber).

Thus, it is important to understand *what a human must teach a robot about the domain of a task.* Some background knowledge (e.g., time as duration, units of time, and time relations) may be best taught or acquired once for many domains, but much human knowledge is domain specific. Learning domain-specific knowledge leads to a whole new set of research questions:

- How does the human know what knowledge the robot (e.g., sub-actions) has so that it can be used to teach new tasks?
- During task learning, what signals indicate the lack of background knowledge and therefore human teaching is required?
- How can existing resources be leveraged to acquire the correct level of background knowledge during teaching?

- What level of granularity should background knowledge be taught by a human?
- How should background knowledge be represented and used for effective reasoning and inference?

Combining Different Forms of Interaction for Task Learning/Teaching

Most previous work on task learning has focused on a single form of interaction for teaching. Except for a limited few (Kirk et al. 2016; Mohseni-Kabir et al. 2018; Niekum et al. 2015; Rybski et al. 2007), techniques that combine language, dialogue, and action demonstration to teach complex tasks are in critical need. As discussed above, different forms benefit different types of knowledge. In addition, as the situation changes (e.g., the lighting situation changes from being good to poor), the form of interaction may need to adapt (e.g., switch from visual demonstration to language instruction). Thus we need to know *how to seamlessly combine and adapt different forms of teaching to enable the most effective teaching*. Is combining and adapting a problem for human teachers or a problem for robot learners? The answer is both.

Teaching Humans How to Teach Robots

After working with a robot, an experienced human teacher (in our scenario involving Mia, this would be David, the employee from the robotic manufacturer) should be able to discern which form of interaction is necessary to teach a specific kind of knowledge to meet specific circumstances. Experienced human teachers should know when to provide a particular kind of feedback (i.e., reward or punishment) so that the robot can learn from such feedback and adjust its behaviors to maximize future rewards. Experienced human teachers may also apply scaffolding, intentionally vary the situation, and design different experiences for the robot to learn the task and aspects associated with the task.

Thus, similar to the setting in human skill learning, human teachers' behaviors and experience have a massive influence on the success of robot task learning. *How, then, should we train a new generation of human partners/teachers, so that robots can be effectively taught through their collaborations?*

Enabling Robots to Engage Proactively in Learning

We cannot expect that every human partner will be capable of identifying and employing the most effective means to teach the appropriate kind of knowledge. Thus a robot needs to be able to share the burden of selecting effective strategies. A crucial issue, as yet unstudied, is: *How can a robot be made to be aware of its own learning situation—one in which it is capable of communicating to the human its limitations and proactively requesting the right kind of teaching from the human?*

Capabilities to Perceive the Environment and Human Inputs

The ability to perceive the environment and human inputs as well as to infer current task-related states and communicative states is fundamental to ITL. A robot must be able to recognize task-relevant objects in the environment, the change of the environment caused by an action, task demonstration from humans, as well as verbal and nonverbal human communicative behaviors. It must also be able to infer human intent, interpret instructed actions and their involved objects, and derive task structures by grounding language to perception.

Visual Perception

Performing or learning tasks inevitably requires an understanding of objects and environments integral to the tasks. This includes objects, their properties, fluents (i.e., attributes which can potentially change), and relations, as well as an understanding of external actions and how they may have changed the perceived state of the physical world. As humans can perform actions to teach robots and apply nonverbal modalities (such as deictic gestures, iconic gestures, and gaze directions) to facilitate communication, the robot should also have the capability to recognize the state and actions of its human partners.

Acquiring perceptual capability has been the main research goal for the computer vision community. Most of the learning algorithms for perception are trained offline and rely on large training data for object recognition, activity recognition, and so forth. Recent years have seen significant progress on recognition of common objects from static scenes (e.g., images) (Grauman and Leibe 2011). However, in a dynamic scene, such as would be encountered in task learning, object tracking and human action recognition still face many challenges (for reviews, see Aggarwal and Ryoo 2011; Sargano et al. 2017). In addition, during task learning, it is likely that neither relevant computer vision models nor sufficient data are available. Thus, it is critical for the robot to continuously acquire new models for object recognition through interaction with its human teacher. The teacher can use language to provide the name, the object type, and related properties to a perceived object in the environment, and the robot needs to learn a generalized model efficiently (e.g., for object recognition) that can be applied in new situations. Key research questions include:

- How can a robot learn reliable models based on a small number of examples with limited human supervision during interaction?
- How can it transfer and adapt models learned from previous experience to a new situation (e.g., transfer learning), perhaps with limited human intervention?

Language Understanding

Language plays an important role in ITL. From a human's linguistic utterance, the robot needs first to understand the underlying intent of the teacher (e.g., whether it is to teach the robot a new step or to correct the robot's current understanding of a learned step or action). When a referring expression is involved, the agent needs to understand what entities, from the interaction discourse or the shared environment, are being referenced. When the utterance describes some task steps, the agent needs to understand what actions are specified and what participants are involved (e.g., agent, patient, instrument, source, destination). The robot also needs to be able to extract any information from the utterance that specifies preconditions, effects, and constraints (e.g., temporal orders) associated with actions and tasks. To help achieve the above-mentioned abilities, recent advances in natural language processing—particularly in syntactic parsing, semantic processing, and discourse processing—can be applied (Jurafsky and Martin 2008). In the event that the robot cannot successfully understand human utterances, dialogue can be applied to clarify human intent and disambiguate different interpretations of linguistic expressions.

In situated interaction, language communication is often accompanied by other nonverbal modalities, such as gesture. Deictic gestures (e.g., pointing to objects in the environment) and iconic gestures (e.g., waving hello or indicating an action or a particular type of object) are vital to an understanding of the teacher's intent. Pointing gestures are essential to task instruction because the array of objects in a task (which may be difficult to describe verbally) lead to the need to point at them rather than rely solely on language descriptions. Matuszek et al. (2014), for example, combine language and gesture to interpret directives in human–robot interaction.

Speech communication is perhaps one of the most natural means of interaction in task learning. Speech recognition has made significant progress over the last decade. More recently, advances in deep neural networks have made it possible for machines to achieve recognition performance on par with human performance. At the time of writing of this article, Google reported a 4.9% word error rate in recognition while human performance is estimated to be around 4% word error rate (Saon et al. 2016). Although encouraging, these results were often obtained based on offline benchmark data. Thus, it is not clear whether the same performance can be attained in a real-time, interactive, and unconstrained environment. *How can recent advances in speech recognition be successfully applied to real-time interactive systems for task learning?*

Unlike traditional natural language processing, linguistic knowledge must go beyond pure symbolic representations—as in a dictionary or thesaurus—to enable communication with physical robots. The meanings of words need to be grounded to the robot's internal representations that are connected with sensors and effectors. Concrete nouns, for instance, need to be grounded to the types of objects or object attributes perceived from the environment (e.g., color words grounded to color histograms). Adjectives are often grounded to the perceived attributes (e.g., the size of the bounding boxes, the weights of an object) and fluents (e.g., door open or closed, box open or closed). Verbs need to be grounded to the underlying action representations, which can be accessed by the robot's control system to plan and execute the corresponding actions. On one hand, existing knowledge of grounded word semantics will be applied to ground language to perception and action (discussed in the next section). On the other, as new words are often encountered during interaction, they should be acquired continuously through situated interaction (Mohan et al. 2012). When a situation changes (e.g., a change in the environment), the learned word representation may not fit the new situation (e.g., a lighting change in the environment may affect grounded word models for color words). Thus, it is important that word models are adaptable to new situations (Liu and Chai 2015; Thomason et al. 2015).

Grounding Language to Perception

The capability to ground human language to the perceived physical environment is particularly important for task learning. Suppose a human teaches the robot how to boil water by demonstrating to the robot how to achieve this task through step-by-step instructions: "pick up the pot, fill the pot with water, boil the water…" To learn how to perform this task, the robot must first understand what perceived objects are involved in each step of instruction by grounding the arguments of action verbs, such as the noun phrase *the pot*, to the perceived objects in the environment.

This task of grounding language to perception of the environment has received an increasing amount of attention (Krishnamurthy and Kollar 2013; Matuszek et al. 2014; Mooney 2008; Tellex et al. 2011, 2014; Yang et al. 2016; Yu and Siskind 2013). Most previous approaches first process language and vision separately, and then integrate the partial results together. In a dynamic scene with ongoing activities, computer vision algorithms still have difficulty reliably recognizing and tracking objects and actions; this leads to a bottleneck in grounding language to vision. Recent deep learning approaches directly fuse raw features from language and vision and have achieved state-of-the-art empirical results on applications such as caption generation from images/videos and visual question answering. These approaches, however, require a large amount of training data. *To integrate language and vision in the context of ITL, what would be the optimal architecture?*

Another line of recent work has explored causality modeling for action verbs (Gao et al. 2016). Here the idea is that knowledge of how concrete action verbs (e.g., cut, slice, pick up, etc.) might alter the world can drive visual detection. For example, from the directive "*slice the cucumber*," knowledge about expected changes to the cucumber will provide high-level guidance to look specifically for grounded objects with relevant features (or the change of features) in the visual scene. Recent work has also explored commonsense physical knowledge about objects that are implied by action verbs (Forbes and Choi 2017). For example, *"he threw the ball"* implies that *"he"* is bigger and heavier than *"the ball."* This kind of implicit knowledge can potentially provide additional cues to ground language to perception.

Capabilities to Act and Communicate

Enabling a robot to learn new tasks requires capabilities to carry out task-related actions as well as actions that facilitate communication. These capabilities span a wide range, from navigation and manipulation to communication.

Task-Related Actions and Grounding Language to Action Representation

A robot's action capabilities can be based on manually designed and tuned controllers as well as policies learned from human demonstrations or through reinforcement learning. In some robotic applications, it is essential for the robot already to possess all of the action capabilities needed to complete a task. For example, previous work in the robotics community aimed to translate natural language instructions to robotic operations (Kress-Gazit et al. 2007; Spangenberg and Henrich 2015), but they were not designed for learning new actions or tasks. In other cases, tasks and actions can be learned simultaneously. For example, Mohan and Laird (2014) developed a system where a robot can learn a hierarchical representation of a new task based on linguistic interaction with the human. Similarly, Liu et al. (2016) applied grammar induction to learn a hierarchical and/or graph representation for a new task from a human's language instructions and visual demonstrations.

To support action learning from language instructions, recent work has begun to explore the connection between semantics of concrete action verbs and action planning (Misra et al. 2016; She et al. 2014) and explicitly represented grounded verb semantics as desired goal states of the physical world as a result of the corresponding actions. Such representations are learned based on example actions demonstrated by the human. For example, a human may teach the robot how to "boil water" by issuing step-by-step language instructions which the robot knows how to perform: "move to the kettle, grasp the kettle, move to the stove…" By following these steps, the robot will experience the change of the physical world. By capturing the differences between the goal state and the initial state, the robot is able to acquire the semantics of the verb frame "boil (water)." Once acquired, these grounded representations will allow the robot to interpret verbs/commands issued by humans in new situations and apply planning to execute actions. One limitation of previous work is that the algorithms were mainly developed based on simulations (e.g., simulated Baxter robots). Except for a few (e.g., She and Chai 2017), uncertainties from the environment were largely ignored. However, the world is full of uncertainties at various levels: from motion planning to perception and language grounding. To extend task learning from language instructions to the physical world, it is paramount to address *how to integrate uncertainties at multiple levels together, so that new actions associated with concrete action verbs can be learned.*

Verbal and Nonverbal Communicative Action

Separate from its task-related actions, a robot will need to perform communicative actions to facilitate its learning/teaching interactions. In situated interaction, both verbal and nonverbal modalities are available for the robot to communicate to its human partner. Some example communication abilities include:

- generating speech and deictic gestures to confirm understanding of instructions or refer to objects in the environment (Fang et al. 2015);
- generating gaze direction, communicative head gestures (e.g., nodding and shaking head), or facial expressions (confused or confident face) to respond to human input at different points in the interaction (Holroyd et al. 2011); or
- displaying visualizations of learned concepts to enable humans to inspect them.

In particular, the embodiment of a physical robot can take advantage of nonverbal modalities (e.g., gaze and gesture) for efficient communication. The robotics community has learned from psychologists that gazing at others and at objects in the environment are quintessential human behaviors. Gaze that is used to convey information to a collaborator is referred to as *social gaze*. Gaze at a collaborator functions to gather attention from the other, to indicate social presence, and to indicate attention to the individual (e.g., turn-taking via gaze aversion). Gaze at objects serves to indicate what one is paying attention to, is about to point at, what one intends to do next, or to indicate that what another has focused on should now be the object of mutual gaze. Collaborators use gaze information to assess how well their partners comprehend their collaborations as well as to assess the collaborators' level of continued engagement (Rich et al. 2010). Every one of these abilities is valuable in task learning, as they enable the assessment of how the learning is progressing, whether the learner is looking in the right direction, and what the teacher intends for the learner to do. Gestures also have similar effects in coordinating interaction, establishing shared attention, and providing feedback. Proxemics, which models the stance of individuals to others and how they approach one another, can be significant in tasks because where the learner stands in performing a task may be crucial. *How to generate verbal and nonverbal communicative behaviors effectively to facilitate task learning remains an important focus for future research.*

Capabilities to Manage and Coordinate Interaction

Managing interactions between humans and robots is critical to support task learning/teaching. At any point in the interaction, robots need to decide what to do next based on interaction history, current situation, and learning goals. These decisions can be made by following simple decision rules that are manually crafted or interaction policies that are learned from experience.

Interaction Management and Active Learning

Decades of work on dialogue modeling are relevant for ITL. Different approaches have been developed, for example, driven by intention and collaboration (e.g., Grosz and Sidner 1986; Rich and Sidner 1998), based on information states (Larsson and Traum 2000) or interaction policies learned from reinforcement learning (Kaelbling et al. 1996; Young et al. 2013). Despite recent progress, dialogue modeling remains a significant challenge. Dialogue models need to be able to accommodate interruption, turn-taking, and other dialogue behaviors, which neither the intention-based nor information state approach have successfully addressed, but are essential in task instruction.

Specifically, to learn new tasks, active learning has been shown to be an important component that contributes to effective interaction management. Most work on task learning assumes a learner that passively receives information from the teacher. However, humans are often suboptimal in their teaching when the learner is passive. One line of work explores active task learning whereby the learner actively requests specific information that it evaluates as most useful. Active questioning enables much more efficient learning. For example, Chao et al. (2010) and Cakmak et al. (2010) demonstrated that an active learner (both human and robot) which requests labels (positive/negative) for specific instances of a task goal outperforms a passive learner taught by examples selected by naïve human teachers. In particular, Cakmak and Thomaz (2012) identified three types of queries that can be used by a human/robot student as part of active task learning:

1. Demonstration queries asking for a full or partial demonstration of the task
2. Label queries asking whether an execution is correct
3. Feature queries asking about the relevance or invariance of specific aspects of the task

Recent work by She and Chai (2017) extended this question–answer style of interaction and applied reinforcement learning to acquire an interaction policy that allows the robot to handle noisy environment and learn new verbs and corresponding actions. To improve ITL, we need to know how to engage in a full range of interaction that can incorporate active learning with other communicative goals (e.g., clarification and disambiguation) to acquire more reliable models of skills.

Extra-Collaborative Effort and Transparency

In human–human task learning, human teacher–learner partners often share similar perceptual capabilities as well as basic commonsense knowledge to support their collaboration.

In human–robot task learning, however, there are huge discrepancies in background knowledge between humans and robots. The robot, for instance, often does not have sufficient background knowledge to learn a new task. Furthermore, although they may be co-present in a shared environment, humans and robots have mismatched capabilities in reasoning, perception, and action: their representations of the shared environment and joint tasks can be significantly misaligned. A significant challenge involves the lack of common ground and discrepancies in the human's mental model of what a robot knows and is capable of doing. Previous work (Chai et al. 2016) has shown that to bridge the gap and strive for a common ground of shared representations between humans and robots, extra effort is needed. This extra-collaborative effort in interaction not only has implications in algorithms for language grounding, but also affects interaction management.

Transparency plays an important role in achieving common ground and promoting accurate mental models during interaction. For example, Thomaz and Breazeal (2006) demonstrated that natural transparency mechanisms, like gaze, can steer the human's behavior while demonstrating a task. Pejsa et al. (2014) used facial expressions to provide transparency about dialogue uncertainties. Alexandrova et al. (2015) employed interactive visualizations of learned actions to enable teachers to verify tasks that are learned from a single demonstration and correct any mistakes they detect. Guha (2016) used pointing to communicate the robot's understanding of a referenced object, and Whitney et al. (2016) used heat map visualizations and facial expressions to communicate uncertainty about its inference. Recent work by Hayes and Shah (2017) allows a robot to automatically generate verbal description of its learned policy (i.e., which actions it takes in which contexts).

To enable common ground for effective task learning, there are many research questions to pursue:

- How can an agent make its internal representations (e.g., causal-effect relations) transparent to the human?
- How can an agent explain its autonomy or decision so that the human can better understand the agent's capabilities and limitations?
- What are the mechanisms to manage interaction so that it can encourage a human's collaborative behaviors and simultaneously create more collaborative behaviors from the robot?

Conclusions

To fully support teaching robots new tasks through interaction, many challenges and open questions remain as discussed above. While the scenarios in the introduction focused on in-home settings, teaching robots new tasks is applicable in many situations, especially ones with highly structured environments. Already robots are being trained by people in ad hoc ways to work in manufacturing assembly lines (e.g., Guizzo and Ackerman 2018). Robots working in warehouses are largely programmed by hand, but it is not difficult to envision the need for them to be taught tasks by human coworkers. The same applies to robots in the service industry (e.g., hotel helpers).

One key challenge in task learning, which we did not discuss, is evaluation—a critical and difficult issue in interactive systems because many confounding factors are involved. In the context of ITL, the following questions arise:

- How do we know that the task has been learned?
- What additional metrics are needed to evaluate the success of task acquisition beyond traditional metrics for evaluating interaction (e.g., efficiency and task completion)?
- What are reasonable baselines and upper bounds, for example, learned fron human–human interaction?
- How do researchers conduct longitudinal studies and evaluation?
- What kinds of products are available that might make longitudinal evaluation (e.g., putting robots in people's house) possible?

While our focus in this chapter has been mainly on task learning where humans serve as teachers and robots serve as learners, it is not difficult to imagine that a well-trained and capable robot could also teach humans new tasks. In the intelligent tutoring world, computer programs have been teaching humans in various ways for more than three decades. Virtual agents teach humans all sorts of tasks, from turbine engine operation (Rickel and Johnson 2000) to negotiation (Gratch et al. 2015) to cross cultural communication (Johnson and Zaker 2012). The idea that robots might teach humans has received relatively little attention, perhaps in part due to the lack of capabilities. Robots are not yet teachers, but for many tasks (e.g., from doing experiments to manipulation of heavy equipment), the physical form of a robot will be useful in ways that computer programs and virtual agents are not. As robots become more capable, a reversal of the teacher/learner role is foreseeable and will bring further research challenges and opportunities.

10

The Essence of Interaction in Boundedly Complex, Dynamic Task Environments

Wayne D. Gray, John K. Lindstedt, Catherine Sibert,
Matthew-Donald D. Sangster, Roussel Rahman,
Ropafadzo Denga, and Marc Destefano

Abstract

Studying the essence of interaction requires task environments in which changes may arise due to the nature of the environment or the actions of agents in that environment. In dynamic environments, the agent's choice to do nothing does not stop the task environment from changing. Likewise, making a decision in such environments does not mean that the best decision, based on current information, will remain "best" as the task environment changes. This chapter summarizes work in progress which brings the tools of experimental psychology, machine learning, and advanced statistical analyses to bear on understanding the complexity of interactive performance in complex tasks involving single or multiple interactive agents in dynamic environments.

Introduction

The shape of a gelatin dessert cannot be predicted from the properties of gelatin, but from the shape of the mold into which it was poured. If people were perfectly adaptable, psychology would need only to study the environments in which behavior takes place (Simon 1992:156).

[B]ehavior cannot be predicted from optimality criteria alone without information about the strategies and knowledge agents possess and their capabilities for augmenting strategies and knowledge by discovery or instruction (Simon 1992:157).

Auch zum Zögern muß man sich entschließen. [Even the hesitation you have to decide.] (Lec 1962)

Simon had it easy. Much of the world he studied moved in discrete steps, as in the Tower of Hanoi (Anzai and Simon 1979; Simon 1975, 1989) or Chess (Gobet and Simon 1996; Simon 1989; Simon and Chase 1973; Simon and Gilmartin 1973). In choosing such tasks, he revealed to us the foundation on which cognitive science could be built and forced us to consider not just the world in our head, but how the capabilities and capacities in our heads were shaped by the mold of the world in which we found ourselves.

The gelatin mold analogy that Simon so liked leads us, perhaps unwittingly, to think of the world as static and ourselves (and maybe the next generation of our creations) as the only active agents. Yet if we stop for too long, it will rain and we will be forced to seek shelter. If we keep working hard, we will become tired and hungry and need to seek food and a place to rest. Indeed, perhaps we are less proactive and more reactive to changes in our task environments than we would like to believe.

In this chapter, we discuss our recent work with single and multiple agents seeking to accomplish complex tasks through a series of sequential choices made in dynamic task environments. In all cases, the choices are presented by an active task environment and the goal is to use our cleverness to deal with that environment and survive as long as possible. In all cases, even doing nothing requires a decision to do nothing.

Although many of the other chapters in this volume discuss humans interacting with robots, or the more general framework of two interacting agents learning from each other (Mitchell et al., this volume), we see the essence of interaction as not defined primarily by biology (or the lack thereof) or agency. Instead, we propose that the essence of interaction lies at the intersection of (a) the skills and abilities of one or more individual agents, (b) the definition of the task, and (c) the nature of change in autonomously dynamic task environments. A key point of emphasis is that intelligent agents learn from interaction with their environments just as they do from each other. The examples in this chapter serve to highlight this insight and its importance in a broader conceptualization of interactive task learning (ITL).

Background

It seems to me therefore that mental training in schools, in industry and in morals is characterized, over and over and over again, by *spurious limits*—by levels or plateaus of efficiency which could be surpassed. The person who remains on such a level may have more important things to do than to rise above it; the rise, in and of itself, may not be worth the time required; the person's nature may be such that he truly cannot improve further, because he cannot care enough about the improvement or cannot understand the methods necessary. But sheer absolute restraint—because the mechanism for the function itself is working as well as it possibly can work—is rare. (Thorndike 1913:181)

The task environment and the properties of the human cognitive, perceptual, and motor systems act as soft constraints on human behavior (Gray et al. 2006). With all else equal, the human cognitive system tends to select the fastest way it can to get a job done, with the result that no one modality is privileged and the mix of methods selected are sensitive to the burden placed on cognitive, perceptual, and motor components of human cognition. However, nothing is simple about human cognition. Once a method is acquired and used, it receives the benefits of knowledge compilation (Anderson 1987), which may make it faster and more efficient than a newer but unpracticed (and thereby "uncompiled") method even though, with practice, the efficiency of the newer method would surpass the old. Unfortunately, this is a fairly common, human situation. If the unpracticed new method is slower or otherwise less efficient than the compiled older method then, as Thorndike observed, it can be exceedingly difficult to entice people to "care enough about the improvement" to put in the time and energy needed to acquire the more efficient method.

A sterling example of this is the time and effort that people who are visually guided typists (a.k.a. "hunt and peck," "eagle fingered") need to spend if they wish to become touch typists. Indeed, the arduousness of this transition, together with the drop in performance while learning the new method, is the main source of recidivism (Yechiam et al. 2003). This pattern of an "easy" but suboptimal method interfering with the acquisition of an initially more difficult, but ultimately faster method has also been shown to be the case for people who first acquire simple menu-based methods for computer-based tasks and are then taught faster scripting-based methods (Cockburn et al. 2014; Yechiam et al. 2004). Fu and Gray (2004) coined the term "stable suboptimal performance" in the context of a study in which an expert architect was shown to have imposed the sequence of steps from his long-established paper and pencil drafting practices onto his current architectural CAD/CAM system.

Our recent work builds on the *soft constraints hypothesis* (Gray et al. 2006) and the concept of *stable suboptimal performance* (Fu and Gray 2004) to explore the elements of extreme expertise in complex, interactive behavior in dynamic task environments.

Framing the Work: Plateaus, Dips, and Leaps

In studying the behavior of people who become expert performers, we must look beyond group measures and focus on the behavior of individuals; that is, at their explorations, failures, and successes as they strive to become experts. For example, we (Destefano 2010; Destefano and Gray 2008) had people play the complex game of *Space Fortress* (Donchin 1995; Mané and Donchin 1989) across 31 sessions of 8 games per session (248 games total). Averaging the data across hours and across players produces the classic performance curve shown in Figure 10.1a: with few exceptions (Anderson 1987; Fitts 1964; Newell and Rosenbloom 1981), performance improves steadily with practice. Unfortunately, as shown in Figure 10.1b, this smooth average represents no player's actual performance. Although each of our 9 players shows improvements over time, these improvements are not smooth: each individual curve is

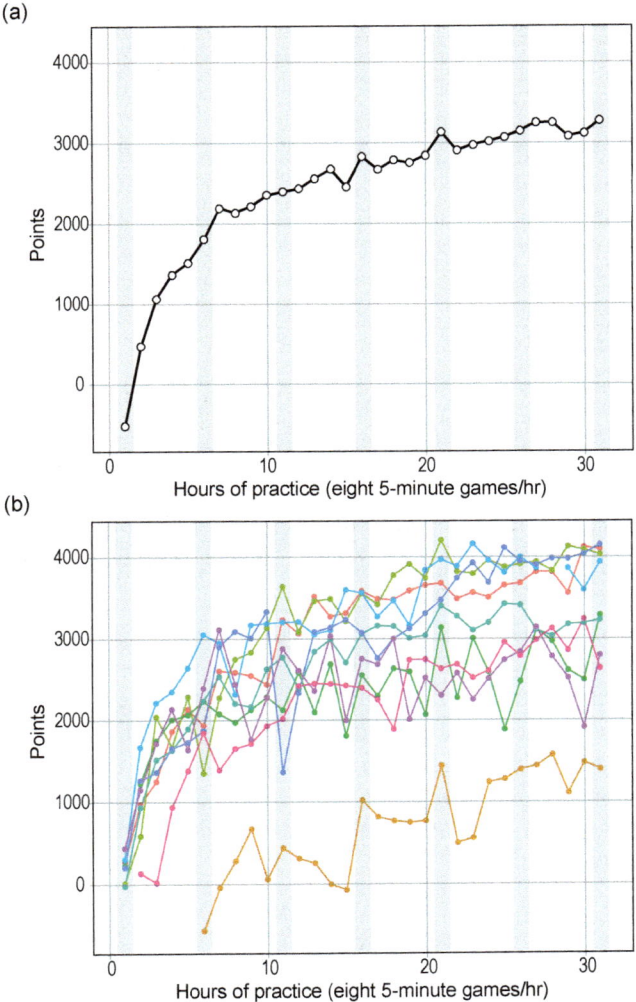

Figure 10.1 *Space Fortress* skill acquisition curves. (a) Mean performance per hour for all 9 players across 31 hr. (b) Actual scores for each individual player. To keep the plots compact, the early games for the lowest scoring players (b) are truncated for hours 1–4. As is clear, the mean performance shown in (a) does not represent the progress of any individual player (b). From Destefano and Gray (2016).

more notable for its plateaus, dips, and leaps (Gray and Lindstedt 2017) than for smooth and steady improvement with practice.

Following Thorndike (1913), we distinguish between "spurious limits... which could be surpassed" and limits due to "the mechanism for the function itself" (Gray 2017). We call the former plateaus and the latter asymptotes. Hence, a new method for completing an old task results in the overcoming of a plateau, whereas an improved tool or a general enhancement of some sort to a brain area raises an asymptote.

The plateau versus asymptote dichotomy is often clear in hindsight when we can show that individuals performing at different skill levels are doing different things. A paradigmatic example is the distinction in high jumping between the Scissors and Straddle versus the Fosbury Flop techniques (see Figure 10.2). In the 1960s, performance in high jumping appeared to be topping out (in our terminology, it was thought to be asymptoting) as only incremental increases (measured in millimeters) in world records were being realized due, primarily, to a larger participant pool and better physical training. Then Dick Fosbury came along in 1968, "flopped" and smashed world records. This made it clear, in hindsight, that prior high jump performance had plateaued due to the method being used, not asymptoted due to an inherent limit in how high humans can jump.

There are three types of activity capable of moving human performance off a plateau: (a) method invention, (b) method development, and (c) practice (Gray and Lindstedt 2017). We argue that method invention and method development are often (but not necessarily) signaled by dips in individual performance, and that the implementation of a successful new method may be signaled by a performance leap that takes behavior well beyond the incremental improvements available through regular practice.

(a)

(b)

Figure 10.2 Techniques in high jumping: (a) scissors and straddle and (b) Fosbury flop. Figure used with permission from Carlos Lopez.

Techniques: Changepoint Analysis

With the plateaus, dips, and leaps (PDL) framework guiding our work, we have attempted to develop tools to help automate the identification of change-episodes; that is, periods in which the learner discovers or invents new methods in individual performance (Destefano and Gray 2016; Gray and Destefano 2016). Although our current techniques are either not sufficiently intuitive or not sufficiently automated, we can provide an example of what we are trying to do.

The essence for our use of changepoint analysis lies in comparing multiple performance factors within the same individual at the same moment in time. This requires detailed data collection with timestamping. Figure 10.3 plots three factors (two features and one score)[1] for one *Space Fortress* player across each of the 248 games that he played.

Figure 10.3 uses the *intuitive changepoint analysis method* (Gray and Destefano 2016) in which a slope is computed for each factor (feature or measure) of interest. For this figure, we then computed the running slope across each of the five games (i.e., games 1–5, 2–6,..., 244–248) of *Space Fortress* for that factor.

The horizontal line in each of the three plots is the normalized slope across all 248 games for that factor and that player. It is always plotted at zero. The other lines plot the running slope for each 5 games as deviations from the overall running slope. Hence, upward sloping lines represent an increase in a factor whereas downward sloping lines represent a decrease. In general,

- when all three factors move down at the same time, the player is probably asleep or distracted,
- when all three follow each other up and down, it is hard to conclude anything, but
- when some move up at the same time that others move down, that is interesting.

Hence, in Figure 10.3 the two gray bands were added by the analyst to highlight periods of interest; namely, periods in which some of these three factors were moving up while others were moving down.

The leftmost gray band (games 71–77) highlights periods of discrepancy between dips and leaps in Fortress Kills and those in Mine Kills. During this period the player discovers the following:

- If you kill the Fortress fast enough (the leap in FortKills between game 71–72), mines will never appear.

---

[1] For *Space Fortress*, there are four scores that the player sees as s/he plays each game. We also collect data on approximately 30 features of game play.

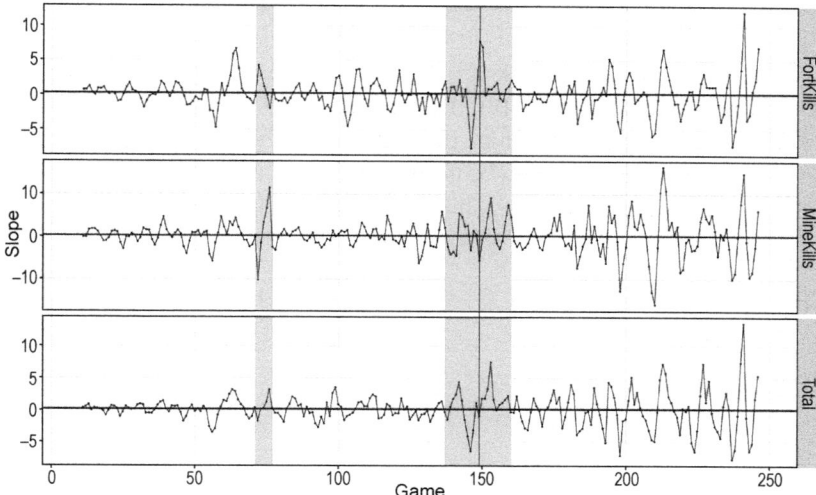

Figure 10.3 A plot of one player's data based on two features—number of Fortress Kills (FortKills) and number of Mine Kills (MineKills)—and one score (TOTAL). See text for detailed discussion (Gray and Destefano 2016).

- By preventing the mines from appearing, your mine kills dip drastically (MineKills between games 71–72) as do the total points earned (TOTAL at game 72).
- A player thus invents, implements, tests, but ultimately rejects this strategy.

Games 137–160 are also highlighted (the vertical black line through the middle of this period is merely a visual aid for the reader). During this period, our player discovers and implements a strategy that was new to us:

1. Shoot the Fortress as quickly as possible to increase its vulnerability to "9" (without shooting it so quickly that its vulnerability resets).
2. Wait for the mine to appear.
3. Manage the mine as a "normal" mine. (*Space Fortress* has two different types of mines which need to be killed in different ways.)
4. Killing the mine gives you points and also increments Fortress vulnerability by "1" making it eligible to be killed.
5. Finally, double-shoot the Fortress as quickly as possible.

Incrementing vulnerability this way saves you the cost of "one" shot while giving you points for destroying mines.

As can be seen in this gray-banded area, across this period of thirteen *Space Fortress* games (each dot is a separate five-minute game) the total score per game fluctuates wildly. As the score stops decreasing and begins to increase (game 146), so does the number of Fortress Kills. At game 149, the number of

Fortress Kills peaks whereas the number of Mine Kills plummets and the total score returns to its average. This is a true invention. Although we programmed this version of *Space Fortress*, we were not aware that what this player invented in this sequence of games was even possible. Indeed, while several of our players discovered the strategy used in the leftmost gray band (see Figure 10.3), very few discovered this one.

In summary, applying changepoint analysis to these data supports our interpretation of dips and leaps as sometimes signaling periods of discovery, invention, and change. There is no way for knowledge compilation or other known practice-based cognitive processes to account for these discoveries.

Applying Machine Learning Insights to *Tetris*

A related line of research compares similarities and differences between performance by humans and machine learning models (Sibert et al. 2017). The most recent work in this thread compares the machine learner "tortoise" with the human "hare" (Sibert and Gray 2017).

Reinforcement Learning Modeling for Minds and Machines

To handle the pattern-matching component of placing a new *Tetris* piece (or zoid) in the pile of existing pieces, we turned to the feature learning method of cross-entropy reinforcement learning (see Sibert et al. 2017). Janssen and Gray (2012) discuss three parameters of reinforcement learning which could be modified to better match human learning and practice: when, what, and how much to reward. The "what" component was interesting to us because all *Tetris* machine modeling research we could find reinforced the "number of lines cleared." These models played a lot of episodes (often over 300,000) and cleared a lot of lines (often over 100,000). At that time the best player we had in our lab played for 506 episodes (one zoid per episode) and cleared close to 200 lines. We wondered if playing for lines versus playing for score would produce different weights in our feature sets and affect performance. Both of these things happened. As Sibert et al. (2017) show, the different feature rates were learned for the two objective function conditions and, when judged by total score per game, the lines model rapidly plateaued at around 100,000 points whereas the score model peaks around 200,000 (see Figure 10.4).

Figure 10.4 shows that the lines controller (which is the one always used by the machine learning community) produces a flat function across its many generations, whereas the score controller shows plateaus, dips, and leaps. At first this finding seems like an interesting mystery. On further thought, it seems completely understandable and provides a bit of an "ah-ha" moment.

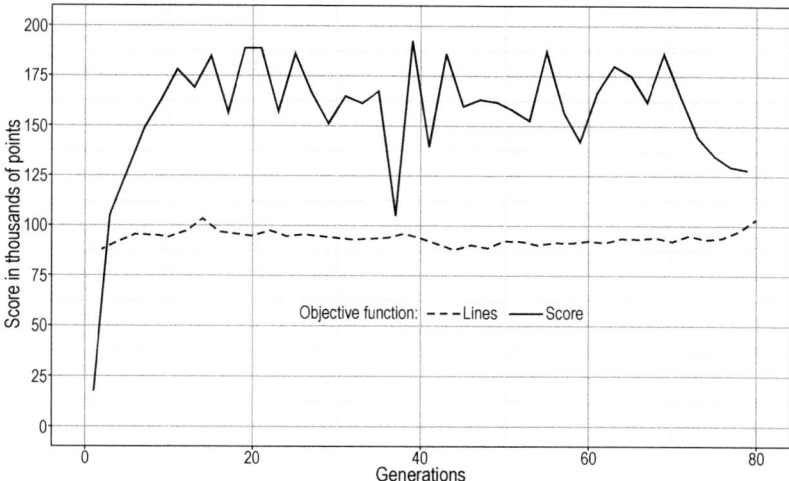

Figure 10.4   Learning curve of models. In each game and generation, the total number of zoids available is 506. At the end of each generation, before the next set of 100 controllers is generated, the new starting controller played 30 test games (for details, see Sibert et al. 2017). The mean performance for these 30 games, for each of the 80 generations of training trials, is plotted here. Adapted from Sibert et al. (2017, fig. 3).

The lines model is playing to clear the maximum number of lines it can and it does this extremely well. With 506 zoids, the maximum possible is 202 lines cleared. Most of these clears are 1-line clears—not 2-, 3-, or 4-lines. Hence, when the y-axis plots score (as we do in Figure 10.4), this model's score is about as high as it can be by clearing one line at a time.

In contrast, the score model is clearing more 2-, 3-, and 4-lines. As the score function rewards multiple line clears (e.g., clearing 4-lines at once yields 7.5 times as many points as clearing 1-line, four times), the score model learns to maximize points by maximizing the number of simultaneous lines cleared. However, the zagging line for score in Figure 10.4 shows that this is a risky maneuver. Doing multiple line clears requires allowing the average board height to become higher than for 1-line clears. This is a dangerous trade-off because when the board becomes too high and the stack of zoids reaches the top of the screen, the game ends.

A Necessary Digression: *Tetris* Technicalities

There are several key differences between machine and human play of *Tetris*. As the level increases, the time it takes a zoid to drop 20 lines keeps decreasing. At level zero, an unhampered zoid would fall 20 lines in 16 sec (to the bottom of the board). At level nine it drops the same distance in 2 sec. By level 16, the drop takes 1 sec. At level 19, the fall takes 0.66 sec. For human play, the drop rate tops out at 0.33 sec to fall those same 20 lines at level 29. Clearly, such rates of fall present a significant challenge to human perception, action, and decision making. To motivate humans to play faster and faster (and some of the champions at the annual Classic Tetris World Championships do play beyond level 29), the number of points awarded for clearing lines "escalates" as a function of drop rate.

In contrast to humans, the machine models (MM) evaluate the goodness of all possible placements in a blink of the eye and "move" the piece to that location instantly. There is no difference for them between play at level zero, level 30, or level 12,000. If we began rewarding the model at level zero rates (the "base" score) and escalated the scoring system beyond level 30 (the highest level reached by humans) all the way to level 12,582 (the highest level reached by our models), the model's escalated score would be 34,847,635,540 points.

With such escalated scores, it becomes misleading to compare MMs at high levels of play to each other and certainly to humans. Hence, we have adopted the convention of presenting base scoring and escalated scoring. When we are discussing the long games played at levels only achieved by the MMs, we always use base scoring, which rewards *Tetris* play at all levels the same as for level zero. However, when we are within human levels of play, we report both escalated and base scores.

The Tortoise and the Hare

The Sibert et al. (2017) study left us in a bit of a quandary, if only because the entire corpus of machine learning research on *Tetris* uses lines as their objective function. In their defense, these machine learning studies focused on developing methods for feature search, not on model behavior and, as discussed above, the models do not play under time pressure. From that perspective, we realized that by capping the number of zoids at 506, Sibert et al. (2017) artificially simulated a type of time pressure.

To investigate this issue more thoroughly, we abandoned cross-entropy reinforcement learning and took up Mind Modeling or, to be more precise, MindModeling.org (Glendenning et al. 2016; Gluck and Harris 2008). In this work, Sibert and Gray (2017) turned to a grid search of the feature space to test 3,543,122 models that each played one game. The models played in one of two conditions: tortoise or hare. In the tortoise condition, each of the 1,771,561 models played until it lost. In the hare condition, each of the 1,771,561 models played for a maximum of 506 zoids.

As shown in Table 10.1, the best tortoise/long model (row 1), which was allowed to play to its limits, scored 5,527,820 (base) points. The highest performing model in the best hare/short game condition (row 4) scored 240,900 (escalated) points, clearing 199 lines using 506 pieces. (The maximum number of lines possible to clear with 506 pieces appears to be 202).

Table 10.1   Model scores and lines cleared.

Model	Length	Points	Scoring	Lines
Best tortoise	Long	5,527,820	Base	125,829
Best hare	Long	68,440	Base	2243
Best tortoise	Short	92,700 8720	Escalated base	202
Best hare	Short	240,900 18,900	Escalated base	199

We then allowed the best hare model to play a long game (Table 10.1, row 2), and the best tortoise model to play a short game of maximum of 506 zoids (row 3). The best tortoise model did poorly in the short condition with 92,700 (escalated) points, far below the 240,000 (escalated) points of the best hare model in the short length condition. Similarly, under long length conditions the best hare model performed well above average (68,400 base points), but scored nowhere close to the best performing tortoise model (5,527,820 base points).

Model Behavior

The above comparisons are interesting; however, as we are focused on human behavior, we are not merely interested in the models' various scores (i.e., the "points" and "lines" columns in Table 10.1) but in their behavior. Did the differently trained models show different behaviors from each other and/or in their transfer conditions?

For this comparison, we looked at key behaviors known to separate human *Tetris* experts from novices; namely, the proportion of 1-, 2-, 3-, and 4-lines cleared. As Table 10.2 shows, the tortoise models had proportionally more 1-line clears than the hare models, but the hare ones had more 2-, 3-, and 4-lines than the tortoise ones. More interesting, the tortoise models made almost no 3- or 4-line clears (all were less than 1% of the total), whereas the hare

Table 10.2   Model behavior: differences in the percentage of types of line clears.

Line Clear Type	Best Tortoise	Best Hare
1-Line	85.16	70.37
2-Lines	14.03	18.52
3-Lines	0.78	4.44
4-Lines	0.03	6.67

models ranged from 4.4% to 6.7%. The hare pattern of more 4- than 3-line clears shadows the pattern of our best human players.

Summary and Conclusions

These models are not replicas of human decision making, but they provide insight into how humans make complex, rapid decisions. The Sibert et al. (2017) work showed that models trained to optimize the objective function of points performed differently than those trained to optimize lines cleared. Not only did they find that the points-trained model achieved a higher score in fewer lines than the lines model, but it better predicted performance differences between expert and novice human players. The lines model essentially did not distinguish between human experts and novices. As such, these strong differences in optimal strategies are issues which must be considered when humans partner with robots, such as is intended to be the case in the canonical uses for future ITL. To make this connection very clear, an implication of this research is that the details of the content taught to the ITL agent matter and should likely change depending on the circumstances, regardless of whether that agent is human or machine. Rather than rewarding performers for number of items completed, the reward needs to be framed in terms of number completed per unit of time.

We characterized the points-trained models as riskier than the lines-trained models because the only way to gain more points with a limited number of pieces is to do more 3-and 4-line clears. These require building higher piles than do the lines models. This, however, is a riskier strategy, and we interpret the dips and leaps in Figure 10.4 as showing that risk.

The MindModeling models support these conjectures. The MindModeling models were not reinforcement learning models; hence, there was no objective function. Using "total points" as our sole criterion, we simply selected the model that scored the most points in 506 episodes and the one that scored the most points when given unlimited play. Although the best hare model upped its score 12-fold when allowed to play long, it was nowhere near the level of the best tortoise model. Even more surprising, the best tortoise model, the one that scored more than 5 billion points, delivered a pathetic showing when it went short, a mere 92,700 (escalated) points!

It is worth noting that the tortoise-trained model completely brackets the hare-trained one. The tortoise/long scores 80 times as many (base) points as the hare/long, while the tortoise/short scores only 38% of the (escalated) points as does the hare/short (46% of the base points). These comparisons drive home the oft-missed point that "time" itself is part of Simon's gelatin mold and to a large degree the methods we choose are proportional to our time horizons.

Finding the "I" in Team

Our first two paradigms, *Tetris* and *Space Fortress*, focus on individual performance and individual interactions in dynamic task environments. Our third paradigm is the game *League of Legends* (LoL), which is the most popular game in the wildly popular genre of multiplayer online battle area games (MOBAs). (For further information on LoL, see https://goo.gl/d7mcs8.) Our research on LoL focuses on finding the "I" in team. The term "team" is not rigorously defined in psychological science; many things in which more than one person is involved are often casually referred to as "team tasks" and the group of people who are involved in those tasks are considered "teams." We will not provide a taxonomy of teams but will provide examples of the characteristics we are studying.

In common with *Tetris* and *Space Fortress*, LoL takes place in a dynamic task environment. Also like *Tetris* and *Space Fortress*, for LoL, doing nothing requires a decision to do nothing. In *Tetris* the environment acts on the human agent via the zoid that appears and the frequency and recency with which each of the seven differently shaped zoids occur (Sibert and Gray 2018). As players demonstrate competence by completing levels, the environment literally speeds up so that the time required for a zoid to drop 20 rows began as 16 sec at level zero, sped up to 1 sec by level 16, and topped out at 0.33 sec at level 29 (few players ever get above level 16).

Compared to *Tetris*, in *Space Fortress* the environment is more overtly hostile to the player and vice versa. Mines spawn randomly and are attracted to the player's ship which they try to ram, thereby blowing up the player's ship and themselves. The "Space Fortress," after which the game is named, is less suicidal than the mines, as it primarily defends itself from player attacks by trying to shoot and destroy the player's ship.

In contrast to these games, the lure of LoL is that the environment is not just dynamic and not just hostile; instead, it is intelligently hostile and cleverly dynamic. Although there is more to the LoL environment than just the overtly hostile five people on the opposing team, this is the main feature. They are trying to kill you. Many, if not most, of the players have a lot of experience playing together as a team (e.g., teams that have played together hundreds of times are not uncommon) and with attacking opposing teams with the goal of defeating them.

Of course, there are two major ways of describing the differences between our first two paradigms and LoL. The first, taken above, is to discuss the dynamism and hostility of the task environment. The second is that LoL is a team event; this is the key difference between LoL and the first two games and is, for us, the most interesting feature of LoL. It also is the feature that ties this game even more closely to ITL, because team members are able to learn interactively from each other during game play. We discuss this feature further after a brief introduction to the game.

What Is *League of Legends*?

In LoL, one team (comprised of 3 or 5 players) battles another team of equal size in matches which last about 30 minutes each. In 2012, one billion hours of LoL were played worldwide each month (Kenreck 2012).[2] With LoL as our paradigm, we have adopted big data (Goldstone and Lupyan 2016; Griffiths 2015) approaches to harvest play data from the web. To date we have collected 1.9 million records from 539 thousand matches. Figure 10.5 shows the "gods' eye view" of LoL. In addition to that view, players see a third-person view of their avatar, the avatars of nearby players, and close-up views of the surrounding terrain.

LoL contains elements that are attractive for empirical studies of team performance:

- It is a team-based game with high demands for coordinated action across team members.
- It is highly instrumented, with detailed records kept on many aspects of performance.
- Its view of performance is multifaceted, with many explicit measures both at process and outcome levels.
- It enables various measures of team composition to be extracted or derived from match records, such as the working history of team members.

The Role Structure of LoL Teams

Without providing a tutorial of LoL play, we stress that like other invasion games,[3] a player's role in LoL is largely determined by the position played (Williams et al. 2011). For our purposes, position refers to the combination of the lane a player occupied and the role they fulfilled. LoL players may occupy the (a) top, (b) middle, or (c) bottom lane or (d) the jungle (i.e., the territories in between the lanes) (see Figure 10.5). There are five roles a player can fill: none, duo, duo support, duo carry, and solo. Different roles are available for the different lanes, resulting in sixteen different positions that a LoL player could play.

---

[2] For readers who are unfamiliar with LoL, we advise viewing the five-minute overview on YouTube before proceeding: https://goo.gl/d7mcs8 (accessed Feb. 10, 2019).

[3] See "Teaching Games for Understanding" at http://tgfu.wikifoundry.com/page/Helena+Baert and "What Are Invasion Games" at https://thephysicaleducator.com/game_category/invasion/ (accessed Jan. 31, 2019).

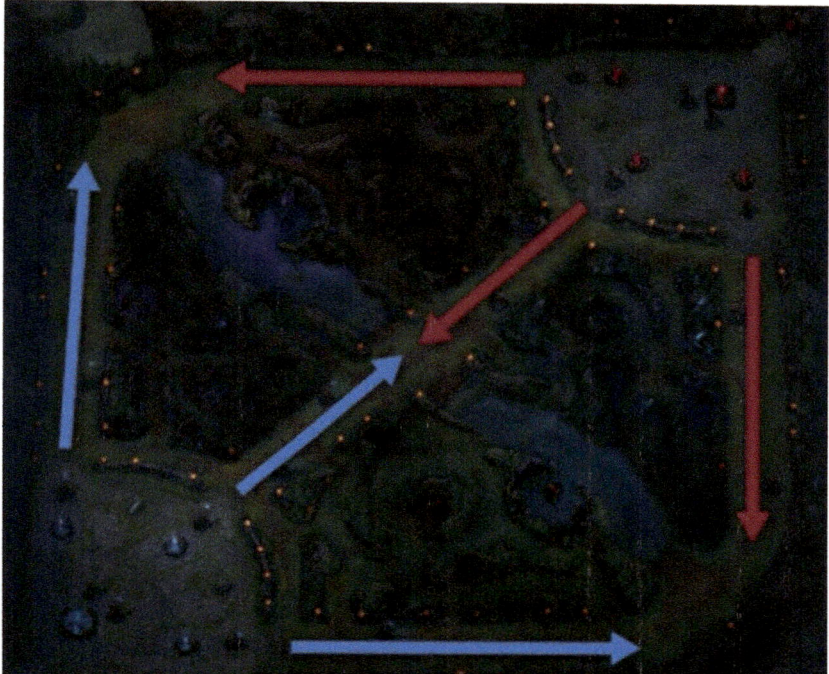

Figure 10.5 Screenshot of a battlefield map for the multiplayer online battle arena game, *League of Legends*. The river flowing from the upper left to the lower right separates the "home" territory of one team from the other. Territories can only be crossed via one of the three lanes (top, middle, and bottom marked here by blue and red arrows). Each team begins in one of the large quarter circles shown in the lower left (blue team) or upper right (red team). These bases include the all-important "nexus" (the blue or red area closest to the lower left and upper right corners). The darker green spaces between the lines are less traversable terrain referred to as the "jungle."

Research Issues in Team Performance of Dynamic, Interactive Teams

"Team" is a loose concept that can be applied to many groupings of individuals. For instance, a shift of telephone directory assistant operators could be called a "team," where the judgment of "team goodness" per shift is simply the sum of the number of calls handled during that shift. However, we would not consider this team to be either "dynamic" or "interactive." For our purposes, examples of dynamic, interactive teams include (but are not limited to) (a) emergency response teams, (b) combat teams, and (c) sports teams for invasion games. That is, the essence of interaction for our definition of a dynamic, interactive team includes interactions among team members which directly contribute to team outcomes.

In talking about team performance, whether the team is composed of humans, robots, or humans and robots (any combination of agent types in the broader ITL framework), we can easily talk about performance at two levels:

- The team: Was it successful? Did it accomplish its mission?
- The individual team member: How well did she, he, or it do? How does the member's performance compare to that of members in other teams or in its own team?

For dynamic teams (including both traditional sports teams and e-sport teams), team outcome can be very simply defined by whether the team won or lost. However, three types of team issues must be considered:

- Individual (Figure 10.6a): Does the team outcome reflect the simple sum of each player's individual competence in carrying out her/his role independent of the other players?
- Teaminess (Figure 10.6b): Is each member of the team doing essentially the same task but to varying degrees? In this model, there are no differences in type of expertise among team members, though there might be differences in degree of expertise.
- Teaminess + Individual (Figure 10.6c): Is there a component of shared expertise but also a component of individual expertise?

An important conundrum for these types of teams is that sometimes a team member's role does not seem to be directly connected to team outcomes. For example, in basketball it often seems as if one or two of the five players are making all of the shots and scoring all of the points. What are the other three doing? And how can we measure how well they are doing it?

Individual Contributions to Team Tasks

The success of our work on *Space Fortress* and *Tetris* has driven home to us how important data on player skill at shifting and focusing visual attention is to understanding differences in expertise among players. Indeed, the more we study and learn from individual experts, the more we believe that an important component of "team member" expertise and/or of "expert teams" is how individual team members divide and overlap their visual attention. Perhaps somewhat counterintuitively, it may be an individual's expertise at attending to other team members that creates an expert team.

Our strongest conclusion, so far, is a methodological one; stated metaphorically, *if you want to understand how a clock works, you have to understand the clockwork.* You have to understand what each part is doing, what parts it interacts with, and the nature of those interactions. We believe we will not understand the "I" in team until we understand the expertise of individual team members. Interestingly, the second edition of the *Cambridge Handbook of Expertise and Expert Performance* (Ericsson et al. 2018) contains 42 chapters and 969 pages, but only one chapter devoted to team expertise. Although that chapter is written by expert team researchers, we have difficulties relating their discussion to the measures and factors that we have found in our

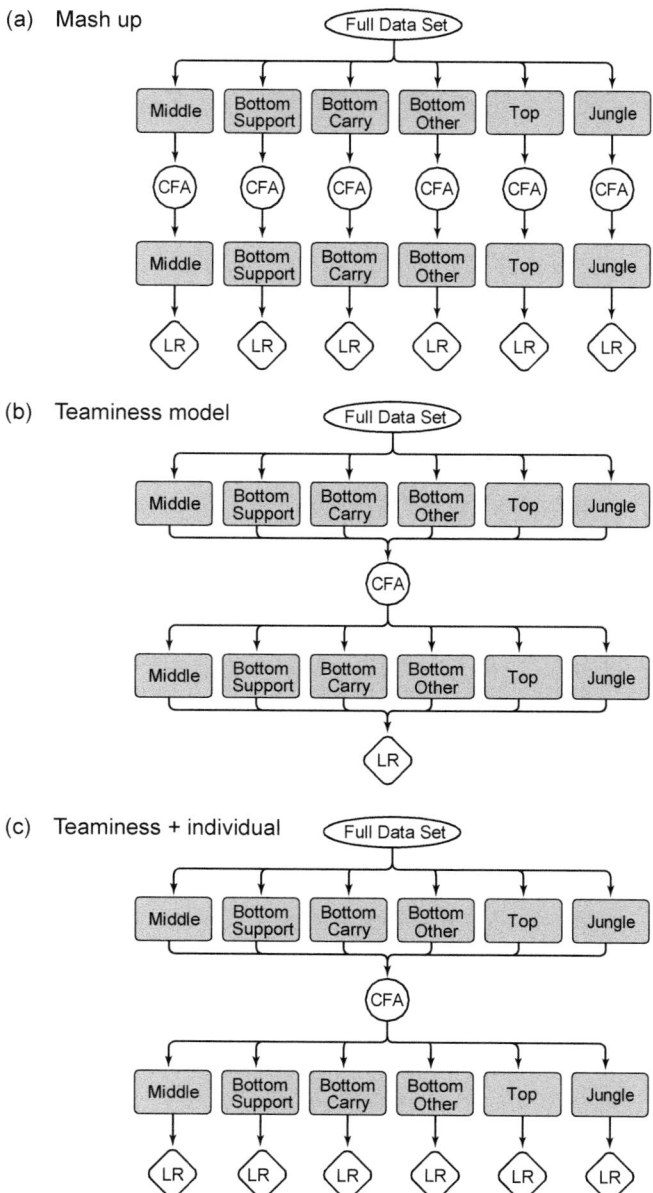

Figure 10.6   Three possible models of team and individual performance in dynamic, interactive team tasks. (a) Mash up: six independent models where each role/position model includes a teaminess and an individual component but to unknown degrees. (b) Teaminess model: one model for the entire team. Each individual is viewed as having more or less of that one model. (c) Teaminess + individual: all data is used to find the best fit for each player position. This model captures both the teaminess component and the individual component.

teams. Perhaps the way forward for understanding team expertise in dynamic, interactive task environments is to follow the "joint action" community (e.g., Knoblich et al. 2011; Sebanz and Knoblich 2009) and focus on the second-by-second interactions among dyads, triads, and tetrads. In this effort, having one or more ITL team members, whose behavior could be programmed as well as learned, might make ITL (as represented by many of the other chapters in this volume) a core topic in team research and, at the same time, make team research the overarching perspective on ITL.

Paradigms for Discovering the Essence of Interaction: Action Games

Our discussion of the essence of interaction started with the plateaus, dips, and leaps framework, which focuses on individual performance over time. The discovery of inventions made by one player after about 9 and almost 20 hr of play highlights the point that the slow-but-steady mechanisms of incremental learning simply cannot account for the dips and leaps seen in skill acquisition.

We then jumped to our machine learning data and used these models to explore the "optimal possible performance" once a certain objective function was adopted. We used the MindModeling system to access large-scale computational resources and perform a grid search across a parameter space covered by 1,771,561 models. That work dramatically demonstrates that "slow and steady" wins the race, but only when the race is long. When it is short, as it is for humans, the best hare model wallops the best tortoise model.

Jumping back to humans, one conclusion from our current work on "finding the 'I' in team" is that solving the "team problem" requires solving the problem of individual expertise. In addition, a methodological revolution is needed in both team studies and studies of ITL, and that revolution might be fueled by the methodological techniques and insights of the joint action community. Finally, if we are talking about intelligent, autonomous, robots and software agents, then there may be few inherent differences in the ways we study human-only teams versus mixed human–machine teams.

Paths Forward?

The central question addressed at this Forum concerned the acquisition of new tasks through natural interaction. Here, we have provided examples of human behavior in task environments that are interactive, dynamic, and require sequential decision making. For such task environments, action games provide a right-sized challenge, and we have provided our answer as to the sorts of methodologies and studies that should be done.

A perennial problem in the psychological sciences is collecting the massive amounts of detailed data required to draw strong inferences regarding the research question of interest. Indeed, a conclusion that we and others draw from the current "crisis" in data analysis (Baker 2016) is that most studies have too few subjects and too few samples of data. If, as a field, we are truly interested in studying human interactive behavior, we must collect detailed data from many subjects. If we are interested in how interactive skill is acquired over time, then we need to run longitudinal studies, sampling expertise from a very large number of performers across a wide range of skilled performance, or both.

This problem with longitudinal or sampling studies of expertise represents the standoff which the human factors community has been battling for decades. We advocate confronting this standoff by taking advantage of the availability of vast numbers of people on college campuses who have acquired almost obscene levels of excellence in action games. Of course, what makes these campus activities so useful is the electronic nature of the tasks and the realistic prospect for researchers to collect detailed data with millisecond precision. Indeed, our current conclusion is that the study of ITL needs to follow the "joint action" (e.g., Knoblich et al. 2011; Sebanz and Knoblich 2009) path of detailed empirical studies (e.g., Vesper et al. 2009) rather than the traditional "team studies" path of eschewing the direct observation of millisecond level interactions among team members.

Acknowledgments

The work was supported, in part, by grant N00014-16-1-2796 to Wayne Gray from the Office of Naval Research, Dr. Ray Perez, Project Officer.

Instruction

11

Task Instruction

Julie A. Shah, Kevin A. Gluck, Tony Belpaeme,
Kenneth R. Koedinger, Katharina J. Rohlfing,
Han L. J. van der Maas, Paul Van Eecke, Kurt VanLehn,
Anna-Lisa Vollmer, and Matthew Yee-King

Abstract

An early concept of interactive task learning (ITL) assumed a human teacher and machine learner. This book broadens the thinking about this relationship by explicitly allowing flexibility regarding the teacher and learner roles. Future ITL systems will be maximally useful and beneficial to the extent that they are effective and efficient *learners* as well as effective and efficient *instructors*. Focusing on task instruction, the primary goal of this chapter is to relate the critical role of instruction in ITL to key existing literature from related areas of research. The general concept of co-constructive task instruction is introduced and differentiated from traditional conceptualizations of fixed instructor and learner roles. Frameworks, models, and methods for task instruction are discussed, and broad connections are made between ITL and structural and adaptive improvements to instruction, historical developments in programming, and the extraordinary challenge that fluid, flexible, co-constructive task instruction and learning places on the vision for ITL.

Co-Constructive Task Instruction

The human and machine roles in interactive task learning (ITL) were viewed earlier as strictly discrete and fixed, with one always serving as the teacher and the other as the learner (Laird et al. 2017a). We anticipate it will be more accurate and useful to view these roles as fluid and dynamic, with the human and computer learning together and co-constructing an understanding of their task.

Group photos (top left to bottom right) Julie Shah, Kevin Gluck, Katharina Rohlfing, Matthew Yee-King, Anna-Lisa Vollmer, Tony Belpaeme, Han van der Maas, Ken Koedinger, Julie Shah, Paul Van Eecke, Katharina Rohlfing, Anna-Lisa Vollmer, Han van der Maas, Matthew Yee-King, Paul Van Eecke, Kurt VanLehn, Tony Belpaeme, Ken Koedinger, Julie Shah, Kurt VanLehn, Kevin Gluck

In co-construction, the aspect of iteration is crucial for the model of the learner and teacher because it changes the notion of intent (Figure 11.1). Accordingly, when the course of interaction is viewed as a possible method of providing feedback and instruction (Figure 11.1a), the initial intent remains stable throughout the entire interaction and is generated by the teacher as s/he guides the learner.

However, the iterative nature of an interaction offers not only the possibility of giving feedback or instruction, it also provides the possibility to adjust and align with the learner's action. In this way, it is possible for the initial intent to be changed or reformulated as the interaction unfolds. Teaching within this framework refers to a sequence of coactions established to achieve a specific goal. A teacher will have knowledge of the tasks that must be completed to arrive jointly at the goal. A knowledgeable teacher is able to constrain the actions of a learner such that the intended goal will be reached (Heller and Rohlfing 2017). However, in this framework, the strict role division between learner and teacher vanishes: individual actions that must be performed can be taken by both partners as they each contribute to reaching the joint goal.

The process of *co-construction* is temporal and embeds learning and future interaction for both human and machine. Co-construction is a process through which mutual understanding or "common ground" can be achieved between

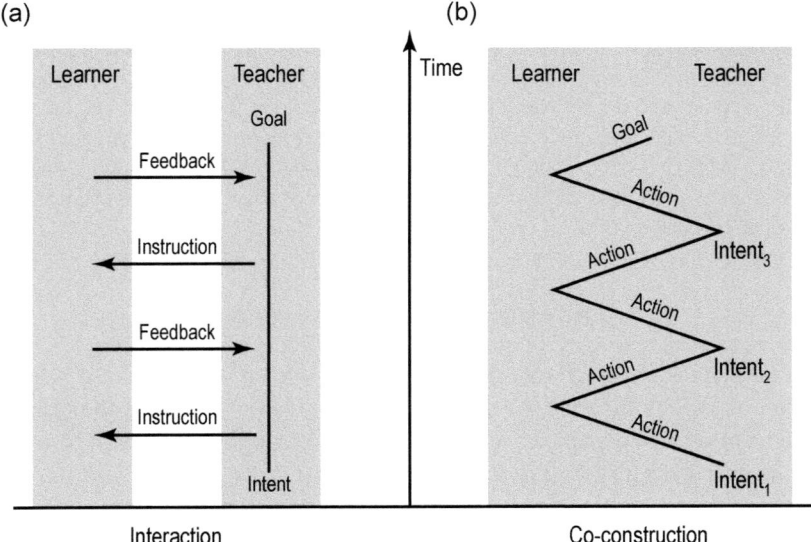

Figure 11.1   The iterative nature of an interaction is depicted: (a) the teacher has an intent, provides instructions to the learner accordingly, and receives feedback; (b) both agents co-construct a joint goal. While in the approach focusing on the feedback (a), roles of the agents are separated, they are both responsible for the intent in the co-constructing approach (b).

the teacher and learner. Common ground, in this sense, is neither fixed nor established once in an interaction, but is subject to constant change and evolution. Importantly, it is based on the history of interaction.

At the beginning of the learning process, the language channel between human and machine (gray arrows, Figure 11.2) may be quite limited. Nevertheless, the agents can make progress through ITL via shared interactions in the world, as when the teacher demonstrates, the learner practices, and the teacher adjusts to the learner's multimodal action and reformulates the goal(s) for their joint actions. Both agents may make coordinated use of language, gestures, and actions which, while not immediately understood, embed the understanding of language into shared actions toward specific goals (bottom arcs).

As the learning process progresses, the agents become increasingly experienced in working together and adjust to one another. As a consequence of relying on their joint experiences in perceiving and manipulating the world/task environment (Figure 11.2, right bottom arcs), the two partners develop new, more conventionalized forms of language interaction (represented by the thicker arcs at the upper right, Figure 11.2). Task- and communication-relevant knowledge inside the machine and human also grow (not shown).

Team training practices can facilitate and accelerate the process of humans and robots learning to work together (Gorman et al. 2010; Nikolaidis et al. 2015; Ramakrishnan et al. 2017). The cross-training approach is designed to improve team adaptivity through practice by requiring team members to switch roles with one another. Through this process, teammates are able to build a shared mental model of the task and collaboration strategy, and they can thus better anticipate one another's actions (Nikolaidis et al. 2015). In another team training method, perturbation training, a team experiences slight disturbances (perturbations) during the training process that are intended to help team members learn to coordinate effectively under new variants of the given task. This method is well-suited to heterogeneous teams, as each member practices his

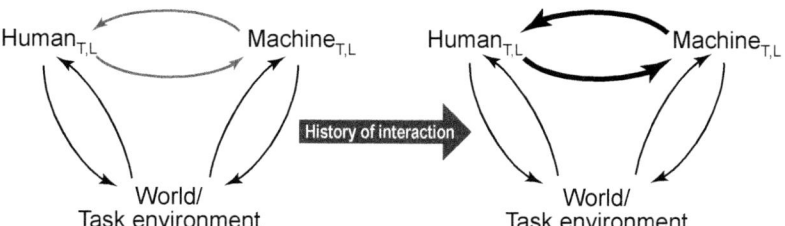

Figure 11.2 For human and machine, co-construction embeds learning and future interaction. Agents cannot rely on common knowledge or language; they must demonstrate individual understanding through actions (left: stronger links to world/task environment are necessary). With recurrent interaction, agents can increasingly rely upon joint experiences, grounded with language (right: stronger links between agents indicate increasing knowledge of each other).

or her own role on the team during the training process; it also does not require switching roles among team members, as in the cross-training method. Results from recent human studies (Gorman et al. 2010) and human–robot studies (Ramakrishnan et al. 2017) indicate that perturbation training yields high levels of human–human and human–robot performance on novel tasks.

Illustrative Example for ITL

In human–robot interaction, many factors have an impact on teaching. The human teacher's expectations for and knowledge of the robot learner are shaped by the robot's appearance, (mis)conceptions acquired from media and general knowledge, any prior human–robot interactions, and the *in situ* behavior of the robot (Hegel et al. 2009; Pitsch et al. 2013; Vollmer et al. 2014). In humans, a basic understanding of the learner can be derived from their outward appearance. For example, we infer that infants do not have the verbal capabilities of adults. By contrast, software agents often have no physical experience, and two robots with the same appearance are not necessarily equipped with identical software and capabilities. Thus, during interactions in which a machine learns from a human user, part of the student model that we normally would assume is available to the teacher (i.e., its existing knowledge and skills) may be unknown to its human instructor.

Consider a scenario in which a robot learner is being taught to clean windows by a human. The robot learner is a full-size humanoid robot with the following characteristics:

- It can perceive 3D motions of manipulative action demonstrations with sensors/cameras.
- It knows that both end-effector position and trajectory may be important for goal attainment.
- It understands the concept of "careful" performance (with regard to the amount of force applied).
- It is able to generalize a shown movement to a certain extent, such that movements performed with one object can be carried out with different objects.

The co-construction interaction unfolds as follows:

Teacher: Takes the cleaning agent, and says, "Look, I will show you how to clean this window now. You have to be really careful; it is quite old." T sprays the cleaning agent on the window.

Teacher: Takes the orange cloth, places it on the window, and says: "It's important not to push against the window too hard," and moves the cloth in straight lines from top to bottom. At the end of the demonstration, T moves the cloth around the edge of the window and says, "Done. Do you want to try it?" T puts the cloth on the windowsill and begins to spray the cleaner on the window again.

Learner: Says, "Okay," attempts to grasp the cloth, and fails. (End effector not suitable to grasp this object).

Teacher: After observing two failed attempts, T says, "Oh, sorry. Don't worry. You can use the sponge as well," and gives the robot the blue sponge.

Learner: Moves its arm and places the sponge gently onto the window, removes it again, then sets it down on the windowsill at the exact place the teacher put the cloth.

Teacher: Says, "Oh no, no, It's about the movement I showed. I'll show it again." T picks up the sponge and demonstrates the downward motion again on the window (a simpler demonstration: slower, with, more pauses, while monitoring the learner's attention), this time with the sponge. T sets the sponge down on the windowsill and sprays the window again.

Learner: Says, "I think I've got it now," picks up the sponge, places it gently on the window and performs a few up-and-down movements, then sets the sponge down on the window sill. "Done."

Teacher: Says, "Great, you repeat these movements until there is no cleaning agent on the glass anymore. This is how you clean a window."

Learner: Says, "Got it."

Co-Constructing Intentions

The co-constructive view of learning has certain implications for intentions that are usually considered to drive the partners' actions. In a co-construction process, partners interact iteratively, in turns, to reach joint agreement on their intentions and goals; consequently, modifications to movement and intent occur as the process unfolds. Offering the learner a different object with which to perform the task following multiple failed attempts is an instance of such adaptation. It requires the teacher to have perceived prior failed attempts and understand what the problem is.

In the window-cleaning example, the fact that the robot's gripper (end effector) is not suitable to grasp the cloth represents knowledge about the learner. This could be a part of the domain knowledge that is initially unknown to both the teacher and learner, which they will discover together over the course of the interaction. As another example of domain knowledge, consider a scenario in which the robot has learned about the goal state of the window-cleaning task and, during its turn, discovers a spot on the window that will not come off when cleaned using the sponge. The robot then suggests, "There is a spot of what I think is paint. I cannot remove it with the sponge. We need a different tool." At this moment, the robot leads the interaction and communicates novel information to the human, who, accordingly, learns about the spot from the robot. In such a case, task knowledge has changed: it now concerns the removal of paint from the glass. The tutor might then show the robot how to remove the spot of paint with a razor blade.

This illustrates that intentions are generated over the course of interaction as the domain knowledge evolves for the partners.

Jointly Clarifying and Completing the Goal

Similar to the dynamic view of intentions, the co-constructive view of learning implies that partners are working jointly toward a goal. In the window-cleaning task above, when the teacher cleans the window for a second time, the teacher monitors the learner more closely over the course of the demonstration. If the teacher observes, at the beginning, that the robot's eye gaze is shifting from the relevant object (the sponge) to the goal position on the windowsill, the teacher might stop the movement, shake the object, or say "Robot! Look!" to regain the robot's attention. Once the robot's gaze is again on the object, the teacher's movements become exaggerated to ensure that the robot continues to follow (track) the movement.

This interaction demonstrates that human teachers are very flexible; they adapt and adjust dynamically to the learner, taking the history of their interaction into account. In addition, both the teacher and learner adapt to the environment. For instance, the teacher might intend to produce the same movement with the sponge as with the cloth, but during the production/demonstration of the movement, the teacher observes that the sponge covers a larger area of the window and that fewer up and down movements are necessary. Consequently the teacher will produce the movements further apart from one another, deviating from the spacing of movements when using the cloth. In addition, the teacher has a certain goal intent and an intended action production (i.e., demonstration), but the realized movement may not coincide with either of these initial intents.

In the above case, the final production of the robot is actually not satisfying the requirements of the goal state. The teacher, thus, understands that the robot learner has not yet understood the goal state and clarifies the goal verbally.

If the modality of language is unavailable (e.g., if the robot simply lacks speech recognition, or there is a construction site outside the window and speech recognition is difficult in noisy environments), or the robot simply does not know what it has to do to remove all of the cleaning agent from the window, the teacher might include an additional iteration at the end of the interaction, modifying movement and intent to emphasize removal of the cleaning agent.

Frameworks for Task Instruction

ICAP Framework

The interactive, constructive, active, and passive (ICAP) hypothesis predicts the effectiveness of various types of instruction (Chi 2009; Menekse et al. 2013; Chi and Wylie 2014a; Chi and Menekse 2015). It defines four classifications of observable student behavior:

- *Interactive:* "We operationalize *interactive* behaviors to dialogues that meet two criteria: (a) both partners' utterances must be primarily *constructive*, and (b) a sufficient degree of turn-taking must occur. We do not restrict who the partners can be, provided that the criteria are met. Examples include a learner talking with another person who can be a peer, a teacher, a parent, or a computer agent…" (Chi and Wylie 2014a:223, italics in the original).
- *Constructive:* "Our taxonomy defines *constructive* behaviors as those in which learners *generate* or produce additional externalized outputs or products beyond what was provided in the learning materials" (Chi and Wylie 2014a:222, italics in the original).
- *Active:* "Learners' engagement with the instructional materials can be operationalized as *active* if some form of overt motoric action or physical manipulation is undertaken" (Chi and Wylie 2014a:221, italics in the original).
- *Passive:* The passive mode of engagement involves "learners *being oriented toward* and *receiving* information from the instructional materials without overtly doing anything else related to learning" (Chi and Wylie 2014a:221, italics in the original).

Suppose a learner, for example, was watching a video of an instructor demonstrating how to wash a window, as in the example described above. If the learner just watched the video without pausing it or doing anything else, then the activity would be classified as passive (i.e., paying attention). If the learner rewound the video to view portions of it over again or mimicked the actions of the teacher, then the activity would be classified as active (i.e., manipulating the given information without extending it). If the learner explained the actions by saying to itself, "The trajectory is intended to cover all the glass with minimal overlap," then the activity is classified as constructive (i.e., generating information not mentioned in the video). If the learner works with a second learner to co-construct some additional information, then the activity is classified as interactive (i.e., transactive collaboration). As an example, suppose one learner asks, "As long as we wipe every bit of glass, can we progress from bottom to top instead of left to right?" while another learner says, "Maybe that would drip dirty water on the clean glass." That is, they both construct information not presented in the video and build off each other's contribution.

This is exactly the kind of co-construction discussed earlier, except that the ICAP hypothesis allows two learners to be involved rather than a learner and a teacher. When ordered according to effectiveness for learning, the hypothesis ranks the four behavior modes as follows: interactive > constructive > active > passive. This is equivalent to: collaborate > generate > manipulate > pay attention.

According to the ICAP hypothesis, students learn the most when they collaborate and the least when they are merely paying attention. The ICAP hypothesis is consistent with a very large number of experiments. For example, Menekse et al. (2013) incorporated a topic in materials science as the target knowledge. They randomly assigned students to four groups, corresponding to the four ICAP modes. The students first took a test to determine how much they knew about the given topic prior to studying it (all four groups were about the same). Students then engaged in one of the following actions:

- They read a text passage (paying attention group).
- They read and highlighted important sentences within the text (manipulating group).
- They individually interpreted a graph that described the information contained within the text passage without access to the text (generating group).
- They interpreted the graph jointly with a peer without access to the text (collaborating group).

The groups then took a test to see how much they had learned. Test results were consistent with the ICAP hypothesis; that is, the collaborating group retained the most knowledge, followed by the generalizing, manipulating, and paying attention groups. The ICAP framework also hypothesizes cognitive processes that underlie the four modes and explains why they exhibit the observed ordering of effectiveness.

KLI Framework

The knowledge-learning-instruction (KLI) framework specifies three interacting taxonomies of kinds of knowledge, learning processes, and instructional methods (Koedinger et al. 2012). One fundamental underlying claim of this framework is that different instructional "treatments" are optimal for different kinds of "content" knowledge goals because of how those treatments best support the particular learning process relevant for that knowledge goal. In other words, "content-treatment interactions" are a common occurrence, whereby a particular instructional treatment (e.g., studying lots of examples) aids or accelerates learning for certain types of knowledge content (e.g., skills that implement multistep flexible procedures, such as math or science problem solving) but slows the learning of other types of knowledge (e.g., second-language vocabulary).

The bottom of Figure 11.3 depicts examples of three broad classes of knowledge: facts, rules (i.e., skills), and principles. On the left are three broad categories of learning processes: memory and fluency, induction and refinement, sense making and understanding. Also on the left are associated instructional treatments from "optimal scheduling," which enhances memory, through "worked examples," which enhance induction, to "accountable talk" (i.e., a collaborative dialogue prompting technique), which enhances sense making. The cells indicate cases in which the treatment (rows) was compared to a matched instructional control with regard to instruction of the indicated knowledge content (columns): +, 0, and − indicate a positive, null, or negative effect, respectively, of the treatment over the control on student learning outcomes. The bottom-left cell, for example, indicates that students learned Chinese vocabulary better when adaptive "optimal scheduling" selected their tasks, compared to receiving tasks in a fixed order.

The KLI framework provides empirical evidence for content-treatment interactions as well as a theoretical analysis to predict and explain when and why such interactions may occur. It is related to and largely consistent with the ICAP theory; however, one important difference is that ICAP does not consider as many types of knowledge as KLI, but instead focuses on more complex forms of knowledge. As a consequence, the claims made for the ICAP framework may only be relevant to "principles" in KLI framework terms. For

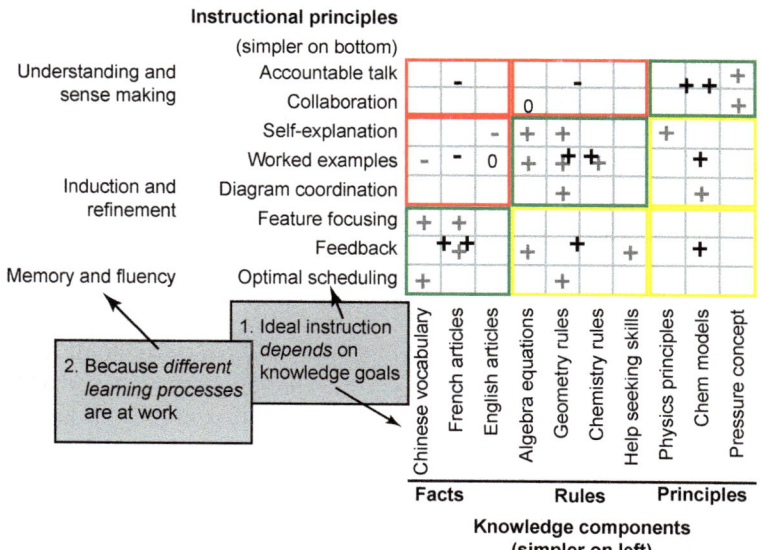

Figure 11.3 The KLI framework suggests "content-treatment interactions." Effectiveness of an instructional treatment (see "worked examples" row) depends on which learning processes it facilitates (e.g., induction) and inhibits (e.g., memory) and whether those learning processes are necessary (see "geometry rules" column) or not necessary (see "Chinese vocabulary" column) to meet the knowledge content goal.

example, according to KLI, forms of instruction that promote active rather than constructive learning may produce better outcomes when the goal involves less complex forms of knowledge, such as facts or skills/rules.

KLI and ICAP are directly relevant to ITL when a robot or software agent is the teacher and a human is the learner. Both frameworks indicate which instructional methods are most likely to work given the target knowledge content. KLI and ICAP may also be relevant when a robot or softbot is the learner, suggesting which instructional strategies and learning processes/mechanisms may be most effective for particular task domains.

Models for Task Instruction

Here we discuss the models that a teacher can hold with respect to an ITL learner, including models of the student, the domain/task, and a model of the pedagogical approach. We also discuss challenges related to communication with regard to models, as well as open modeling questions.

Overview of Models

The teacher in an ITL system (human or machine) must be able to assess the student's current learning state and make decisions that move the learner from one state to the next in a way shaped by the goal. This requires the teacher to maintain models of both the *student* and *domain*. These models may be created or evolve throughout a learning and co-construction process, wherein the teacher and learner develop an understanding of their shared world and/or each other. The teacher must also maintain and possibly evolve a model of a *pedagogical approach* in order to shape interactions.

The student model itself includes three components, the first of which is a model of the *learning state*, or the "knowledge overlay." This model incorporates a representation of correct knowledge that the student has acquired throughout the learning process and a representation of missing knowledge. The teacher maintains a model of each student with reference to the domain model, which can be very rich and may be represented as an "overlay" over the domain/task model. The second component is a model of the student's *misconceptions*, and the third is a model (or models) required for communication and interaction. This could include a model of modalities for *communication/ interaction* (e.g., language, depiction) or abstractions over the level of concepts to communicate.

The teacher also holds a model of the *domain/task* that is separate from the student model. The domain model is maintained in the teacher's mind and is representative of what experts know as well as what intermediate and/or novice students know. It can also include conceptual alternatives of the task. The domain model can change in structure or content over time.

Learning State Model

The model of the learning state is analogous to the tutoring system "student model," which refers to information that represents what the student currently knows and has accomplished so far. This includes the history of interaction over time and is not simply a snapshot of the current state. The process by which the student arrives at the current state is important and holds a lot of information for the teacher, who is shaping the interactive experience toward some learning.

This information serves one or more of the following purposes:

- It assigns or recommends a task for the student to perform next.
- It reports the student's current state to the instructor.
- It reports the student's current state to the student.
- It guides the system's selection of feedback, hints, or other scaffolding.

Sometimes the student model contains only simple information. For instance, if the tutoring system's only use for the student model is to meter the progression through a linear sequence of tasks, then the student model need only record where the student is located in the task sequence. If the system's only use for the student model is to decide when the student has correctly accomplished three of the previous four tasks, then it only needs to track how many of the last four tasks have been completed.

However, many systems incorporate student models that are based on assessing what knowledge students have learned so far, rather than just which tasks they have completed. Such systems often divide to-be-learned knowledge into pieces, typically referred to as "knowledge components" (if they are relatively small) or "instructional objectives" (if they are larger). If a tutoring system incorporates knowledge components, it often represents which ones a given student knows using a simple data structure called an "overlay." In its simplest form, an overlay is a single binary variable (mastered/unmastered) per knowledge component; another common representation is a value between 0 (completely unmastered) and 1 (completely mastered). It is traditional (albeit confusing) to refer to a student model that uses only correct knowledge components as an "overlay" model, whereas a student model that includes both correct and incorrect knowledge components is referred to as a "buggy" student model.

Model of Misconceptions

The tutoring agent in an ITL system may also maintain a model of student misconceptions. This model can contain the misconceptions held by multiple students or those of a particular student. If the tutor agent understood the learner's misconceptions, the tutor could potentially be able to exploit this knowledge to enhance the effectiveness of its teaching; this has formerly been studied as the "diagnostic remediation hypothesis." Evidence, however, rejects this hypothesis, at least under certain conditions. Study results from Putnam (1987) and Siler and VanLehn (2015) have shown that human teachers rarely utilize knowledge about students' misconceptions, even when this information is explicitly presented to them. Other studies provide evidence that use of knowledge about misconceptions does not help students to learn algebra (Sleeman et al. 1989). This may be due to the fact that misconceptions rarely result in systematic errors. Researchers have also found that the explicit remediation of misconceptions (or "mind bugs") can have a positive effect when learning about physics (Albacete and VanLehn 2000).

Although evidence of the usefulness of knowledge about misconceptions is mixed, and greatly depends on how systematic the errors are, it should be noted that human teachers maintain several conceptions about how a learner works. This is crucial for adapting language and level of explanation to a target audience (e.g., children of different ages or adults with different levels of expertise).

Task Domain Models

In the expert, or domain, model of ITL, the system stores its knowledge relevant to solving tasks within the given domain. In a subclass of intelligent tutors, sometimes called "cognitive tutors," there is a commitment that the domain model is a "cognitive model," one that contains knowledge able to produce a broad range of possible solutions for domain tasks, including variations in expert solutions as well as in student solution strategies, both correct and incorrect. This cognitive model may also include knowledge about a progression (or "learning trajectory") of approximate conceptions or misconceptions. For example, in human biology, the model may include a misconception that the human circulatory system consists of a single loop involving only the heart, or of one loop involving both the heart and lungs, in addition to the correct conception that there are two loops: one involving the heart and one involving the lungs.

This domain model can provide adaptive tutoring to students at multiple temporal grain sizes of interaction and adaptivity. Using a model-tracing algorithm, the tutor can monitor student task performance in comparison with the task performance the domain model can generate (i.e., a kind of plan recognition in AI terms). Steps taken during a student's performance of the task are correct when they match application of correct knowledge in the domain model and are incorrect when they match application of incorrect knowledge (in which case specific feedback can be provided to explain why the given step is incorrect) or when there is no match (in which case a simple error feedback indication can be provided). When students have difficulty completing a task, they can ask for an instructional explanation or demonstration of the next step, which would provide the means for students to learn how to perform these steps on their own in the future.

Model tracing of each student step yields an "overlay" on the domain model, indicating the probability that each knowledge component is known for each student. We refer to this above as the model of the student learning state. A knowledge-tracing algorithm updates these probabilities as students perform correct or incorrect steps during task performance, and the results are used to adapt future task selection to facilitate cognitive mastering of the correct knowledge components. For cognitive tutors, the tutoring interaction is dependent on the completeness and quality of the domain (or "cognitive") model. Thus, the broadest loop of adaptivity is to use student performance data to identify flaws in the domain model (e.g., common learning challenges that are not independently represented) and to improve it and the associated task design (Aleven et al. 2016).

Pedagogical Approach/Expertise Model

One of the goals of ITL systems is to design a robotic agent capable of learning a task from a human tutor, performing the task, and potentially to teach the task to a different human learner. Task knowledge must, therefore, be combined with didactic/pedagogical knowledge. To realize this goal, one possibility is that the agent reuses the teaching strategies incorporated when the agent itself first learned the task. Alternatively, a didactic/pedagogical model could be included in the agent; the agent would then apply this model to the task knowledge to generate instructions or other tutoring aids.

Model-Based Scheduling

Thus far, the intelligent tutoring system research community has focused primarily on curriculum adaptation (i.e., which content or problem to present next) and support adaptation (particularly with regard to changing the type of feedback provided) based on a student model. These adaptations address the "what" and the "how" of instruction; however, there is a new, promising area of research and development that focuses on the "when" of instruction (Gluck et al. 2019). Here, the emphasis is on personalizing the scheduling of learning events based on performance history. The adaptation is conducted at the level of individual knowledge components, a collection of which would comprise proficiency in problem-solving or task completion.

Personalized scheduling of learning depends on the availability of a computational model of the dynamics of human learning systems (Raaijmakers 2003; Pavlik and Anderson 2005; Walsh et al. 2018). The general idea is to calibrate parameters based on accumulated empirical evidence, then predict forward in time to determine a good schedule for the next study or practice opportunity. What represents a "good" schedule depends critically on whether the bias is toward rapid initial acquisition or longer retention of information. This trade-off relates to a phenomenon known as the "spacing effect" (Cepeda et al. 2006; Benjamin and Tullis 2010). If rapid acquisition is more important, then cramming learning events into a tightly massed sequence is preferable. This will generally accelerate the initial acquisition of that material but will also reduce the duration of retention. If better retention is more important, then spacing repeated learning events over a longer period of time is preferable, up to a point. If temporal spacing is too long early on in the process, then people forget the information more quickly than they learn it, and the potential advantage of distributing the practice over time is lost.

The implication for ITL is that these predictive, adaptive personalization capabilities can be implemented in the machine intelligence and applied in any context in which one of the agents participating in the interaction is a computer.

Communication with Respect to Models and Asymmetries between Human and Machine Learners

One key to a successful interaction and, by extension, to a successful tutoring interaction is that the teacher and learner share sufficiently similar communication processes and concepts; this is often referred to as "common ground" (Clark and Brennan 1991). When the teacher is tutoring, an implicit or explicit assumption is made by the teacher that the learner can correctly interpret the teacher's communicative signals. This requires a shared set of communicative signals to be interpreted in a manner considered successful by the teacher (or another external observer).

Systems for tutoring humans are designed based on assumptions of the students' knowledge base (e.g., what the learner knows, what the learner does not know, common ground, language, how the learner learns). When designing a machine to teach a human, we leverage substantial knowledge about how humans communicate and learn. However, when the learner is a machine, the validity of many of these assumptions is weakened or disappears. When a human teaches a machine, it is more difficult to make *a priori* assumptions about the types of communication available. Systems capable of constructing and building upon a model of language are crucial. Depending on the complexity of the learned task, a significant effort may be required to design communicative modalities that allow instructions and knowledge to be transferred from a human teacher to an artificial learner. Forbus et al. (2017) demonstrated a system that integrates visual processing, spatial representations, and conceptual knowledge; for the system to learn successfully from a human tutor, it needs to establish a common set of communicative signals (language and visual sketching, in this case) and requires access to the ontology of the world.

Similarly, when a machine teaches a human, the machine requires a model of the human learner, but it is challenging to encode all the relevant information. Asymmetries also exist between the representation and use of the learned knowledge. A human may learn to perform a task and use this as a basis to teach the task to another learner. A machine may be able to teach based on its learning mechanism, or teaching must be addressed separately.

Human–robot interaction exaggerates the problems associated with asymmetry, as in the correspondence problem. First identified in the context of imitation learning (see, e.g., Nehaniv and Dautenhahn 2007), the correspondence problem relates to the difficulty of mapping actions from one body to another. If a human demonstrates a skill to an agent with a dissimilar body plan or dissimilar actuation (e.g., a robot arm with five degrees of freedom that is learning to pour a drink by observing human demonstrations), how is the learning agent to map this demonstration to its own capabilities? The correspondence problem also holds in ITL: while the interactive element allows skill transfer to be scaffolded more gradually, the correspondence problem still requires a solution.

Methods for Task Instruction

There are myriad methods of instruction that are natural to human–human interaction, but whether emulation of these methods can or will support ITL is an open question. These instruction methods include *signal-focused methods*, which involve the modulation of signals such as speech and gesture, and *directing/attending behaviors*, such as pointing as well as other methods such as coupling of language with action. All instruction methods involve a persistent process of *monitoring, adjustment*, and *feedback* based on the projected cognitive process of the given student. This persistent process is important for the teacher's choice and adjustment of feedback type, including timing considerations.

Signal-Focused Methods

Developmental studies have addressed some of the scaffolding methods that people use to improve understanding of action and speech: motherese, motionese, gesturese. Collectively known as the "eses," these methods may be of fundamental interest to ITL. As techniques used to make instruction more effective and efficient with children, they may generalize to make instruction more effective and efficient with machines:

- *Motherese*, parentese, or child-directed speech refers to modifications in verbal behavior (Dominey and Dodane 2004). These modifications can be performed on all linguistic levels, including prosody (speaking with higher pitch and long pauses), semantics (making references to here and now, regular use of repetition, lower levels of abstraction), and grammar (shorter, simpler sentences).

- *Motionese* concerns the performance of actions. As reported by Rohlfing et al. (2006), child-directed actions are performed with rounder (smoother) movements and more frequent pauses.
- *Gesturese* refers to how caregivers modify the frequency, type, and duration of their gestures when they talk to children. Grimminger et al. (2010) reported that gesture frequency increases when teachers issue instructions for difficult tasks, or when they instruct children at risk of delayed language development (compared to typically developed children).

Learner feedback is crucial in these signal-focused methods and must be kept in mind when applying them to interactions with a machine. As demonstrated by Lohan et al. (2012), it is important for a robot to behave in a contingent manner and react in a timely manner to the communications initiated by the tutor. Fischer et al. (2011) indicate how tightly the "eses" are tailored to the learner: if the robot reacts to the tutor's instructions using visual behavior alone (e.g., eye gaze), then the tutors will modify their behavior in this modality. Consequently, behavior modifications can be observed in the form of "motionese" (visually perceptible), not in "motherese," as this would most likely require a verbal response from the robot. This research suggests that if a robot indicates sensitivities toward interaction in the form of a contingent behavior and particular modalities of communication, then tutors can provide beneficial input in the form of parentese, motionese, and/or gesturese.

Showing, Pointing, and Depicting

When children are already experienced with regard to interaction, their learning can be guided through "social cues" embedded in a sequence of particular actions. For instance, a pointing gesture is embedded as a social cue within a referential frame (e.g., eye gaze, pointing to an object, labeling it). Such sequences (pointing, showing, depicting) engage children to guide their attention and actions toward a specific goal.

When showing an object or performing a task, language and action often are synchronized such that they reinforce each other. This intermodal redundancy foregrounds key information (Bahrick et al. 2004) and creates meaningful "acoustic packages" (Hirsh-Pasek and Golinkoff 1996) to facilitate comprehension and learning. It is a practice that is generally useful in communication and is especially helpful with young children. Among infants who cannot yet understand semantics, labels or words were found to highlight the commonalities between objects and situations, facilitate object categorization, and override the perceptual categories of objects (Rohlfing and Tani 2011). Conversely, when action is provided concurrently with speech, it seems to embody the meaning of language, even in young infants (e.g., Nomikou et al. 2017).

Directing, Performance Modeling, and Peer Performance

Musical instrument instruction is an interesting practice to consider in the context of ITL, as it involves humans executing physically demanding tasks in the real world, with multidimensional success criteria. The ontology of feedback on music instrument instruction, provided by Yee-King et al. (2014), indicates the diversity of success measures in music instrument learning (e.g., timbre, groove, and articulation). Musical instrument tutors employ three key techniques that can be applied in ITL scenarios:

1. *Directing* involves the student playing a piece of music, with the tutor providing verbal and gestural directions during the performance, similar to a conductor. For example, the tutor might say "build up the speed now," or "careful with the fingering for the right hand." The information is highly contextual and can be directly acted upon by the student to improve performance. This approach could apply to an ITL scenario in which a machine is learning to set a table: "careful with the placement of the fork, as there's no space for the plate."

2. *Performance modeling* involves the tutor playing the instrument and presenting the student with both good and poor examples of how to play sections of a given piece of music. In the context of ITL, this correlates to a tutor providing a set of carefully chosen, labeled examples.

3. *Peer performance* involves students performing a complete piece for an audience of their peers as well as the tutor. At the end of the performance, the peers are encouraged to give feedback to the performers. Peer performance is important for the performer, as it takes place in a context where failure has a high cost (embarrassment) and closely mirrors the actual context in which music is commonly performed. It is also important for the listeners, as they learn to discern between good and poor execution. Both performer and listeners gain a wider view of the performance: What effect did it have on an audience? How does the sound change when performing in a concert hall? How does a person's performance respond to the pressure of being observed? This raises the intriguing idea of one robot washing a window while others critique its performance.

Monitoring, Adjustment, and Feedback

Feedback is especially important, perhaps essential, to almost all instructional interactions. Feedback can be provided/received at any time during a task: at the beginning to inform upcoming action on the basis of prior performance, in real time during task performance, and at the end when the task is complete. Many details need to be considered when providing and receiving feedback (e.g., timing, synchrony of cues, task structure, conventional social norms).

Feedback received by a teacher or learner can trigger a process whereby the agents interact to repair incorrect inferences or clarify ambiguities. Feedback received by the learner can also trigger a feed-forward process, whereby the learner is able to apply feedback received from a teacher to future attempts at completing a task, in the absence of the teacher's guidance (Hattie and Timperley 2007). The feed-forward process maps across several desirable aspects of ITL as follows:

- It results in an efficient use of instructor time, since feedback is reused.
- It signifies a transition on the part of the student from being unconsciously incompetent to consciously incompetent: students can detect their own mistakes and use previously received feedback to rectify them.
- The student switches his/her role to that of his/her own instructor.
- It includes metacognitive aspects, wherein the student becomes aware of his/her own learning process.

During human–human interactions for learning and teaching, particularly between parents and infants, monitoring the learner is important. Teachers modify their behavior online in a moment-to-moment co-construction toward a goal that is shared with the learner, with interactional loops between teacher and learner, such as between the teacher's hand movements and the learner's eye gaze (Pitsch et al. 2014). Study results have shown that these processes translate to human–robot interactions, with the robot as the learner and the human as the teacher, under certain conditions (Vollmer et al. 2009; Pitsch et al. 2013). These conditions include a certain appearance and social behavior (such as feedback or contingency) on the part of the learner that forms a social interaction interface for the human to employ during teaching. It has been shown that the way in which an action demonstration is instantiated and how the action is structured (in terms of highlighting what about it is important when teaching the action to a robot) strongly depend on the learner's feedback (Vollmer et al. 2014). In a case involving a robot learner, these interactional loops with the adaptive behavior of the teacher could be exploited for the benefit of learning. The robot could, through its feedback, elicit certain adaptations to the teacher's behavior, allowing for testing of a hypothesis about, for instance, the importance of a certain part of a demonstration.

The lessons gained through human–human learning/teaching interactions are most clearly relevant to robot learning, although some of these teaching methods have analogous forms in the virtual world of graphical user interfaces within which software agents learn. The signal-focused and directing and attending methods discussed above may not be implemented in the same way in a graphical user interface, but similar implicit communicative functions can be achieved in other ways. Analogous to a robot finger pointing at (or an eye gaze directing attention toward) an important element or feature within an object or scene, features may be highlighted through other means, such as bolding, circling, or underlining. Such "feature focusing" has been used to enhance human learning; for instance, to help teach Chinese symbols by highlighting semantically meaningful components (known as "radicals") within them. It has also been used to aid machine learning of grammatical structure in algebra (Li et al. 2015) by highlighting elements that interrelate (e.g., highlighting "–3" in "–3x"). Do other aspects of human–human learning/teaching interactions have relevance within the virtual world of agent learning? This is an interesting question that awaits further exploration.

Learning by Teaching

When the primary goal is for a (human) student to learn a task, studies have demonstrated that it is often more effective to assign a human the role of a teacher, who then must tutor a computer agent by demonstrating, for example, how to solve algebra equations while providing feedback about the attempts (Chase et al. 2009; Matsuda et al. 2013). In these "teachable agent" applications, the computer agent takes a quiz after being instructed by the human, and the agent's performance provides students with feedback on the ways in which they succeeded or failed to teach the agent. This method of learning proves to be an effective "ego-protective buffer" whereby students' knowledge gaps are revealed by their agent's performance rather than their own.

When the primary goal is for an artificial agent to learn a task, it may also be valuable to swap human and computer roles so that the agent becomes the teacher and the user becomes the learner. One such scenario is the use of the ITL agent to help teach or "entrain" humans to use the language understood by the ITL agent. Consider a situation in which the human says, "Start up my mail" to the agent, rather than "Open the email app." In response, the agent might say, "Well, show me," and the user subsequently succeeds in teaching the agent via demonstration. Key to this example, the learning agent then switches into the teaching role and explains the procedure it learned back to the user, using terms the agent understands (e.g., "First, I opened the email app. Then, …"). Upon observing this explanation, the user may have a better chance in subsequent attempts of using language patterns that the agent understands. In addition, the agent has had the opportunity to expand its vocabulary.

Closing Thoughts

Structural and Adaptive Perspectives on Improving Instruction

Instructional improvements based on education technology (including ITL) can be broadly classified as adaptive, structural, or both adaptive and structural (VanLehn 2016). Instructional improvement is adaptive if the content and structure of the new instruction are the same as in the baseline instruction, but the new instruction interacts with learners differently based on their performance. An instruction improvement is considered structural if the structure, or plan, of the instruction differs significantly from the baseline instruction. Consider the following examples:

- Purely structural: A new, three-week module in a middle-school science class is revised to focus on accessing data obtained from a real radio telescope.
- Purely adaptive: An organic chemistry practice system contains the same large set of problems as before, but the system now monitors the students' successes and failures, and recommends the best problem to address next.
- Both structural and adaptive: A collaborative learning system has students working in small groups to build an economic model. The system monitors their interactions and sends advice covertly to group members who are not speaking up enough or who are dominating the group dynamics.

Regardless of whether one is designing instruction that is adaptive, structural, or both, there are several choices to consider, including demonstrating, telling (through verbal instruction), and providing feedback (e.g., as accomplished through depicting, showing, pointing to, or coupling language and action). Instructional strategies include numerous choices with regard to structuring the environment, choosing learning examples, comparing or contrasting tasks or problems, making sequencing or curriculum decisions, and determining how elaborately to expose the steps and whether, when, and how to provide reward or feedback (Koedinger et al. 2013).

Historical Perspective on Programming and Implications for ITL

In the early days of programming, FORTRAN, COBAL, and LISP were the only languages available. Many people thought that redesigning the languages would make programming significantly easier. This led to ALGOL, PASCAL, and eventually myriad languages currently in place, which may be a little easier to use than the first languages, but certainly did not vastly simplify programming. Similarly, natural language programming, graphical programming languages, and programming-by-demonstration did not have significant impacts, as they proved useful only within a limited range of programming problems.

Perhaps the biggest advance in software engineering has been the recognition and systematic development of agile programming methods (Beck et al. 2001). Prior to this recognition, the orthodox method of programming (now called "waterfall") involved three phases:

1. Writing detailed specifications in natural language, and sometimes diagrams (e.g., flow charts; UML).
2. Converting the specification into a giant program.
3. Testing the program.

In contrast, the agile method divides the overall function of the desired program into many small pieces, and follows the following three-step approach to develop each piece:

1. Write a specification of the new piece of functionality, called a user story.
2. Add that function to the program.
3. Test the new function while also testing that the preexisting functions remain intact.

Nowadays, the waterfall method is typically used when developing a program that has been previously developed but needs to be repurposed for a new situation. Under these circumstances, the complete functionality of the program is well known, so the specifications can be accurate and complete. Most other programming is performed using agile programming, because it allows the ultimate users of the program (often called "clients") to use the emerging program as functions become available, and to provide new user stories or revisions to old user stories. In other words, agile programming is more interactive.

The ITL vision is similar to these historical features of programming methods in many respects. First, it is interactive: like agile programming, the interactivity of ITL is probably the most powerful simplification of the overall job of creating engineered activity. Second, ITL combines natural language and demonstration as ways of expressing user stories. A skeptic may note that natural language and demonstration provided only limited simplification of traditional programming and that perhaps an implication of this is that they may have limited importance for ITL, as well. However, it is important to keep in mind that the traditional programming model of machine as a tool (rather than a partner) and implementation for niche specialization (rather than generalized interactive learning of new tasks) inherently limited the potential utility of natural language and demonstration.

One way to understand ITL's potential is to consider its scope. It is clearly not intended to replace traditional programming for large systems; ITL will most likely replace scripting, writing simple programs of perhaps a page or two. More specifically, it may fit best in the middle of a continuum of ways that users can express their intentions to computers. On the simple end of this spectrum are menus that allow a user to choose from a small set of predefined behaviors. Next, in terms of simplicity, are forms that have a set of menus, type-in boxes, buttons, and other controls to allow users to express somewhat more complicated intentions. To indicate intended behaviors that go beyond those readily designated by form-based user interfaces, we currently write scripts, demonstrate the intended behavior to a macro-writer, or both. This capability is likely to be replaced by ITL. The next step up in terms of complexity will likely require traditional programming.

The same complexity continuum applies even when all agents are human, such as when an employer describes intended behaviors to employees. On the simple end, a short phrase suffices to explain the boss's intentions (e.g., "Hold my calls.") On the more complex end, modern organizations use formal languages, often called *business process languages*, or procedure manuals. An ITL system would most likely incorporate intended behaviors "taught" to human workers through a short email message or narrated demonstration.

Based on these analogies, it seems likely that ITL systems may address a "sweet spot" within the continuum of complexity, resulting in many potential applications. Moreover, these applications are poorly served today because the scripting/macro-demonstration process is manually intensive.

Similarities and Differences between ITL and Human/Animal Teaching and Learning

To clarify similarities and differences between machine and human/animal learning, consider the following examples. The first involves a "cat flap" (i.e., a portal that allows a cat to enter or exit a house at will): through a system of rewards (and some patience), most cats learn to open and pass through the cat flap. This is an example of operant conditioning. The second involves standard computer programming: a computer is initially unable to perform a given task (e.g., play chess), so the user provides it with a set of instructions. Initially, the instructions are of poor quality, and the computer is still unable to perform the task. Responding to this failure, the user continually updates the instructions until the computer succeeds. This scenario is both interactive and involves the learning of a task.

Note the cat flap example does not involve intelligent behavior or reasoning; it is a primitive form of teaching (operant conditioning). Once learned, however, the cat would be quite flexible with regard to its new task; that is, confronted with a new cat flap or the same cat flap in another door in the same house, the cat would most likely succeed in going through the flap. Also, the cat can learn many other tasks through operant conditioning without interfering with its ability to maneuver through cat flaps. In contrast, the computer programming example involves hardly any learning by the computer, whereas the task for the teacher is immense.

Humans and animals have many advanced, fine-tuned, and mutually aligned learning mechanisms. Human–human ITL begins with learning through reinforcement in the first months of life, followed by demonstration and imitation, and later on through exploration, instruction, and reasoning. It is quite clear that our current AI systems and robots lag far behind in this respect: they may be very good at applying a single learning mechanism (e.g., learning through examples or parameter optimization), but the broad integration and application of multiple learning mechanisms in complex, real-time task learning and performance remains elusive. We aspire to create artificial systems that combine diverse mechanisms for co-constructive task learning and instruction.

12

What Do Human Tutors Do?

Kurt VanLehn

Abstract

People teaching an agent or robot might use the same methods that they use when tutoring a human student. Because teaching agents and robots is a central topic of this Ernst Strüngmann Forum, this chapter reviews research that characterizes human tutoring. Most of this research was done to improve the design of computer-based tutoring systems, which were assumed to be inferior to human tutors. However, it turns out that human tutors and a certain class of tutoring systems actually behave quite similarly, and their effectiveness is about the same. This chapter begins with a description of prototypical human tutoring behavior before discussing some common hypotheses about human tutoring behavior, which turn out to be unsupported by studies. It concludes with an attempt to synthesize these descriptions and apply them to the goals set forth at this Forum.

Introduction

Our discussion at this Ernst Strüngmann Forum focused both on a person teaching a bot (a robot, softbot, or agent) and a bot teaching a person. To understand both approaches, we need to have a general understanding of how a person behaves when teaching another person. Thus, in this chapter I review what is known about human–human tutoring.

I use the term "tutor" to refer to an adult domain expert who works with a single student, but note that in some literatures, "tutor" refers to someone working with a small group of students while in others, *peer* tutoring indicates a situation where one student teaches another. Here I review only studies where the tutors are adults who have expertise in the knowledge to be taught, and they are working one-on-one with a human student. This seems the most likely analogue to a human teaching a bot or a bot teaching a human. Further, I will discuss both expert and novice tutors. Both are domain experts, but an expert tutor has considerably more experience *as a tutor* than a novice.

I will refer only to studies that have sought general properties of human tutoring. These have appeared mostly in the cognitive science literature or in the intelligent tutoring systems literature. In contrast, many disciplines have studies that address the effectiveness of human tutoring compared to ordinary instruction in that discipline (e.g., Bausell et al. 1972; Marston et al. 1995; Scruggs and Richter 1985; Shanahan 1998; Wasik 1998; Wasik and Slavin 1993) or the occurrence of discipline-specific tutor behaviors. As an example of the latter, Juel (1996) analyzed dialogues of human reading tutors to find out, among other things, how frequently they asked tutees to sound out words.

What do human tutors do? This very question requires a description as its answer. Although some of the research reported here formulates and tests specific hypotheses, most of the research is qualitative and descriptive. Thus, this review is qualitative and descriptive as well. I begin by describing, in broad strokes, what human tutors do. Thereafter I analyze what a few exceptional tutors sometimes do, but most tutors do not do, before comparing the effectiveness of human versus computer tutors. Finally, I attempt a synthesis of learning and tutoring, and speculate on its implications for instruction of bots.

A Prototypical Session with a Human Tutor

A prototypical tutoring session can be viewed as a setting in which participants work through a sequence of *tasks*. A task could be solving a problem, studying an example, or answering a question. A task could even involve the student reading a text, while the tutor watches and helps.

A key fact is that most human tutoring supplements classroom instruction. Because the tutees are also taking a course in school, the student's teacher (i.e., not the tutor) usually determines the curriculum, so students often arrive at a tutoring session with a list of tasks to do (i.e., their homework) or a topic that requires further explanation. If they bring only a topic, then the tutor often selects tasks from a standard resource, such as a textbook, rather than inventing new tasks on the spot. In short, tutors prefer not to do on-the-spot instructional design, so they seldom invent new tasks. In this respect, human tutors play a very different role than a human instructing a bot. Presumably, bot instructors would need to invent their own tasks, because a school for bots with homework assignments seems highly unlikely!

Once the tasks to be accomplished during the session have been determined, the tutor takes control. After analyzing 72 one-hour tutoring sessions, Graesser et al. (1995:500) concluded that "the students never set the agenda for the tutoring session. Thus, the tutors carried the burden of setting the agenda, introducing subtopics, and proposing problems to solve." Typically the tutor works through one task after another. What the tutor does during each task depends on the nature of the task. Let us consider several common tasks.

Reading

Let us assume that the student is asked to read a short text, such as a one-page description of mitosis. Although the tutor could stop the student after each paragraph to ask questions (Chi et al. 1994), tutors typically let the student read silently until the end of the reading.

Answering Complex Questions

Now let us assume that the student is given a complex question to answer. Such questions are commonly assigned as homework. Tutors sometimes interrupt other tasks to ask the student complex questions. The students' behavior appears to be similar regardless of whether the question is a task in itself or an interruption of a task.

Graesser et al. (1995) characterize human tutoring of such questions as exhibiting a five-step pattern:

1. The tutor asks the student an open-ended question, such as "What happens when a computer boots up?"
2. The student gives an initial answer: "It starts up the windows."
3. The tutor gives some brief feedback: "Yes. Good."
4. The tutor then conducts a subdialogue to extract a better answer from the student:
    *   Tutor: "What is the software called that has those windows?"
    *   Student: "Windows"
    *   Tutor: "I meant the type of software."
    *   Student: "Do you mean operating system?"
    *   Tutor: "Yes. So when you boot the computer, it starts the operating system. From where does it get the operating system?"
5. The tutor typically ends the dialogue with an evaluative comment or question:
    *   "So do you understand 'booting up' now?"
    *   Student: "Yes"

Although the five-step frame is common, novice tutors have a tendency to cut it short and replace step four with a brief lecture (Glass et al. 1999; Kim et al. 2005).

Studying Worked Examples

Studying a worked example (i.e., a multistep process needed to perform a task or solve a problem) is another task type. Consider the following example in physics:

Problem: Suppose a roller coaster is at rest on a track 200 meters above ground. The track tilts, and the coaster zooms down the first drop. Its descent is so steep and the wheels are so well lubricated that friction can be ignored. What is its speed when it reaches ground level?

Solution: This problem can be solved by applying conservation of mechanical energy.

At the start of the fall, the coaster's mechanical energy is entirely gravitational potential energy, because its velocity is zero, so

$$E_1 = m \times g \times h.$$

At the end of the fall, the coaster's mechanical energy is entirely kinetic energy, so

$$E_2 = m \times v^2.$$

By conservation of mechanical energy,

$$E_1 = E_2$$

so

$$m \times g \times h = m \times v^2$$
$$g \times h = v^2$$
$$v = sqrt (g \times h) = sqrt (9.8 \times 200) = 44.7 \text{ m/sec.}$$

Each line in the solution can be considered a step in solving the problem. If students explain each line thoroughly, they will learn well from studying the example (Chi et al. 1989). Although human tutors can be taught to prompt for such explanations after each line, most prefer to lecture instead (Chi et al. 2001).

Solving Problems

Tutors often spend most of their time helping students solve multistep problems, like the physics problem explained above. Their behavior varies with the competence of the student in a pattern widely known as *model-scaffold-fade* (Collins et al. 1989). When students lack the competence to solve a problem themselves, then tutors "model" how to do it by executing all or most of the steps. As they do a step, tutors usually explain it and then ask: "Do you understand?"

Once students gain even a little competency, tutors will let them solve the problem while giving them several kinds of "scaffolding." Forms of scaffolding include the following:

- The *prompting* of steps, especially difficult steps, through hints given before the student has even attempted to do the step. This is a natural means of scaffolding the task performance of bots.
- Giving *feedback* after a student's initial try, by indicating whether the attempt was correct or incorrect. The tutor may elaborate on this feedback and explain why the attempt was correct or incorrect. When giving negative feedback, tutors seldom say "wrong" or some other explicit negation. Instead, they use circumlocutions or even just pause

a moment before starting in on an explanation (Fox 1991, 1993; Rose et al. 2003). This particular habit, of giving negative feedback in a subtle, face-saving way, may be problematic when humans tutor bots.

- Providing *hints* when the student asks for one, or when it appears that the student is struggling to enter a step but has not yet asked for a hint.
- Asking students open-ended *questions*. Here a tutor might ask the student to explain a correct step or a concept related to it. In one study of tutored problem solving, 18% of the tutor turns were open-ended questions (Rose et al. 2003).

As students gain competence, the tutor removes or "fades" the scaffolding. Prompting is often the first form of scaffolding to disappear, but feedback may be withdrawn as well. Tutors can often determine, after just a few minutes of tutoring, how much scaffolding their tutee needs (Siler and VanLehn 2015).

When a problem is finished, tutors will sometimes review the solution with the student (Cho et al. 2000; Katz et al. 2003). For instance, if the student makes an incorrect step during the tutoring session, and the tutor believes it was due to a serious misunderstanding, the tutor may simply correct the step for the student and then, during the post-solution review, discuss the step and the misunderstanding.

Novice and expert human tutors vary in how they tutor. Novices tend to do more modeling (explaining) whereas experts tend to let the student do as much of the work as possible (Glass et al. 1999; Wood et al. 1976). Novice tutors sometimes give too much prompting, which reduces errors and learning (Kim et al. 2005; VanLehn et al. 2003).

What Human Tutors Seldom Do

The behaviors described above strike many people, including some participants at this Forum, as surprisingly mundane. People often think that human tutors engage in much more sophisticated behaviors, such as the ones listed below. Although at least a few tutors performed as hypothesized, research has shown that their behavior was not common enough to account for the effectiveness of human tutoring.

Detailed Diagnostic Assessments

In a few early studies of human tutoring (Collins 1977; Stevens and Collins 1977; Stevens et al. 1979), tutors debugged students. These studies involved tutors who went through a script that asked the student about the causes of heavy rainfall in various parts of the world. When a tutor uncovered a student's misunderstanding, the tutor conducted a misconception-specific subdialogue that resulted in removal of the misconception. This tutoring strategy is often called "diagnose and remediate." This strategy, however, appears to be uncommon, and is perhaps even unique to inquiry instruction, which is the instructional technique studied by Collins and Stevens. Let us consider how prevalent the diagnose-and-remediate strategy is.

Although human tutors in other domains usually know which *correct* knowledge components their tutees have not yet mastered, tutors rarely know about their tutees' *misconceptions, false beliefs,* and *buggy skills* (Chi et al. 2004; Jeong et al. 1997; Putnam 1987). Moreover, human tutors rarely ask questions that could diagnose a student's specific misconception (McArthur et al. 1990; Putnam 1987). Thus, although tutors can try to uncover misconceptions, they rarely do so, and thus rarely know what misconceptions their tutees harbor.

Several studies asked whether human tutors could remediate when given a diagnosis. When human tutors were told which *correct* knowledge components were not yet mastered by their tutees, their behavior changed and their tutoring became more effective (Wittwer et al. 2010). On the other hand, tutors typically do *not* change their behavior nor become more effective when given detailed diagnostic information about their tutee's misconceptions, bugs, and false beliefs (Sleeman et al. 1989). Siler and VanLehn (2015) found that human tutors who worked with the same student for an extended period, and could thus diagnose their tutee's strengths, weaknesses, and preferences, were *not* more effective than when they rotated and had little familiarity with their tutees.

Human tutors do not seem to infer an assessment of their tutee that includes misconceptions, bugs, or false beliefs, nor do they seem to be able to use such an assessment when it is given to them. Instead, they often infer an assessment of which *correct* conceptions, skills, and beliefs the student has not yet mastered, and are able to use such an assessment when it is given to them. In short, tutors attempt to fill in the missing correct knowledge of their tutees, but do not attempt to remove their incorrect "knowledge." In this respect, human tutors operate just like computer tutors that use an *overlay* model of the student, where the overlay model represents only the presence or absence of correct knowledge (VanLehn 1988, 2008).

Adaptive Task Selection

Another hypothesis is that human tutors select tasks adaptively. That is, they give the student a task that is just what that student needs at that time. Studies suggest, however, that when human tutors are not working off a set of tasks brought to the tutoring session by their tutee, the human tutors select tasks using a *curriculum script*, which is a sequence of tasks ordered from simple to difficult (Chi et al. 2008; Graesser et al. 1995; Putnam 1987). Human tutors use their assessment of the student's mastery of correct knowledge to regulate how fast they move through the curriculum script. For example, tutors typically move to the next problem type after students correctly answered two or three problems of the current type (Putnam 1987).

Sophisticated Tutorial Strategies

A common early belief was that the power of human tutoring lay in their use of sophisticated strategies, such as Socratic irony (Collins and Stevens 1982), wherein students who give an incorrect answer are led to see that their answer entails an absurd conclusion. Other sophisticated strategies include reciprocal teaching (Palinscar and Brown 1984), inquiry (Collins and Stevens 1982), or authentic anchored cases (Goldman et al. 1993).

Studies of human tutors in many task domains with many degrees of expertise indicate, however, that these sophisticated strategies are rarely used (Cade et al. 2008; Chi et al. 2001; Cho et al. 2000; Core et al. 2003; Evens and Michael 2006; Fox 1991, 1993; Frederiksen et al. 2000; Graesser et al. 1995; Hume et al. 1996; Katz et al. 2003; McArthur et al. 1990; Merrill et al. 1995; Merrill et al. 1992; Ohlsson et al. 2007; VanLehn 1999; VanLehn et al. 2003).

A related belief is that human tutors sometimes let students make mistakes so that the students can practice finding them. However, tutors rarely let this happen (Fox 1991, 1993; Frederiksen et al. 2000; Merrill et al. 1995). If they ignore a student's error, it is often because the error is trivial and would have no impact on subsequent problem solving (Merrill et al. 1995).

These are statements about frequency and not about capability. For instance, from the Collins and Stevens (1982) studies it is clear that human tutors can, and sometimes do, use sophisticated strategies. Chase et al. (2015) gave human tutors the job of scaffolding an invention task and observed that they frequently employed more sophisticated tutoring practices than typical tutors. The demands of the task may have a large impact on the type of tutoring observed.

Questions Asked by Students

Human tutoring allows mixed initiative dialogues, as students can ask questions or even change the topic. This contrasts with most computer-based tutoring systems, where student initiative is highly constrained. For instance, although students can ask a typical tutoring system for help on a step, they are not able to ask other questions, nor can they cause the tutor to veer from solving the problem. On the other hand, students are free to ask any question of human tutors and to negotiate topic changes with the tutor.

Analyses of human tutorial dialogues show that although students take this initiative more often than they do in classroom settings, the frequency is still low (Chi et al. 2001; Core et al. 2003; Graesser et al. 1995). For instance, Shah et al. (2002) found only 146 student initiatives in 28 hours of typed human tutoring (i.e., tutor and student communicated over a chat connection), and in 37% of these 146 instances, students were simply asking the tutor whether their statement was correct (e.g., by ending their statement with "right?") or establishing common ground (e.g., "Did you mean condition two?"). That is, there were about 3.3 nontrivial student questions per hour. Participants were medical students being tutored as part of a high-stakes physiology course, so apathy is not a likely explanation for this low rate. With high school and undergraduate students, Graesser et al. (1995) found that 71% of the student questions were trivial, so there were only 7.7 nontrivial questions per hour. In short, even though students can ask nontrivial questions of human tutors, they seldom do.

Motivational Comments

Human tutoring is often believed to increase the motivation of students. Episodes of tutoring that aim to increase students' motivation certainly do occur in human tutoring (Cordova and Lepper 1996; Lepper and Woolverton 2002; Lepper et al. 1993; McArthur et al. 1990), but their effect on student learning is unclear.

For instance, consider praise, which Lepper et al. (1993) identified as a key tutorial tactic for increasing motivation. One might think that a human tutor's praise increases motivation, which increases engagement, which increases learning. However, the effect of human tutors' praise on tutees is actually quite complex (Boyer et al. 2008; Henderlong and Lepper 2002). In Kluger and DeNisi's (1996) meta-analysis of hundreds of feedback interventions, praise had a small negative effect ($d = -0.17$). Kluger and DeNisi (1996:275) conclude:

> The debilitating effects of praise on performance received some direct experimental support both in the laboratory and in the field and were explained, respectively, by a model of self-attention (Baumeister et al. 1990) and by control theory (Waldersee and Luthans 1994). These findings are also consistent with a review of field studies (many of which did not qualify for the meta-analysis) that concluded that "praise may not be widely effective as a reinforcer."

There are many interventions that increase motivation, such as convincing students that their intelligence is malleable rather than fixed (Dweck 1986). The effects are often quite large compared to the brevity of the intervention (Lazowski and Hulleman 2016). However, most of these studies involve interventions with large groups of students, such as a whole class. It is unclear what the impacts would be if these interventions were implemented by human tutors (or computer tutors). This is clearly an area where more research is needed.

Habits of Tutors That Bots Should Be Designed to Handle

Human tutors, especially novice human tutors (Glass et al. 1999), utilize certain habitual dialogue to support the self-monitoring and self-correction processes in a student. Looking toward future encounters between a human tutor and a bot, designers need to be aware of this and reflect it in the design of bots.

Self-Monitoring: "Do You Understand?"

Tutors often ask students: "Do you understand?" Such questions often occur at the end of the five-step frame (see above), when students have finished a reading, or when they have finished studying an example (Glass et al. 1999; Graesser et al. 1995).

A student's answer reflects their ability to monitor their own understanding (self-monitoring). Chi et al. (1989) divided their students into Good and Poor, based on learning gains, and found that while the self-monitoring statements of Good students were 45% negative and 54% positive, the self-monitoring statements of Poor students were 15% negative and 85% positive. This suggests that the Good students were good at self-monitoring and thus accurate at reporting their degree of understanding, but that the Poor students were not good at self-monitoring and their default was to assume that they understood. Graesser et al. (1995) found the same pattern when self-monitoring statements were elicited by the tutor's question. These findings reflect the general challenge that many learners have with self-monitoring (Glenburg et al. 1982).

Because bots are likely to get such questions from tutors, they should have good self-monitoring capabilities. Moreover, they should be able to act like Chi et al.'s (1989) Good students, whose negative self-monitoring statements were quite specific (e.g., "I'm wondering whether there should be acceleration due to gravity?") compared to the Poor student's self-monitoring statements (e.g., "What should I do now?").

Negative Feedback That Saves Face and Encourages Self-Correction

Although human tutors are quite accurate at spotting errors in the students' work (Evens and Michael 2006; Merrill et al. 1995), they often will not give explicit negative feedback. For example, Graesser et al. (1995) found that tutors directly acknowledged student errors only 24% of the time. Fox (1993) and Lepper et al. (1993) found that human tutors prefer dialogue filled with pauses, prompts, and questions which encourage the student to notice and repair the error themselves. In addition to giving the student practice in self-correction, they hypothesized that the circumlocutions minimize the impact of negative feedback on motivation and allow the student to "save face."

Merrill et al. (1995) found that the proportion of explicit versus implicit negative feedback depended on the type of error. If the error was clearly an unintentional slip, such as a typo, and the students did not themselves detect it, then the tutor explicitly indicated the correction. If, however, the error was plausibly due to a misunderstanding or lack of domain knowledge, then the tutor often used more indirect dialogue to allow the student to do as much of the error detection and correction process as possible.

This is not to say that human tutors fail to give negative feedback. McArthur et al. (1990) found that *every* error was followed by some kind of remedial dialogue. Merrill et al. (1995) found that 95% of student errors which were not immediately caught by the student were followed by some kind of remedial dialogue by the tutor.

Looking toward human tutor–bot interactions, a bot can count on human tutors giving negative feedback on almost every error, but the *way* in which tutors give the negative feedback might make it hard for a bot to recognize it as such.

The Effectiveness of Human Tutors

It is often thought that human tutors are the most effective form of instruction on the planet. For instance, Bloom (1984) claimed that human tutors produced a two-standard deviation effect size, which is extremely large. However, a review of experiments comparing human tutors with computer tutors suggests that human tutors are only slightly more effective than step-based tutoring systems (VanLehn 2011); that is, tutoring systems which can give feedback and hints after each step in the student's problem solving. Both human tutors and step-based tutors were around 0.75 standard deviations more effective than no tutoring.

Discussion

So much research has been done on human learning, tutoring, and instruction that it is easy to lose track of the bigger picture. Here, I will summarize my "big picture." I include the four most important factors that distinguish, in my opinion, human learning from instruction, and then use them to explain why human and computer tutoring are so effective. Finally, some implications are considered for instruction in the context of interactive task learning.

The "Big Four" Factors Affecting Human Learning

The first of the "big four," *engagement*, is arguably the most important factor when explaining the success or failure of students in classroom instruction. In laboratory studies, students are almost always engaged in doing the tasks assigned to them, but in classrooms, students often disengage and "go off task."

The second most important factor concerns the *qualitative nature of the learning curve*. A learning curve is a quantitative display of students' progress in learning a specific piece of knowledge. The horizontal axis corresponds to the episodes where the student tried to apply that piece of knowledge. The vertical axis is either the error probability or the duration of the episode. The curve descends in an exponential or power-law shape. That is, as people practice the same thing over and over, their errors decrease and their speed increases. However, learning curves only look nice and smooth when they represent averages over hundreds of students. The errors and durations of a single student's episodes are hardly smooth. This is because students' behavior and thinking varies qualitatively as they practice. Many theorists like to use the following stages as an approximation of the qualitative changes that underpin the learning curve (Anderson 1982; Fitts and Posner 1967; Koedinger et al. 2012; VanLehn 1996):

- *Sense making*: During the initial phase, students are just trying to understand information presented to them about the target knowledge. During the first attempt to apply the target knowledge, they might have to refer back to the text or example that comprised their original instruction. They may also need help in finding out what they need to know or in interpreting it. Such an episode can take minutes. If the second attempt immediately follows the first and the task is almost the same, then the episode might go quickly and smoothly, but the result may be incorrect. There is high variability in errors and duration during the sense-making phase.
- *Refinement*: During the middle phase, students understand the core or basic elements of the target knowledge, so their learning covers the fine points, or what programmers call "corner cases." Thus, students make errors and struggle on unusual tasks, but perform smoothly and rapidly on tasks that are similar to those previously experienced. Variability in errors and duration is less in the refinement phase than the sense-making phase.
- *Fluency building*: During the final phase, students fully understand the target knowledge, so their errors are only unintentional ones. As they practice, the probability of such slips decreases and their performance speed increases.

Different theorists have used different terms to describe these stages; the terms above are from Koedinger et al. (2012). The important point is that the cognitive and overt behavior of learners changes qualitatively as they practice. Instruction that is optimal at one time may not be optimal at a different time. This explains why model-scaffold-fade is so popular and effective. It is an effective way to make instruction adaptive.

The third important factor in human learning is the *type of behavior students engage in when learning a task*. This is well captured by the ICAP framework (Chi 2009; Chi and Wylie 2014b; Fonseca and Chi 2011; Menekse et al. 2013). The acronym stands for four types of student behavior, listed below from least to most effective:

- *Passive:* The student is paying attention to the instruction, but does not move or do anything overtly.
- *Active:* In addition to paying attention to the instruction, the student is displaying overt (visible) behavior that suggests she is evaluating and

selecting portions of the information presented. Highlighting phrases in text, taking verbatim notes during a lecture, or answering multiple-choice questions are often active student behaviors.

- *Constructive:* To qualify as behaving constructively, students must generate information that is not contained in the instruction. That is, they are constructing/generating *new* information (e.g., inferences, examples, judgments).
- *Interactive:* This type of behavior occurs only when two or more students are working together; it is not an individual behavior. Interactive students are co-constructive. That is, both students are constructive but, critically, each student's construction builds upon the information that their partner generated.

The ICAP framework aptly predicts learning gains during the sense-making phase of learning and perhaps also during the refinement phase. However, it probably does not make the right predictions for fluency building. ICAP is also ordered by learning gains, not by duration. Passive activities generally go faster than the others. Interactive activities are probably the slowest, given that all four types of activity "cover" the same knowledge. Thus, for the very first exposure of students to a piece of knowledge, a passive activity might be a better choice than an interactive one just because it takes less time. Thus, ICAP may often be compatible with model-scaffold-fade. Although ICAP is a classification of student behavior, it is often used to classify instructional tasks. Thus, for example, a task is classified as interactive if the instructions for doing the task ask that students behave interactively. This does not mean that all students will behave interactively, which brings us to the last major factor.

The fourth major factor influencing human learning is *feedback* that compares the student's actual behavior to the desired ICAP behavior, pointing out discrepancies and suggesting ways to improve the student's behavior (Hattie and Timperley 2007; Narciss 2007; Shute 2008; VanLehn 2016). The feedback may be immediate or delayed, and it might be given directly to the student or conveyed via a teacher. The feedback might come from the task itself, as students discover what doesn't work or uncover contradictions in their beliefs.

Another type of feedback, although it is not usually recognized as such, is deciding when a student has passed a module, a course, or a grade level based on assessing the student's competence. When this feedback loop is applied to modules or units of a course it is called "move on when ready," "gated instruction," or "mastery learning" (Bloom 1984). Such feedback is another way in which instruction can be adaptive.

What Makes Tutoring Effective

Tutoring is effective because it addresses each of the "big four" factors listed above:

- Tutors (especially human tutors) *keep students engaged.*
- Tutors can determine from the student's behavior *where they are located on the learning curve* for each piece of target knowledge.
- Tutors can modify episodes within a task along the ICAP dimension so that the episode becomes *appropriate for the student's position along the learning curve.* For instance, if the tutor determines that a piece of target knowledge is completely unfamiliar to the student, the tutor may explain it to the student rather than watch silently as the student tries to construct it for themselves.
- Tutors can *provide feedback* that helps students approach optimal behavior for the task. During the refinement and fluency phases, tutors can encourage students to take over the feedback loop and self-regulate their behavior. Tutors that are allowed to control when a student passes a module can use *mastery learning* so that students move on to the next modules only when they have achieved mastery of the current module.

Implications for Teaching Bots

Discussions at this Forum usually assumed that someone instructing a bot would first explain how to do a task, then monitor the bot's initial performance of the task, and help the bot debug itself if there were errors. How well this works depends on both the human instructor and the bot. The discussion above suggests what our expectations of humans should be for each phase:

- *Initial explanation*: Human tutors love to give explanations, but human students rarely master a concept or skill from just an explanation, so it is appropriate to assume that bots will probably need all three stages (i.e., initial explanation, monitoring, and debugging) to master new knowledge. It is worth pointing out that some people think that if they explain something very clearly to students, then the students will understand it perfectly and that will suffice. Even professors who have been lecturing for years harbor this belief. Such people may become frustrated if the bot makes mistakes or asks for clarifications even after receiving a "perfectly clear" initial explanation.
- *Monitoring performances*: Human tutors are quite good at comparing a student's performance to what they would do in the same situation and spotting differences. They almost always give negative feedback but sometimes in an implicit way, which makes it hard for a learner to recognize that an error has been made.
- *Debugging*: Human tutors and learners are not good at debugging. Tutors are not good at asking diagnostic questions, nor are they able to use diagnostic information when it is given to them. Students, especially poor learners, often do not know when they fail to understand. Students seldom ask nontrivial questions that might help them debug

their knowledge. Instead, when errors occur, the tutor simply teaches the correct knowledge again, and this seems to suffice for most students (Sleeman et al. 1989).

The strength of human tutoring lies in monitoring learners. Bots can trust their human tutors to point out errors, albeit in an oblique way sometimes. Since human tutors are not good at debugging, the bot may need to invent its own test cases or deep questions, and use them to infer on its own how to fix the errors pointed out by the tutor.

Humans can do this as well, and this behavior is referred to as self-regulated learning (Zimmerman 2008). Self-regulated learners constantly test their performance against desired performance, note discrepancies, and correct their errors. In the expert-novice literature, this is often called deliberate practice (Ericsson and Lehmann 1996). This is the kind of student that every tutor would be proud to teach because the tutor merely has to point out the occasional failure to exhibit perfect performance, and the student does the rest. If future ITL systems are designed for deliberate practice, tutoring bots may turn out to be infinitely more satisfying than tutoring humans.

Acknowledgments

Preparation of this chapter was supported by the ES Forum, NSF grants IIS-1628782 and DUE-1525197, grant OPP1061281 from the Bill and Melinda Gates Foundation, and the Diane and Gary Tooker Chair for Effective Education in Science, Technology, Engineering and Math.

13

Strategies for Interactive Task Learning and Teaching

Katrien Beuls, Luc Steels, and Paul Van Eecke

Abstract

A strategy is a way to make decisions that come up when handling a task. It requires a problem solver able to address routine cases and a set of diagnostics and repairs to handle, in a flexible way, unusual or unforeseen situations. Between humans, interactive task learning and teaching appear to involve strategies at three levels: (a) the execution of a task with available knowledge (task strategy), (b) interactive learning to expand the available knowledge and thus become a better problem solver in the future (learning strategy), and (c) interactive teaching or tutoring to help others learn (teaching strategy). This chapter examines the general architecture that is needed to build artificial agents that can play either the role of teacher, by carrying out teaching strategies, or the role of learner, by carrying out learning strategies that benefit from these teaching strategies. Focus is on artificial teachers that interact with humans or artificial learners as well as on artificial learners that interact with human or artificial teachers. We argue that the use of a meta-layer is of primary importance for understanding and implementing strategies and point to operational examples from an implementation of this hypothesis in the domain of second-language teaching.

Introduction

Humans invent and acquire an amazing number of day-to-day tasks and routinely execute them throughout life. Some of these tasks, such as medical surgery, manufacturing, and business procedures, may be more professional in purpose but they rely, nevertheless, on the same cognitive capacities as "commonsense" tasks, such as cooking or cleaning a room. Some task execution procedures have been designed or analyzed explicitly; they have been written down and are taught through explicit education. Other procedures, however, are acquired through interactions with people who have the requisite knowledge to carry out the task.

Our discussion here centers on the cognitive process and interaction patterns by which tasks are learned and taught in an interactive-situated manner, at a sufficiently deep level so that we can emulate these processes in artificial systems. The notion of a strategy plays a central role in this endeavor, and the expertise needed for task execution, learning, and teaching relies on domain knowledge and strategies, builds up a context model, and performs goal/subgoal decompositions. We begin by examining conceptual issues and then discuss an application, the *Spanish Verb Tutor*, in an intelligent tutoring system that is designed to assist second-language learning.

What Are the Components of Task Execution and Task Learning?

It is possible, and indeed common in artificial intelligence, to view learning as a particular kind of problem solving (Mitchell et al. 1983; Simon 1996). There is often more than one way to generalize or specialize a concept or inference and often more than one possible extension of a strategy or different hypotheses about domain models. Just as nontrivial tasks require consideration of different possible avenues to tackle a task and the heuristics to make a decision, these different options must be evaluated based on learning heuristics to derive the most likely option to pursue. Given this perspective, we can ask whether the same components found in the kind of problem solving involved in the execution of tasks are also found in learning (Steels 1990).

Task Expertise

From diagnosing failure in a mechanical system, to writing a computer program, or deciding whether an insurance claim is valid and assessing how much needs to be paid, a task requires coming up with a series of actions to address the peculiarities of that task in a given context. These actions must satisfy the *goals* and *subgoals* required by the task.

Some tasks are so routine that there is a ready-made solution at hand. Here, however, we are interested in tasks that require problem solving. This involves obviously a fair amount of facts and inference rules about the domain (captured in a *domain model*) and facts about the particular case in which the task is to be executed (captured in a *context model*). These facts are formulated in an ontology that defines the categories with which to describe the objects and properties that make up these models. In addition, problem solving requires a *task strategy* for making use of the domain model to expand the context model and the goal/subgoal decomposition to achieve the task.

We refer to the set of domain ontologies, domain models, and task strategies as the *domain knowledge* of a particular task domain. The contents of domain knowledge are relevant across different tasks in the same domain, whereas the context model and goal/subgoal decomposition are specific to a concrete task context. Elements of domain knowledge typically have several degrees of generality. The ontology and domain models consist partly of concepts and facts of which a great deal is valid across domains, whereas some of it is more specialized. Similarly task strategies range from very specific strategies, which work to solve only a limited set of tasks in a very specific contextual setting, to more generic approaches, such as "divide and conquer" or "simplify and extend" (Polya 1945).

It is common to add an additional component to a task execution agent, a *meta-layer*, so that there are two levels of problem solving (Figure 13.1):

1. Basic processing where available domain knowledge is applied and possibly leads to acceptable performance.
2. Meta-level processing, which becomes active when basic processing fails, in which case the problem solver moves to a higher level to apply the available knowledge much more flexibly, for instance, by relaxing certain constraints so that an inference schema that does not fit completely with the current situation can still be applied, or by ignoring gaps in the input and continue to find the best possible solution.

These strategies constitute metaknowledge about the task domain, and architectures to represent and operationalize these strategies have been studied intensely since the 1980s (e.g., Rosenbloom et al. 1986). Meta-level strategies can also invoke learning expertise to expand the domain knowledge and consolidate adaptations. The different components of task expertise are summarized in Table 13.1 (Steels 1990).

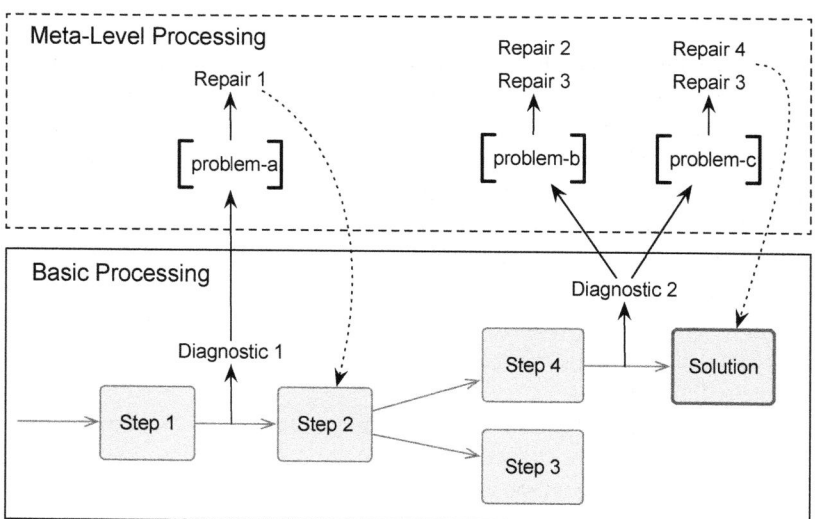

Figure 13.1 Diagnostics and repairs operate on top of basic problem solving, which traverses a search space of possible steps toward a solution. Diagnostics signal problems (e.g., problem-a, problem-b). A diagnostic can be handled by a range of repair strategies. The repair strategies are compared, the most plausible strategy is chosen for execution, and after the repair the basic layer is evoked again to continue or restart from an earlier step in problem solving.

Table 13.1 Components of expertise, found both in task expertise and learning expertise.

Task General	Task Specific
Ontologies	
Domain models	Context models
Strategies	Goal/subgoal decomposition

Learning Expertise

Can we find the analogues of goals, ontologies, domain models, strategies, context models, and goal/subgoal decomposition in learning?

The expansion of knowledge in a particular task domain (the domain ontologies, domain models, and task strategies) is the *learning goal*. For example, when an apprentice receives instruction to become capable of operating a complex machine, this learning task becomes an explicit goal, and the learner will need to decompose this goal into many learning subgoals. Of course, a lot of learning also takes place in the background, triggered when there are opportunities for generalization or specialization or when new domain facts arise. In this case, we would say that these learning goals are continuously active.

There are many *learning strategies* and among them there are important individual differences. A learning strategy is a series of steps that are followed to acquire some aspect of task expertise. For example, the learner could chunk some of the inference steps into a more compact single rule to speed up problem solving later, and thus handle more complex problems; alternatively, the learner could detect similarities between a set of objects and introduce a category to be able to represent inferences that apply to the class of objects of the same category or seek additional information from external sources. Many of these learning strategies have been operationalized in the symbolic machine learning literature and classified in terms of dimensions such as the degree of supervision (unsupervised/supervised), the learning objectives (similarity, classification, discrimination), the nature of the domain model (decision trees, weighted networks, inferences rules), the nature of available data, and the availability of automatic mechanisms cooperating with deliberate strategies (Mitchell 1997).

The *learning ontology* provides the building blocks for formulating *learning models* that capture general facts and experiences about learning, in particular which strategies are most effective for certain learning tasks. For example, the acquisition of part of a particular domain ontology might be based on decision-tree learning, whereas the acquisition of heuristics might be based on exploring a search space and storing information on major decision points based on their success in a task; deblocking an impasse in a specific task domain might be effective by relaxing constraints so that inference rules could apply more flexibly, whereas in another domain it could be by relaxing some supposed constraints on the task itself. The ontology is also required for building the rich *learning context models* needed in learning. These learning context models represent information about a specific learning episode so that the learner can make decisions on how to handle it.

How Do Task Execution and Task Learning Interact?

Much recent research in machine learning, particularly in statistical and neural network learning, assumes that there is a strict separation between a learning phase, based on cycling through a large amount of data, and an execution phase (Bengio 2009). However, this is certainly not the only form of learning. Interactive task learning is situated within a concrete context. It requires incremental learning strategies, which absorb information from a concrete case and are implemented in tight interaction with the problem-solving processes required for task execution. How can such a tight interaction be achieved?

A plausible solution is to assume that the *meta-layer architecture*, introduced earlier for implementing more flexible problem solving, can also be used for invoking learning strategies. At the meta-level, the agents reason about gaps in the domain knowledge and how they can be filled by invoking an appropriate learning strategy. After the learning strategy has taken its course, processing can move back to the task execution layer and the newly available knowledge can then be used to pursue problem solving and task execution further.

The move back to the meta-level gets triggered when a learning opportunity arises. This can occur, for instance, when there is an error in performance, signaled through feedback from the environment (e.g., an action did not generate the expected consequences), when there are impasses during the execution of a task (e.g., the learner gets stuck trying to achieve a particular domain goal), or when an opportunity is sensed for integrating new knowledge into the existing body of domain knowledge (e.g., by generalization or specialization after handling a concrete case).

How Does Teaching Expertise Relate to Task Learning and Execution?

Although learning and teaching are often seen as two different activities that can or should be studied separately, we argue that it is better to view them as two sides of the same coin, just as language understanding and language production are intimately intertwined and the same competence intervenes for both. Those who are better learners tend to be better teachers because they have more successfully developed metacognition skills, such as diagnosing and repairing knowledge gaps (Veenman et al. 2006). Conversely, learners can become better by acquiring skills associated with teaching, such as setting learning goals, timing, and motivation (Zimmerman 2010). We argue that learning and teaching share general characteristics with problem solving, which implies that the same fundamental components (as discussed above) can be expected. Indeed, in the educational literature, teaching expertise has already been viewed as a problem-solving/decision-making process (Shuell 1990), and there have been extensive efforts to document teaching strategies and study their effectiveness in many domains. Moreover, there are considerable differences in the strategies that different individuals use to teach, and it is possible to become a better teacher through practice and instruction. Thus, teaching expertise can be approached in the same way as task expertise (Steels and Tokoro 2003).

Some teaching strategies involve the systematic presentation of domain knowledge and the introduction of ways to exercise and test whether this knowledge has been acquired by the learner (see VanLehn, this volume). Traditional top-down teaching strategies that dominate classroom teaching exemplify these types of strategies. Other teaching strategies employ a more active, learner-oriented approach. They assume that the teacher monitors carefully what the learner already knows, or what kind of errors the learner is making, and intervenes with examples, challenges, and corrections that zoom in on these issues. The learner-oriented approach is a more natural form of teaching and is practiced by parents, caregivers, or peers in natural learning settings. It is particularly this form of interactive teaching that interests us, even though it is much more difficult to capture in computational learning environments. More sophisticated teaching strategies use a model of what the learner already knows to plan possible exercises or gauge whether new material can already be presented (Amaral and Meurers 2007; VanLehn 1988).

The implementation of teaching strategies requires the same components as for task execution and learning: an ontology to describe teaching situations, domain models that now focus on capturing knowledge about teaching, a context model that represents the current teaching episode, teaching goals and subgoals, and teaching strategies. Teaching expertise can either be in the driver seat of an interaction (e.g., in a classroom situation where a teacher has explicit teaching goals and then uses strategies to present new material or come up with appropriate exercises) or invoked when discrepancies are discovered between the behavior of the learner and the behavior that the teacher was expecting.

Application in Intelligent Tutoring Systems for Second-Language Learning

The insights and proposals in this chapter derive partly from our technical work on language speaking, understanding, learning, and teaching (Beuls 2014; Steels and Hild 2012). Speaking and understanding language can be seen as very complex tasks that have all the characteristics of other commonsense tasks. Speaking is similar to a planning or design task: it requires that a (communicative) goal is decomposed into various subgoals and the speaker needs to find the best way to translate some aspect of meaning into words and syntactic structures that conform to the conventions of the language. Understanding is similar to a plan recognition task: the listener must grasp the purpose of each word or syntactic structure and interpret it within the present context.

Young children, being native language learners, rely heavily on experience; they memorize stereotyped patterns and only gradually systematize them. They do not get explicit instruction, and their knowledge of the underlying structure of their native language remains implicit and cannot be verbalized (e.g., Dabrowska and Lieven 2005). Second-language learners, by contrast, are given instructions on the verbal paradigms, phrase structures, and many other aspects of the language that they are trying to learn, and they then gradually internalize these rules so that they become routine (cf. multiple contributions in Robinson and Ellis 2008). They also need a lot of practice and experience, which seems contradictory to the pedagogical aim of second-language learning: learning is to be achieved at an accelerated pace with less input than natural learning, even though this is not always successful. Language is a good test bed to explore experiential and symbolic knowledge acquisition and ways in which they interact.

In this context, we briefly introduce a concrete example of a second-language teaching application designed for learning Spanish verbs (for details, see Beuls 2013). This application, the *Spanish Verb Tutor*, illustrates the three strategy levels (task execution, learning, and teaching) as well as the use of meta-level processing.[1]

The task for the human learner is to engage in language games that approach, as much as possible, real-world situations in which verb conjugation is relevant. Two events are visually shown on a time line containing a past, present, and future time sphere. The time line may also contain the actor in the event, described as the subject of the verb (see Figure 13.2). In a game, the human learner goes through a series of situations and has to act either as a listener or as a speaker.

When playing the listener role, the learner gets a sentence with a conjugated Spanish verb and the task consists of choosing which event is in the correct location on the time line (see Figure 13.2a). As a speaker, the learner selects an event on the time line and then enters the correct verb form to meet this temporal condition (see Figure 13.2b). If the required skill level is too high, the learner can choose to skip the lesson. After each interaction, the learner receives feedback on the answer and an explanation of the type of error.

The *Spanish Verb Tutor* is implemented using two agents (see Figure 13.3): a Teacher Agent (in charge of teaching) and a Learner Agent (in charge of

---

[1] For a demonstration of the system, see https://www.youtube.com/watch?v=5BWDVGUjEEs (accessed January 23, 2019).

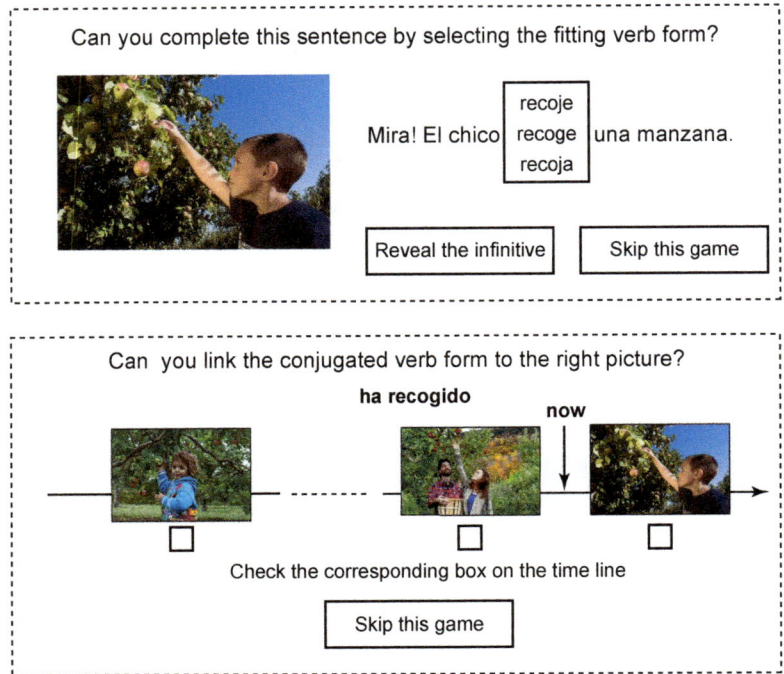

Figure 13.2   Screen shots of the *Spanish Verb Tutor*, illustrating (a) comprehension and (b) production language tasks.

learning a model of the human learner). The Learner Agent is used by the Teacher Agent to personalize the teaching experience of the human learner. Each agent has two components.

The first component involves task execution. Both the Teacher Agent and the Learner Agent have *active expertise* in the sense that they are able to play the language game autonomously. The Teacher Agent, however, plays the game equipped with a fully competent, accurate understanding of the grammar of the *Spanish Verb* system, whereas the Learner Agent models a human learner's competence in Spanish verb conjugation, which is necessarily partial and erroneous. The language behavior of the agents is implemented using fluid construction grammar (FCG) (Steels 2017), but it could be implemented in any other formalism with similar functionalities. To play the game, *flexibility* is required from both agents. The Teacher Agent will be confronted with ungrammatical or partial input from the human learner and requires flexibility to respond properly. The Learner Agent needs flexibility to parse sentences produced by the Teacher Agent, when the learner's own grammar (being a model of the grammar of the human learner) is unable to answer correctly. Flexibility is implemented using the meta-level functionalities embedded in the FCG formalism (Van Eecke and Beuls 2017), which uses operators that return a description of what constraints had to be relaxed.

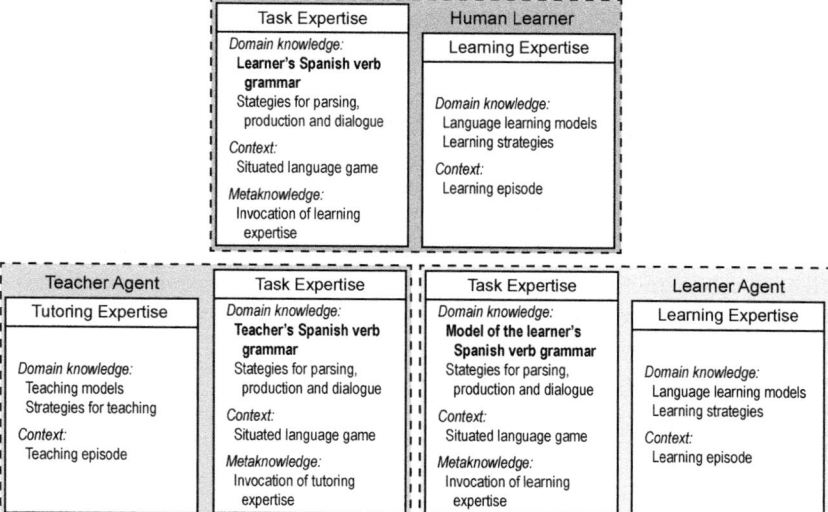

Figure 13.3 The main architecture of the *Spanish Verb Tutor* consists of two agents that interact: the Learner Agent, which is an active model of the human learner, and the Teacher Agent, which can take on the role of teacher. The Learner Agent has task expertise to predict how the human learner will perform in a game, and it has learning expertise to learn the grammar of the human learner to become better at predicting his behavior. The Teacher Agent has tutoring expertise as well as task expertise to formulate exercises or correct human behavior in the task. Teaching strategies either get invoked in a top-down manner or through meta-level operators.

The second component permits the Learner Agent and Teacher Agent to implement learner expertise and task expertise, respectively. Learner expertise operates on a meta-level compared to the task execution level.

Recall that the goal of the Learner Agent is to approximate, to the greatest extent possible, what the human learner already knows, so that teaching strategies can make use of this model to personalize feedback, pinpoint new material, and formulate the most productive exercises. When the Teacher Agent launches a new language game interaction, the Learner Agent carries out the interaction in parallel with the human learner and then compares its response with the human response. If the response is different, the Learner Agent will use learning strategies to implement its model of the human learner, including, and in particular, the acquisition of the user's error constructions.

The goal of the Teacher Agent is to drive the interaction with the human learner. Based on previous interactions, the Teacher Agent invokes a new language game and uses its task expertise to come up with a solution. So, the exercises given to the learner are not ready-made, and hence boringly predictable for the learner: they are actively constructed, taking into account what the human learner knows. When the response of the human learner does not fit with the teacher's own response, the teacher invokes a teaching strategy that attempts to address the mismatch, explains the relevant grammatical knowledge, and then provides new exercises.

The *Spanish Verb Tutor* demonstrates how a meta-layer architecture can be used to create a second-language tutoring system, one that is fully personalized and capable of teaching language through situated interactions that approach normal use of verb phrases; namely, coming up with descriptions of situations or identifying correctly where an event is located on the time line.

The performance of the Teacher Agent was evaluated on the lowest and highest L2 learner level in the Spanish Learner Language Oral Corpus (SPLLOC). The Teacher Agent (FCG) achieves an accuracy of 58% (percentage of isolated corrected forms that equal the gold standard correction). Our system outperforms the standard MS Word grammar checker by almost 30%. The evaluation was run on individual word forms that were not embedded in a sentence (see Beuls 2014).

Conclusions

In this chapter, we have provided strong analogies between the expertise needed for task execution, learning, and teaching. Each of these relies on domain knowledge and strategies, builds up a context model, and performs goal/subgoal decompositions. We also emphasized the role of meta-level operations in the form of diagnostics that detect needs and opportunities and repairs that handle them. In our practice in building teaching agents in the domain of second-language learning, we have found that meta-level operators can provide a smooth interaction between task execution and learning as well as learning and teaching.

Acknowledgments

This paper was written to provide background for discussions at the Ernst Strüngmann Forum on Interactive Task Learning. We are indebted to Kevin Gluck and John Laird, the conveners of this meeting, and to Julia Lupp, director of the Forum, for creating a most stimulating context to reflect on the issues discussed here. We also thank the many participants of the Forum for their feedback and fascinating discussions. For the writing of this paper, Katrien Beuls and Paul Van Eecke were funded by the Chist-Era ATLANTIS project and Luc Steels by a fellowship from the Institute for Advanced Studies in Catalunya (ICREA) as well as support from the Institute for Evolutionary Biology in Barcelona. The development of the *Spanish Verb Tutor* was funded by the Flemish government agency for Innovation by Science and Technology (IWT from 2009 to 2013).

14

Creativity and Feedback

Designing Systems to Support Student Learning and Improve Instruction

Arthur Still, Matthew Yee-King, and Mark d'Inverno

Abstract

This chapter provides a historical perspective on the concept of creativity and its relationship to the development of education theory during the first half of the twentieth century. In the early twentieth century, creativity had a very specific meaning, which expanded in the mid- to late twentieth century into a more general, and in our view less useful, meaning. These two perspectives are linked to two conflicting educational theories, represented by Edward Lee Thorndike and John Dewey. Dewey described learning as a natural part of being an inquiring human being in a social and physical world, whereas Thorndike's view was more reductionist, based on stimulus–response connections. The Thorndike's theory gained prominence and still dominates today, over the Deweyan theory, due in part to the ease with which it can be experimentally tested.

Ideas are developed into a two-part manifesto to inform teaching practice and the development of education technology. The first part delineates the conditions for creative feedback in social learning and encapsulates a Deweyan educational approach. The second part describes the characteristics of education technology that can be used to experiment with creative feedback and social learning, and establishes how we can begin to validate experimentally the Deweyan theory of education.

Interactive task learning considers the challenge of interactively training bots to carry out a task. This chapter is most relevant to medium-term and future tasks for bots within a social context involving humans and bots, and may offer subjective or dynamic evaluation criteria. Bot instructors working with these types of tasks may benefit from considering the complexity and nuances of creative feedback.

History of Creativity and the Education Wars

Dewey introduced the concept of "creative intelligence" (Dewey et al. 1917) early in the twentieth century. His use of "creative" in this expression dates back to discussions on the creative processes among a number of writers and artists at the end of the eighteenth century, especially William Wordsworth. Before then, in the eighteenth century, the word had been defined in terms of its product, be it great art or science (Engell 1981). However, Wordsworth referred to the experience as a special kind of activity, one that is unusually focused, mindful and purposeful, and often results in the production of significant art, poetry music, and science. Such creation was seen as analogous to natural growth, rather than as the result of a Godlike mental act. As experience anchored in this notion of growth, it is logically independent of any concrete product, as was apparent in Ralph Waldo Emerson's use of "creative reading" (Emerson 1837/1962) and Matthew Arnold's reference to "creative criticism" (Arnold 1914). This tradition culminated in the work of Dewey and other pragmatists. For them, experience was a flow rather than a succession of moments, as was expressed in William James's "stream of consciousness" and in Dewey's famous reflex arc paper (Dewey 1896). Here, Dewey treated the stimulus and response of the traditional reflex as abstractions of ongoing activity, and meaningless as isolated units.

This rejection of atomism went with the questioning by Charles Sanders Peirce and James of a strictly deterministic view of the physical world. They proposed, instead, that chance variation may be an inherent property of the universe (Hacking 1983). This view was taken further by Alfred North Whitehead during the 1920s and presented in his process philosophy, which he worked out with formal precision in *Process and Reality* in 1929. Whitehead used the term creativity to refer to the principle of novelty in nature, which is ultimate in his system (Whitehead 1979:21). The adjective "creative" had come to refer to a process of growth and production, for which Whitehead introduced the noun "creativity."

After reading Whitehead's philosophy and writing his major book on creative activity, *Art and Experience* (Dewey 1934), Dewey began to use the word "creativity" during the 1940s. From the start of his career, Dewey insisted that human beings are essentially social and argued against the individualism inherent in much of psychology. Nevertheless, what is creative is always individual. It is a capacity to vary: this capacity is cultivated by good education and suppressed by education that consists strictly of "learning lessons" (Dewey, in Boydston 2008:5):

> The emphasis James places upon the individual quality of human beings and all things is, of course, central in his pluralism. But the adjective "individual" is often converted into a noun, and then human beings and all objects and events are treated as if they were individual and nothing but individual. The result is that identification of human beings with something supposed to be completely isolated which is the curse of the so-called individualistic movement in economics, politics and psychology. I find the actual position of James to be well represented in a remark he quotes from a carpenter of his acquaintance: "There is very little difference between one man and another; but what there is, is very

important." It is this element which is precious because it is that which nobody and nothing else can contribute, and which is the source of all creativity. Generic properties on the other hand are replaceable, and express the routines of nature.

Creativity in the Mid- to Late Twentieth Century

After Whitehead, "creativity" became popular in the United States and flourished during the 1940s as part of the jargon associated with advertising, where it reached a state of Pollyannaish vagueness (Osborn 1948), much loved and beyond criticism, far from the precision of Whitehead. From there it was seized upon by Joy Paul Guilford, one of the leading lights of mental measurement, who invented a new entity in the brain or mind he called "creativity." This was a measurable power which he defined as the "the generation of novel and valuable ideas." This power was subjected to all the rigors of psychological measurement, but the word's imprecision remained. As Liam Hudson (1966:100) stated a few years later:

> This odd word [Creativity]...applies to all those qualities of which psychologists approve. And like so many other virtues...it is as difficult to disapprove of as to say what it means. As a topic for research creativity is a bandwagon on which all of us sufficiently hale and healthy have leapt athletically aboard.

Nothing much has changed. Creativity can be a behavioral process (Simonton 2003) as well as a trait or product, sometimes both at the same time (Eysenck 1995:231):

> There are two major definitions of creativity, and these are quite different in many ways. "Trait creativity" is conceived as a latent trait underlying creative behavior..."Achievement creativity" is defined in terms of novel and socially useful/acceptable products.

The situation remained so bad that Mark Runco, the most prolific of academic experts on creativity, confessed in his recent textbook *Creativity* (Runco 2007) that he had considered giving up the word altogether. Then he read Bill Bryson's *Short History of Nearly Everything*, which told him how much ambiguity exists across all sciences, so why not in the science of creativity? Less cavalier about this, George Mandler (1995) suggested that the word should be treated as an umbrella term or chapter heading that gathers together several disparate topics giving different meanings to the word. The implication is that creativity is not a useful scientific concept at all, any more than the word "cricket" is a good cricketer, or Physics is a useful concept within Physics.

The transition of the word creativity, from a well-defined to a less well-defined status, is unusual in science. The French historian and philosopher of science, Gaston Bachelard, pointed out that the start of a science is often marked by a change from an imprecise but poetic common language for dealing with an aspect of the world, to a system of concepts with exact meanings (Tiles 1984). But in the case of "creativity" exactly the opposite has happened.

The meanings offered by Whitehead and Dewey have been replaced (after Guilford made a "science" of creativity) by a prolonged semantic inflation, as the word has gained yearly in popularity since 1950. Our use of creative in "creative feedback" below returns to Dewey and Whitehead. If that means turning back the clock, then so be it. If we are right, the future of scientific creativity lies in its past.

Dewey and Thorndike on Education

Having explored creativity and creative intelligence, we shall now return to the past and recount a story about the development of education practice and pedagogy in the early twentieth century. The chief protagonists are John Dewey and Edward Lee Thorndike, whose views on feedback and education reflect quite different views on the nature of life, and what it is to be human. Both were egregious figures. Dewey was a pragmatist through and through, and managed to be America's greatest philosopher, most notorious educationalist, and most neglected psychologist. Dewey did not like measuring people, whereas Thorndike, Guilford's mentor in the science of mental measurement, followed Descartes in believing that the world is there to be measured. Thorndike pioneered experimental studies of reward and punishment in education, and recognized that social control follows from scientific measurement. He was an enthusiastic supporter and contributor to eugenics, the attempt to breed better human beings. He also, as we shall see, carried out one of psychology's most famous experiments in 1898: cats in a puzzle box.

Thorndike's "Ideas"

The way of thinking behind Thorndike's experiments is based on the stimulus–response (S–R) model, which treats organisms as independent individuals that can be studied outside of their customary environments, similar to laboratory "preparations" in physiology. In the simplest form of the model, S–R connections in the brain are strengthened when followed by reward, weakened when followed by punishment. These connections are mediated by ideas in the brain, which were treated as "atoms of the mind," and Thorndike speculated that "the vague gross feelings of the animal sort might turn into the well-defined particular ideas of the human sort, by the aid of a multitude of delicate associations" (Thorndike 1901:63). This is Thorndike's connectionism, and it has provided the main framework guiding studies of learning throughout the twentieth century. The Thorndikean icon to match Dewey's reflex arc paper (Dewey 1896) is Thorndike's puzzle box experiment, in which cats confined in a strange box learned to escape by pressing a lever. He claimed that the gradualness of their learning showed that they learned through trial and error rather than reason, with the correct response being rewarded and strengthened (Thorndike 1898).

It is important to note, however, that the cats were in an environment that was frightening and alien to them, nothing like their usual settings where they could follow their interests and curiosity. There was none of the familiar feedback on which the security of their ordinary living depended. In a similar, but less drastic way, the human environment for laboratory experiments, to say nothing of examinations and tests of IQ or creativity, is often unfamiliar to the subject or student. Success in tests depends, in part, on the compliance by the person being tested, a willingness to sit down for a long period of time and answer questions that are at best tedious and often seen as ridiculous. They are tests of compliance as well as intelligence. Unlike ambitious Western students, cats are not compliant: they hate being forced into cages, and Thorndike was mistaken to ignore compliance and treat his experiments as pure tests of reason. Later work on animal intelligence has gone well beyond Thorndike (Seligman and Hager 1972) to take into account the animal's normal setting, as well as the pervasive feedback underlying the active life of animals, both of which were largely ignored by Thorndike and laboratory psychology, more generally.

The Thorndike tradition has dominated education for over 100 years. Its basic assumption is that learning takes place in the head, through ideas, connected together as a result of reward and punishment. Apart from their role in connectivity, Thorndike's ideas are largely passive (Thorndike 1913).

Dewey's "Ideas"

Dewey (1916:16) also used the concept of idea but for him, ideas were dynamic:

> Ideas...are anticipations of possible solutions. They are anticipations of some continuity or connection of an activity and a consequence which has not as yet shown itself. They are therefore tested by the operation of acting upon them. They are to guide and organize further observations, recollections, and experiments. They are intermediate in learning not final.

Ideas for Dewey, therefore, are not self-sufficient atoms to be combined together to form knowledge, nor are they the same as intentionality (already familiar by 1916 through the work of Brentano and Husserl), defined as "the power of minds to be about, to represent, or to stand for, things, properties and states of affairs" (Jacob 2003). For Dewey, ideas point to things outside themselves, like intentions, but also to the actor or thinker herself, and to potential actions upon the environment. The relationship is mutual or threefold, like teaching itself, which he believed should be based on inquiry and discovery rather than "lessons" and adds (Dewey 1916:160):

> This does not mean the teacher is to stand off and look on; the alternative to furnishing ready-made subject matter and listening to the accuracy with which it is reproduced is not quiescence, but participation, sharing, in an activity. In such shared activity the teacher is a learner, and the learner is, without knowing it, a teacher.

This mutualist or triadic relationship is characteristic of the anti-dualist philosophy of pragmatism, and it is present throughout the work of Dewey and his predecessors; for instance, in Peirce's concept of a sign (Peirce 1902/1935), what James views as "pure experience" (James 1975), or more recently James Gibson's affordance which "points two ways, to the environment and to the observer" (Gibson and Walker 1984). It is also present in Russian "activity theory," as in Lev Vygotsky's concept of "word meaning" (Vygotsky 1967).

The Outcome of the Battle

Thorndike's S–R model was elaborated by Clark Hull into the mechanisms of behavior theory, in which stimuli were worked upon by internal processes and transformed into responses (Cordeschi 2002). Hull recognized that his terms would be improved with a more sophisticated language, and in this way anticipated the replacement of his S–R model by the input–output models of information processing. Like Thorndike's connectionism, both models focus on internal mechanisms in an isolated organism. In that form, the same opposition between Thorndike and Dewey has continued, with Dewey joined by the Russian activity theorists, starting with Vygotsky. As Aukrus (2007:47) stated in a recent commentary that introduced computer-supported collaborative learning (CSCL):

> Broadly speaking, there are two main traditions within the learning sciences: cognitive psychology and the situated/sociocultural perspective. The former is based on an information processing perspective...and the latter on American pragmatism...and Soviet psychology....In CSCL studies, methods and techniques from both traditions are used and sometimes blended (e.g., interaction analysis). However, within each tradition there are unique interpretation of key concepts, methods and empirical design...

Knowledge Objects

What are the essential characteristics that distinguish a Thorndikean from a Deweyan approach to learning and teaching? In Thorndike's S–R models, the focus is on setting up the connection, and this piecemeal approach is suggested by a list of twenty-five principles of learning presented by Arthur Graesser (2009) in the inaugural editorial for the *Journal of Educational Psychology*. Even when social and cultural context is taken into account, the unit of analysis is always the item to be learned. In Dewey's learning, by contrast, the main goal is further inquiry, and therefore further learning. As one of the founders of CSCL, Carl Bereiter's concept of "knowledge objects" allows for a more Deweyan principle (Bereiter 2002). For Bereiter, a knowledge object is a concept that becomes real for a student who develops a passion for it, whether it is evolution, existentialism, artificial intelligence,

Charles Dickens, or numbers. It is a love object which (ideally) will effortlessly organize, guide, and motivate the social and individual activity which drives further inquiry.

In education, victory has gone to Thorndike: "One cannot understand the history of education in the United States during the twentieth century unless one realizes that Edward L. Thorndike won and John Dewey lost" (Lagemann 1989:185). Thorndike won for at least two reasons. First, experimental studies on feedback are much easier to set up if the independent variable is simply reward or punishment, or information, rather than if it pervades the social situation. Second, teachers are usually obliged to follow a curriculum which allows little room for free inquiry; this forces them back into the default approach with which they are familiar, involving the I-R-E sequence of interaction (initiate-response-evaluation).

Below, we shall revisit the challenges of experimentally investigating Deweyan education theory. This will then lead us to suggest a solution.

A Manifesto

Creative Feedback for Social Learning

Dewey's 1948 definition of creativity was given in the foreword to a book on what we would now call art therapy (Schaefer-Simmern 1961:ix–x). There he describes individuality as the "creative factor in life's experience." It is "the life factor that varies from the previously given order, and that in varying transforms in some measure that from which it departs, even in the very act of receiving and using it. This creativity is the meaning of artistic activity (Schaefer-Simmern 1961:ix). Illustrating this, the first chapter described how a severely withdrawn woman, with an IQ of 49, slowly emerged from her withdrawal through kindness and patience and the opportunity to use paper and colored crayons. At first, following the work of others, she gradually developed her own style. "This creativity is the meaning of artistic activity." She achieved it with the encouragement and creative feedback of others. What was involved in this process? Creative feedback is part of the flow of creative activity, or creative learning, but to some extent a teacher can intervene, in a way that will encourage creative activity through (creative) feedback. We can see this in the work of the nineteenth-century writer Matthew Arnold, as well as the twentieth-century founder of modern counseling, Carl Rogers.

Arnold had an important influence on Dewey. As a young man he was a prolific poet, until he ran out of steam and became one of the great critics of his time, providing feedback to other writers. His experience of joyful creative activity was similar in his work as a poet and critic, and he outlined the conditions under which this (creative criticism or feedback) could be possible (Arnold 1914:35–36):

> To have the sense of creative activity is the great happiness and the great proof of
> being alive, and it is not denied to criticism to have it; but then criticism must be
> sincere, simple, flexible, ardent, ever widening its knowledge. Then it may have,
> in no contemptible measure, a joyful sense of creative activity....And at some
> epochs no other creation is possible.

Rogers acknowledged the influence of Dewey, especially through the teaching of William Heard Kilpatrick (Dewey's pupil and later close colleague). Rogers's views on creative activity and traditional education were similar to those of Dewey (Rogers 1954:250):

> In education we tend to turn out conformists, stereotypes, individuals whose
> education is "completed" rather than freely creative and original thinkers....
> My definition...of the creative process is that it is the emergence in action of
> a novel relational product, growing out of the uniqueness of the individual on
> the one hand, and the materials, events, people, or circumstances of his life on
> the other.

Rogers was the founder of modern counseling and his career, as he saw it, was to help people change from being conformists, imprisoned by the demands of themselves and others, to finding the uniqueness in themselves, which he referred to as "self-actualization." This change was brought about by setting up a therapeutic relationship in which feedback from the counselor enabled the client to let go of self-demands. For this to take place, his famous core conditions were empathy, acceptance or unconditional positive regard, and congruence. Empathy is to reflect the other's point of view, rather than imposing your own; unconditional positive regard is to always accept the other as a worthwhile human being; and congruence is to be honest, not to say what you do not feel. To these core conditions we would add "interest"; that is, the deep interest of both student and teacher in the student's project.

The Nine Elements of Creative Feedback

Elaborating on the views of Arnold and Rogers, and many other writers who followed or anticipated Dewey, we have developed the following principles of creative feedback leading to a manifesto of creative feedback for social learning (d'Inverno and Still 2014), which defines nine characteristics of creative feedback as it applies in a social learning context. We believe that giving feedback to others is a profoundly creative and difficult act—one with many dimensions along which a range of useful skills must be developed. If we consider education as a process wherein people learn to provide effective feedback, we find that the following aspects of that feedback neatly encapsulate the goals of a Deweyan educational approach.

All of these criteria build on the precondition that the tutor or student who is giving the feedback has a genuine curiosity and interest in the student to whom the feedback is being given.

1. *Creative feedback is a social process.* It comes from one social human agent (the tutor or peer learner) who has perceived the feedback object (such as a performance, a proof, an essay or an artwork) to another social human agent (the learner and originator of the feedback object). Note this definition does not preclude students giving creative feedback on their own work which can often provide great insight as long as a sufficient distance can be taken.

2. *Creative feedback is mindful.* This incorporates at least two aspects: (a) that the person giving the creative feedback is aware of the cultural and individual context of the receiver (such as an understanding of the individual's artistic or scientific goals/methods/audiences, etc.) and (b) that individuals are aware of any personal judgments that are being made and can articulate these if required.

3. *Creative feedback involves community awareness.* If creative feedback occurs in a community of learners (rather than one-on-one), then it should embody community awareness of the creative feedback that has previously occurred as well as the part it plays in a complex and developing system. Giving and receiving creative feedback should be embraced equally for the community to sustain itself. It would be difficult, of course, for communities to thrive if everyone wanted to give more creative feedback than they wanted to receive. Creative feedback creates a self-sustaining, self-organizing system where flexibility and robustness need to be balanced. While each learner may have more or less knowledge about what is required to maintain such a system, it is clear that it can only exist if individuals in the learning environment actively encourage engagement in creative feedback.

4. *Creative feedback is clear.* The language used must be unambiguous and the terms used must be mutually understood. There is no attempt to hide meaning behind technical or ambiguous words and sentences.

5. *Creative feedback is democratic.* Being a tutor or student bestows no special right to giving or receiving creative feedback. One might hope that tutors have more experience and skills in giving, but this is not a prerequisite.

6. *Creative feedback is challenging.* Underpinning any creative partnership is the notion of the challenge that each brings to the other. Creative feedback that provides the right level of challenge is arguably the most sought after feedback. To do so involves "skill in means": a Buddhist concept which holds that feedback should be geared to the level and character of the student, and is always open to the student's needs. The idea of programs and feedback challenging students is a critical part of the design and delivery of any course.

7. *Creative feedback incorporates generosity of spirit and compassion.* It is an act of giving and enabling; that is, the giving of guidelines (not rules) for future exploration and awareness.

8. *Creative feedback allows further discussion and explanation.* This is what makes it a social process between equals, allowing for more detailed and nuanced exchanges, making sure nothing is ever closed off.

9. *Creative feedback is comparative rather than absolute.* No absolute judgment about a feedback object can be made. Comparisons (explicit or implicit) of the feedback object to other existing objects are a mindful tactic involving skill in means. For instance, creative feedback to a jazz piano student from a tutor may simply be to say how close the student's playing is to another well-known jazz pianist, or that the student may wish to listen closer to certain aspects of that person's approach.

The very best tutors are able to give feedback that encompasses many of these qualities. The ability to demonstrate to students just how engaged a tutor is with the students and their work is, we believe, critical to keeping students engaged. This coincides with the first of the "big four factors," described by VanLehn (this volume), that impact human learning (i.e., *engagement*). Moreover, creative feedback requires tutors to be sensitive to the student and provide feedback that is *appropriate for the student's position along the learning curve* (VanLehn's second major factor). Furthermore, through creative feedback, students learn how others experience their work, and this provides students with greater abilities to evaluate their own work. This view coincides with VanLehn's fourth factor: *feedback*.

Application and Critique

To demonstrate how one might critique any education technology, including intelligent tutoring systems and robotic tutors, using the manifestos, we present a case study below: the MusicCircle. Thereafter we discuss how the manifesto for creative feedback applies to interactive task learning (ITL).

MusicCircle: An Online Music Learning Support Tool

Built at Goldsmiths during the FP7 Project PRAISE,[1] MusicCircle is a web-based social network that encourages members to share their creative works and receive feedback from others. This novel peer-to-peer learning approach views "learning to give and receive feedback" as integral to getting better. Users can upload, share, and annotate time-based media in several ways: by uploading a file to the browser, using a smart phone app which also allows recording, or recording directly into the browser. An application programming interface (API) allows software agents access to the full set of uploading, sharing, and annotation features.

---

[1] The Social Enterprise Museifi was set up to make MusicCircle available. For details on MusicCircle and Museifi, see museifi.com.

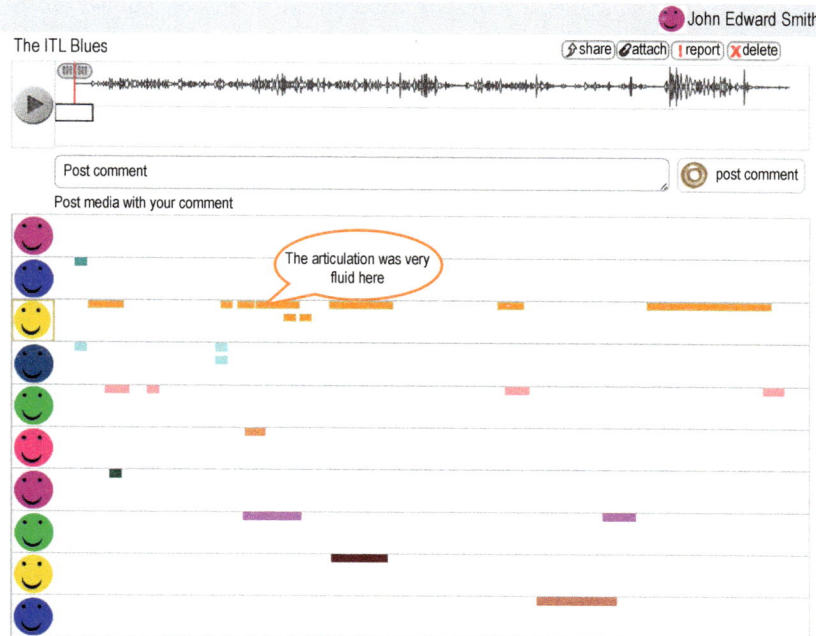

Figure 14.1   Screen shot of the MusicCircle system, showing an audio recording being annotated. At the top is the waveform, then below is the social timeline showing the annotations of various members of the community. This audio recording and the associated dataset is available at http://zenodo.org/record/46232 (accessed Feb. 11, 2019).

A screenshot of the browser-based user interface is shown in Figure 14.1. A key feature of the user interface is the social timeline (Brenton et al. 2014), visible as a set of colored blocks below the audio waveform in the figure. These blocks represent sections of the recording that have been highlighted and annotated. Each person who has created an annotation has their own strip on the timeline. Each annotation becomes a discussion thread visible to everyone who can see the top level media item.

To evaluate how MusicCircle supports the creative feedback manifesto, we enumerate the manifesto's key points and discuss how each point is addressed (or not) by the system. Where possible, we relate these points to VanLehn's discussion (this volume). We present this evaluation as an example of how the creative feedback manifesto can be used in practice to analyze a real system and hold that this approach can also be used to evaluate an ITL system.

1. *Creative feedback is a social process.* This is a core feature of MusicCircle. Feedback is visible to all and can be garnered from all members of a community. This type of interactive behavior relates to the most challenging level of the ICAP framework (see VanLehn, this volume).

2. *Creative feedback is mindful.* Mindfulness in an educational context includes awareness of your own learning as well as that of others (metacognition). Studies of MusicCircle have shown that through its use, students became more aware of their improvement and that other people learned and improved as well. The simple act of recording and reflecting on work in progress allowed this to happen. Mindfulness is social; thus it maps directly onto interactive behavior.

3. *Creative feedback in groups embodies community awareness.* The MusicCircle system includes a set of community guidelines which emphasize the need to be a positive member of the community by considering the feelings of others, for example.

4. *Creative feedback is clear.* Feedback given on MusicCircle consists of an annotation that is mapped to a very specific region of the recording. This encourages clarity in the feedback, and it focuses on the recording, not the person. As VanLehn (this volume) states: "human tutors love to give explanations, but many are not very good at it." This emphasizes the challenge of giving clear feedback.

5. *Creative feedback is democratic.* All members of a MusicCircle community have equal rights to upload, share, and annotate. The user interface also displays everyone's annotations simultaneously.

6. *Creative feedback is challenging.* Once an annotation has been placed, it becomes a discussion thread, and both the annotator and the receiver can discuss the annotation more deeply. Challenging comments are more likely to be received positively if they are tagged to a specific part of the media, not the person. In terms of learning curves, challenging feedback encourages students to operate in the sense-making phase: they are encouraged to gain a stronger understanding of the basic subject matter. The use of a discussion thread may be helpful as a detailed diagnostic assessment, which is a valuable instruction strategy seldom used by tutors (cf. VanLehn, this volume).

7. *Creative feedback incorporates generosity of spirit and compassion.* MusicCircle provides community guidelines related to these aspects, but does not explicitly encourage this kind of feedback.

8. *Creative feedback allows further discussion and explanation.* MusicCircle allows each annotation to become a discussion thread in and of itself. It is also possible to link comments to further media items, either on MusicCircle or other media-sharing platforms.

9. *Creative feedback is comparative rather than absolute.* MusicCircle allows linking to other media items in comments, for comparative or demonstrative purposes. One could extend this to allow multiple versions of the same media item (e.g., recordings of a band practicing a song) to be aligned, thus enhancing comparisons. The current version of MusicCircle does not yet permit aligned comparison. Comparative

instruction is a key technique in ITL, where the performance of the bot is compared to the desired performance.

Creative Feedback and Its Relationship to Interactive Task Learning

Creative feedback can happen reflexively, in which case there is one agent (A). It is also possible in a one-on-one meeting, in which case there are two agents, as well as in a community where there are three or more agents. In any learning scenario, there is at least one agent who receives feedback (the learner, L) and at least one agent who gives feedback (e.g., a tutor or peer in the learning group: the giver-of-feedback, G). In the case of reflexive feedback, A1 = A2 whereas in the other instances, A1 ≠ A2. The effectors of agents, in our view of the world, include being able to make a work (e.g., a performance, a computer program, or a painting) and being able to provide either written (text) or spoken feedback. The sensors of all agents in our model include being able to sense/experience (a) the work, (b) the student who made the work, and (c) the community of other agents, if they are present.

As in the ITL framework presented by Mitchell et al. (this volume), agents have different abilities in terms of their experience of the subject, the experience they have developed in giving feedback, and their ability to empathize with the learner (their goals, their background, their culture, and so on.) This fits exactly with the model of communication that Mitchell et al. (this volume) describe, where aspects of the mental state of any agent can be observed: "To define a learning problem precisely, we say that an Agent A learns to improve its performance, P, at task, T, through experience, E." This provides a strong framework within which we can characterize creative feedback. The task (T) in ITL includes producing a software program, playing a musical instrument, or painting a picture in our world of creative feedback. Performance (P) of these activities is what the agent has to get better at through the experience (E), which is creative feedback. Experience can be in several forms:

- Submitting a program and getting feedback from a tutor or peer learner.
- Submitting a performance online (e.g., to MusicCircle) and requesting feedback from the community to which it is posted.
- Presenting the software program as well as demonstrating its performance and getting feedback on all of this from a tutor and a peer learning group.
- Showing the painting and discussing aspects of it and the process of creating it, and then inviting feedback from a community of tutors and learners.

In all cases, the student wants to get better at programming, playing the piano, or making art by opening themselves up to receiving the creative feedback of others. The working definition of ITL, therefore, applies very precisely to creative feedback; it is an example of ITL. A student (A) improves its creative practice (P) of making a work (T) by communicating to other agents and receiving the creative feedback from other agents who have considered aspects of A, P, and T.

Concluding Thoughts: Can Artificial Intelligence Systems Give Creative Feedback?

The systems described above were set up to support creative feedback and we were essentially agnostic about whether the agent giving feedback on a performance was human or machine. Many of the characteristics of creative feedback place the activity of generating it at the difficult end of the range of tutoring strategies discussed by VanLehn (this volume). For example, community-aware feedback is innately social, and therefore falls at the most highly engaged (i.e., *interactive*) end of student behavior in the ICAP framework (Chi and Wylie 2014a). This means that the task of providing creative feedback is challenging for human beings, and therefore it is interesting to consider how well automated agents might perform in this area.

In conjunction with an EU FP7 Project called PRAISE (Practice and Performance Analysis Inspiring Social Education), we have worked closely with some of the leading AI groups in Europe (e.g., Carles Sierra at IIIA in Barcelona, Francois Pachet who was then at Sony Computer Science Research Labs in Paris, and Luc Steels at VUB in Brussels) to look at the potential of automatic feedback. In this scenario, the automated agents needed to place comments on the audio recordings uploaded by music learners. This is consistent with the expanded view of ITL, in which agents or bots are the instructors, providing feedback on the tasks carried out by the humans. In this case the task is to play a piece of music. We could extend the scenario and say that the teaching bots were carrying out the task of giving creative feedback to music learners and that humans needed to teach these teaching bots, but we did not push the implementation that far.

The challenges with this work were the sensors and processing for the bot. The effectors of the bot allow it to retrieve the data it needed and to post the comment on the platform; this was quickly achieved through a web service API, which provided a programmatic interface to the platform. The sensors of a feedback bot allow it to listen to the audio uploaded by a musician, then to listen to a model recording uploaded by the tutor. The bot could also see the score for the piece of music, which was in a machine-readable format. If the bot were to follow the manifesto of creative feedback, it might consider previous recordings of the student, and possibly the other students who could see the comment, and even previously uploaded comments. Implementing the sensors and processing needed to do the basic audio listening, score reading, and commenting involved a signal processing task and a natural language generation task. Full creative feedback required natural language processing and social network analysis, but we did not implement that. The signal processing extracted the notes and features of the performance and aligned the different performances with the score. The natural language system generated the feedback comments.

The bots were able to generate natural language feedback on specific regions of recordings, but only in quite constrained scenarios with carefully selected inputs and processing algorithms. The technology was not sufficient to operate independently on the live platform in a realistic scenario, due to the difficulty of automatically locating the correct inputs (audio, model, score), and yet not constraining the case study to an artificially simple scenario. In addition, even a simple scenario where students are expected to upload and tag the correct files was challenging to implement from a usability perspective. Further, providing comparative, natural language comments on deeper aspects of the performance, beyond playing the correct notes at the correct speed, was beyond the reach of the available signal processing, score comprehension, and natural language technology.

In summary, we were able to automatically generate natural language comments on musical performances and to place them at the appropriate position in the audio timeline. However, the scope of commenting was limited in terms of instrument types and performance aspects. The bots could not be considered to have generated creative feedback.

The results of our investigation of the capabilities of bots to generate creative feedback stand in stark contrast to recent press reports about the potential of robots to replace human tutors (Bodkin 2017). In this article, Sir Anthony Seldon, Vice-Chancellor of the University of Buckingham in the United Kingdom, is quoted as saying:

> Everyone can have the very best teacher and it's completely personalized; the software you're working with will be with you throughout your education journey...

The article continues to report that

> Experts predict that automated teaching of maths and science will form the vanguard of machine-led learning, but that sophisticated algorithms would soon be devised to teach the humanities.

For creative activity to take place, we need to remember that Dewey's approach needs a physical and social system. We certainly believe that feedback is a creative activity, hence our introduction of creative feedback.

We could imagine taking a Thorndike approach and a Dewey approach to the design of tutors. To bring out, in the simplest possible form, the difference between the mechanistic feedback of a Thorndike-influenced robot (Tbot), and the creative feedback of Dewey (Dbot), let us imagine contrasting robots: The Dbot tries to tap into the student's interests and draws on these to generate inquiry into the topic. Learning for a Dbot is directed toward future learning and is usually initiated by the student. Learning, by contrast, for the Tbot is initiated by the teacher/tutor and is focused on a classroom task, with reward and punishment strengthening correct and weakening incorrect responses. These thoughts are illustrated in the following conversations:

- *Dialogue between a Tbot and a young student of elementary arithmetic*

Tbot:	What is 2 + 4?
Student:	8
Tbot:	Wrong. You need to practice adding numbers up to 10. Let me give you a simpler example.
T-bot:	What is 2 + 2?
Student:	4
Tbot:	So what is 2 + 2 + 2
Student:	6
Tbot:	Excellent.
Student:	Oh I see, 6.

- *Dialogue between a Dbot and a student*

Dbot:	How are your mice?
Student:	Shows Dbot a picture on her iPhone
Dbot:	So here they are peering out of their 2 cages.
Student:	Yes, do you like them?
Dbot:	Lovely, how many are there in each cage?
Student:	2 in that one, 4 in the other
Dbot:	So how many altogether?
Student:	8
Dbot:	Try counting them.
Student:	Student counts the mice in their cages and responds: Oh, there are 6.
Dbot:	So what's 2 + 4
Student:	Oh I see, 6.

We believe that a Tbot could probably work as a robot, similar in structure to a standard teaching machine of 50 years ago (Lumsdaine and Glaser 1960). A Dbot would need a lot more information about the student, and would be unable to capture Dewey's crucial insight that good teaching and social interactions cannot be reduced to rules and mechanisms.

Once again we need to ensure that we understand the limitations of AI and Dewey helps us. Machines for imparting knowledge and machines that take a Tbot approach are all very well, but the best teachers—those who embody a Deweyan approach—will understand that teaching is a social process: one concerned not just with the acquisition of knowledge, but with teaching students by example how to think and how to discuss. It is a process where student and teacher are interested in each other, and where the dialogues are mindful, aware, and clear.

By setting out our criteria for creative feedback based on a Deweyan approach, it is clear that a machine cannot give creative feedback. It may be that the teaching program envisaged by Sir Anthony Selden will yield better exam results in a national curriculum, not just in maths and information technology, but in the humanities as well. This would be what Dewey referred to as "learning lessons," which he distinguished from education. In education, according to our interpretation of Dewey, there is no sharp boundary between the discourse of students and teachers, students among themselves, and students in the world outside school: the knowledge objects absorbed at school permeate the whole of life. This process depends on creative feedback throughout, and would be a move toward a realization of Dewey's belief that "education...is a process of living and not a preparation for future living" (Dewey 1897).

Learning New Tasks

15

Learning Task Knowledge

Dario D. Salvucci, John E. Laird, Franklin Chang,
Kenneth D. Forbus, Parisa Kordjamshidi,
Tom M. Mitchell, Shiwali Mohan, Michael Spranger,
Suzanne Stevenson, Andrea Stocco, and J. Gregory Trafton

Abstract

How does an agent acquire (i.e., learn) knowledge and information about a specific task by interacting with a teacher, so that ultimately the agent is able to execute the task successfully? This chapter reviews critical aspects of the learning process in interactive task learning (ITL). It discusses learning task knowledge through interaction, capabilities that facilitate learning, aspects of interaction that relate closely to learning, and evaluation dimensions and metrics for ITL systems. Given the interconnected nature of ITL, it also explores relationships between learning, knowledge, interaction, and tasks: how tasks influence learning, how knowledge should be represented, and what types of information and communication are needed to facilitate learning.

Introduction

Research into interactive task learning (ITL) focuses on how agents (biological and artificial) acquire new tasks through natural interactions with one another within a shared environment. A core component of ITL involves (a) learning new knowledge, (b) integrating it with existing knowledge, and (c) operationalizing that knowledge to perform novel tasks. This process, in turn, is heavily influenced by the very nature of ITL: a learner continually acquires information from both the teacher and the environment, and such interactions have the potential to fundamentally affect what is learned as well as when and how learning takes place.

Group photos (top left to bottom right) Dario Salvucci, John Laird, Parisa Kordjamshidi, Michael Spranger, Suzanne Stevenson, Tom Mitchell, Greg Trafton, Andrea Stocco, Shiwali Mohan, Ken Forbus, Dario Salvucci, Suzanne Stevenson, Franklin Chang, Shiwali Mohan, Greg Trafton, Tom Mitchell, Andrea Stocco, Franklin Chang, Michael Spranger, John Laird, Parisa Kordjamshidi

We begin by reviewing multiple aspects of learning in the context of ITL and explore two questions in depth: What are the characteristics of learning mechanisms most appropriate for ITL? What overall capabilities of the agent are desired (or even essential) to enable and facilitate interactive learning? We then examine the relationship between interaction and the task itself and propose a set of metrics that may be used to understand and evaluate ITL systems with respect to the dimensions of learning, task, and interaction.

Background to Learning Task Knowledge

Because ITL can be applied in a wide range of domains (e.g., real-world robots, agents in virtual environments, video game agents, and virtual personal assistants), learning mechanisms must be able to respond to different levels of complexity and diverse characteristics inherent to each domain. Such variance means that there is no single process of learning to be followed in ITL. The unique characteristics of a domain will determine what needs to be learned, the degree of difficulty of the learning problem, and the appropriateness of a learning algorithm. Consideration must be given to whether the environment is partially or fully observable, whether the environment is discrete or continuous, whether actions are deterministic or stochastic, and whether the environment has complex dynamics. In addition, the extent of uncertainty in sensing and acting in the environment must be evaluated: the more complex and uncertain a domain is, the more difficult it will be for an agent to learn new tasks quickly.

Currently, ITL agents focus primarily on tasks within a single domain, and designers tailor their learning algorithms to that domain. As we look to expand our research into ITL, we need to assess whether there are fundamental representations and learning algorithms that would work across all domains. Do certain domains require specialized learning approaches and, if so, to what degree? Here, neuroscientific studies of the human brain may provide inspiration for general learning algorithms, as the human brain has demonstrated an unrivaled ability to learn across the lifespan and multiple domains (Cole et al. 2013).

We begin our discussion with a review of characteristics of the task being learned and the knowledge needed to perform the task. This is especially important in the context of the knowledge and knowledge representations that it elicits in the learner.

Tasks and Task Knowledge

The space of ITL tasks centers around goal-oriented tasks, and thus one of the first aspects of the task to be learned is the goal itself—what is being achieved. For some tasks, the goal is straightforward to describe, such as the end state in checkers or chess; for other tasks, the goal might be harder to state formally or even informally, such as writing an essay or giving an eloquent but humorous speech. The goal might also be dependent on the current environment—for instance, when setting a table, the appropriate setting might depend on the types of food and drinks being served.

The actions that can and cannot be performed in the context of a task offer additional constraints. On one hand, there may be actions that are tempting to perform but not available to the learner agent (e.g., going through a locked door); on the other hand, other entirely possible actions might not be realized at first by the learner (e.g., insight problems that require "thinking outside the box"). The former constraints embody limitations of the task and/or external world themselves, whereas the latter constraints may embody limitations (perhaps temporary ones) associated with the learner. There is a more general question of whether we need an explicit metric to assess how well the learner has learned the basic properties of a task and distinguish it from how well the learner performs the task successfully. Though this distinction seems useful and indicative for evaluating the learner in domains such as games, in general, the line between what are "legal" actions and what is simply good performance is not clear cut. Moreover, there is not always a need to learn all the rules to be able to be successful—some situations never occur, and sometimes even a learner who knows a few basic rules can perform well. These challenges are even more difficult in ill-constrained problems, such as writing and telling jokes.

Besides the general performance metric described above, learners must often form their own notion of this metric, and its formation becomes part of the learning process. This metric is important for deciding and learning the sequence of effector actions while pursuing a specific task, as it is useful in evaluating various actions available in a state. The metric may be explicitly defined using general criteria: for example, for some tasks, the time needed to execute actions may be more important than the number of actions taken; this performance metric will cause the agent to prefer a longer sequence of actions that takes less time over a shorter sequence of actions that is more time consuming. While a performance metric might be explicitly communicated, it is arguably more commonly implicitly communicated: for example, the Argentine tango has no explicit goal state, but instead is danced in a counter-clockwise movement around the outside of the dance floor.

In broad terms, then, learning can be thought of as an *iterative refinement* of a number of aspects of the learner's knowledge: the overall knowledge base being extended and refined, including not only the typical notions of skill and factual knowledge, but also iterative refinement of the learner's understanding of the goal, performance evaluation metric, and problem state.

Characteristics of Learning in ITL

There are a number of characteristics of learning in the context of ITL that, all together, distinguish this challenge from others in the broader space of problems in machine learning, cognitive science, and artificial intelligence.

One of the most significant ones is the necessity of learning from a small number of training instances. In contrast to approaches that rely on thousands if not millions of training examples, a human teacher typically provides only a handful of examples from which to learn. While these examples are typically of much higher quality than in a large-scale data set—carefully chosen to illustrate specific points—the paucity of data often necessitates a radically different approach to the learning problem. This idea is related to the work done on "one-shot learning," although in the ITL case, the potential for continual interaction that influences the learner (not to mention the teacher) provides for a richer set of data and experiences.

Another distinguishing feature of ITL relates to the interaction of the learner and teacher—in particular, the possibility that the learner can directly ask the teacher for feedback and guidance. While most learning algorithms operate in batch mode on large data sets, an ITL agent may direct its own learning by prompting the teacher for specific information—for instance, more instructions, clarifications on ambiguities, or additional specially chosen examples. In turn, the learner's ability to ask such questions implies the existence of meta-cognitive abilities on the part of the learner, namely an ability to inspect its own knowledge and to identify gaps or potential sources of errors. This ability is needed to provide the teacher with specific feedback. A simple "dump" of the agent processes would not be enough, or not be informative for a human teacher; instead, the agent must identify specific problems, or explain its own difficulties in a targeted way.

The learner's interaction with the teacher also creates conditions for a "push-pull" alternation between two major types of learning, generalization and specialization. For example, typically the examples provided by the teacher to the learner would be specific in nature (e.g., "grab this cup"), and it would be up to the learner to generalize these procedures to related tasks. In fact, children are typically conservative in their use of new artifacts (Casler and Kelemen 2005) and tend to imitate even causally irrelevant actions in learning new sequences (over-imitation; Lyons et al. 2007). But, if and when any overgeneralization takes place, specialization can be invoked to make instances of the more general rules, and/or augment the general rule with exceptions to the rule.

A teacher can also help the learner to construct hierarchical structures of understanding around their behaviors. Again, although the teacher may start with specific instances, they can also point out higher-level general properties of their examples (e.g., "notice that we always fill the pot with water before turning on the stove"). Recent cognitive modeling frameworks have some of these properties and would allow the learner agent to combine smaller behaviors into larger ones in an incremental manner. Learning about multiple concepts and the way they relate to each other, even in the simplest case, takes us from classical statistical learning to a structured learning paradigm, but this structure is usually expressed in a high-level abstraction and in a relational form (e.g., Kordjamshidi et al. 2015). The type of learning techniques that are used must consider the relational structure of the domain and background knowledge, and to learn from examples in the context of these challenging issues. A natural consequence of these characteristics is that the learning examples may become structured as well—for example, expressed in terms of demonstrations, explanations, and interpretations of the world.

Last but certainly not least, one of the most important elements of ITL agents is a rich endowment of knowledge. Learning almost never happens "from scratch," but rather combines existing pieces of knowledge, large or small, in new ways to achieve a new task. An agent's background knowledge will necessarily be incomplete (if they knew everything already, they would be operating in a static world and no training would be needed). To date, existing large-scale open-source ontologies, such as CYC (Lenat 1995) and YAGO (Suchanek et al. 2007) can be used to provide a backbone for such domain knowledge. The advantage of starting with a large-scale ontology is that it reduces the time to build new systems, since reasonable knowledge representations for predicates and concepts have already been created. In ontologies such as OpenCyc, there is also a context structure, called *microtheories*, that is useful for focusing reasoning and also enables irrelevant subsets of the knowledge base to be removed if desired, to reduce an agent's memory footprint.

Capabilities to Support Interactive Task Learning

People are, of course, currently the best interactive task learners in existence. A deeper understanding of how people learn tasks interactively may provide valuable insights in how to create artificial systems that can learn as flexibly. At the functional level, we will catalog some capabilities that interactive task learners need.

There are a number of capabilities that can be identified for a learner agent in the context of ITL—capabilities that are not necessary per se, but certainly desirable as a part of the learner, with a degree that depends on the task domain. Some of the most important cognitive capabilities are:

- *Theory of mind*: In establishing common ground in communication (see Levinson, this volume), humans draw on "theory of mind" capabilities: knowledge or assumptions about the state of mind of others (e.g., their beliefs, desires, and intentions) and the ability to reason about these (e.g., Wimmer and Perner 1983; Leslie et al. 2004). Similarly, establishing a shared understanding and common knowledge in ITL—necessary to support communication of relevant information between the student and teacher—will require an agent to have some elements of theory of mind, such as detecting goals and intentions, and having beliefs about their human teachers and other agents.

- *Causality*: Understanding causality is fundamental to human rationality (Gopnik and Schulz 2007). This ability will not only improve an agent's ability to reason about the world as it learns and performs its tasks, but it is crucial to the process of determining its human users' intentions and goals as part of interacting and establishing common ground.

- *Self-monitoring*: Another key capability of ITL agents is the ability to monitor their own performance and learning. When being taught a new task, potential collisions with prior learned knowledge (perhaps trained by a different teacher) should be noticed by the agent and resolved in some way (e.g., by inferring the best solution, or by explicitly asking the teacher). Due to the fact that the agent is responsible for its own internals, it needs to monitor its own performance and use that to either ask for additional instruction or formulate and tackle its own learning goals.

- *Retrospection*: An ITL learner will be continually engaged in interactions with an instructor as well as with its environment. These interactions occur in real time and therefore, the agent may not have an opportunity or computational resources to recognize general patterns in the data it is observing or to judge how similar the current situation is to something in the past. In such cases, the capability of retrospection—to be able to recover, reason about, and learn from data at a later time may become crucial. Retrospection has been shown to be useful in category learning (Kuehne et al. 2000), where the agent maintains a set of exemplars from the past and previous generalizations. Each new observation is compared to these sets to discover new patterns in experiential data. Another example is explanation-based learning, which requires an analysis of a complete trace of task performance for generalization, but the complete trace may not be available during incremental interactions with a tutor (Mohan and Laird 2014). While using an explanation-based strategy, the ITL learner can store away its interactive experience and at a later point reason about why the instructed trace was useful in achievement of the goal. This retrospective analysis of its own behavior could be critical to produce generalizations and further learning.

- *Explainability* (or transparency): The ability of the agent to explain elements of its learning can promote generalizability and improve its potential to communicate effectively with the human user. Although this ability may be especially evident in cases where the learner agent then becomes a teacher of others (human or agent), this type of transparency can play a key role in self-monitoring and learning from retrospection, regardless of the end task.

- *Directed attention*: When learning any task, but especially in a busy real-world environment, the ability to focus on task-relevant details—and to ignore task-irrelevant details—is critical to the success of learning and performance. Given the ambiguous nature of most communicative acts, directed attention is also very important to successful interaction by narrowing the search space over what is being referred to or what is intended.

- *Operationalization of knowledge*: One of the most challenging aspects of interactive task learning is the problem of converting the externally specified knowledge—whether from language or examples—into an internally executable form. It is somewhat reminiscent of a run-time compiler, but for task knowledge instead of a computer program. There are several possibilities, including creating a declarative representation of task knowledge that is interpreted by an agent, and compiling declarative task knowledge into more efficient rules or even code. One option is to create a declarative representation of the task knowledge that is interpreted by the agent; for example, in the Northwestern Tic-Tac-Toe program (Hinrichs and Forbus 2013b), task information extracted from natural language and sketching is translated into a simple representation and then interpreted by an interpreter. Another option is exemplified by Rosie (Kirk and Laird 2016), in which declarative task descriptions are automatically converted into a native rule representation, which execute 80 times faster than if they were interpreted. Both of these approaches seem to find support in the neuroscience of how the human brain rapidly learns new tasks (Stocco et al. 2012; Cole et al. 2013). Existing data suggest that tasks are represented in a declarative and unified format (Cole et al. 2013) and that these representations are interpreted "on the fly" or "just in time" by subcortical brain structures, like the basal ganglia (Frank et al. 2001; Stocco et al. 2012). With time, however, tasks undergo significant reconfigurations, with the original declarative knowledge being recoded into more efficient, procedural terms (Chein and Schneider 2005).

- *Handling large amounts of knowledge*: Given the need to learn multiple tasks over time, through multiple means (instructions or examples) and possibly in multiple modalities, an ITL agent should be able to handle and maintain a large amount of knowledge in a scalable and efficient format. In particular, it should be able to use representations that can integrate different task descriptions (from instructions or examples) and different actions (for different tasks), avoiding conflicts between representation and building up a knowledge base (world knowledge, teacher preferences) that can be used to facilitate learning future tasks.

Interaction in Task Learning

Although interaction has arisen as an issue throughout the earlier chapters, it is worth highlighting the critical role of interaction in task learning in its own right. One way to provide context to this problem involves asking a series of questions about the role of interaction in ITL.

First, why is interaction so critical to ITL? Interactive learning can lead to agents that are much more adaptable to the contexts in which they are deployed. However, relying on a human user to train an artificial agent on a range of tasks imposes some degree of burden on the user, and interaction for its own sake is not recommended. Incorporating interaction into a task-learning scenario should be motivated by restrictions imposed by the task goals and environment, such as the variety of tasks that might be important to learn, not all of which (or all variations/parameters of which) are known *a priori*; the lack of sufficient training data; and the difficulty of hand-coding sufficient knowledge into the agent.

We define the goal of interaction in task learning as communicating any and all aspects of the task to be learned, and an important requirement for effective communication is *common ground*. Common ground refers to the knowledge in each agent of which aspects of the situation are shared between the two agents (Clark and Brennan 1991; Clark 1996). For example, a teacher and the robot can both know how to do a task (e.g., making coffee knowledge is in their situational awareness), but they might not know that the other knows how to do this task and that would mean that it is not in their common ground. Hence in their interaction, the teacher might think that she has to teach the robot how to make coffee. Interactive task learning can take place without common ground (e.g., the robot simply memorizes the commands of the teacher), but when common ground is also created between teacher and robot, this interaction becomes more natural because there is more shared information to use in the interaction.

How can interaction best facilitate establishing common ground between learner and teacher in the context of achieving effective learning? One of the main challenges that a participant in an interaction faces in establishing common ground is determining the intentions of the other participant(s) in the interaction. Natural communicative acts between humans (as opposed to artificial symbolic systems) are generally highly ambiguous and/or underspecified (Keysar et al. 1998). Humanlike interaction between a human user and artificial agent will face these same issues. For example, if the human tells her domestic robot "put the fish in the refrigerator," background knowledge will help resolve the ambiguity between "fish" as a pet and "fish" as a food. If she says "it's cold in here," only experience with that person (or further interaction) will enable the robot to decide whether the person prefers a blanket or having the thermostat adjusted. At the same time, humans have expectations about communication, as in the Gricean maxims of quantity, quality, relation, and manner (Grice 1975), and pedagogy (Csibra and Gergely 2009) which guide how they interpret utterances and these kinds of expectations could be incorporated into ITL robots.

Given the current state of the art, we have a long way to go before artificial agents can fully interact naturally with a person, the way people so effectively and efficiently do with each other. It is useful to specify various dimensions of interaction and to consider the level of sophistication of the mechanisms—and the resulting level of naturalness of the communicative capabilities—of the agent along each of those dimensions, as indicated in Table 15.1.

How much should interaction be used in ITL systems? System designers will need to determine an appropriate balance of the burden of effort between the human user and the computational agent—or perhaps the learner agent can determine its own need for interaction and behave accordingly. For the human teacher, issues to consider include: how much can they be expected to effectively elaborate task/background knowledge, especially for implicit knowledge; and how much can they be expected to (learn to) interact in a restricted way. The design of the agent will also need to examine to what degree user needs can be anticipated, how feasible it is to program in the needed knowledge, and what the limits are of the state-of-the-art in the mechanisms and algorithms of interactive learning.

To better understand interaction as part of ITL, we need to consider both the types of knowledge being communicated between agents, and separately, the types of communication and information that package this knowledge for transmission between agents. First, there are a number of types of knowledge that could be communicated as part of the ITL learning process, including:

- Perceptual and attending knowledge (e.g., object identification in the world, perceptual chunking)
- Goals: final state(s), space of possible goals

Table 15.1 Dimensions of interaction: general questions about communicative interaction abilities are listed in column one, followed by the corresponding dimensions along which agents can be assessed and the scale endpoints of those dimensions (columns two and three, respectively).

Dimensions of an Agent's Communicative Abilities:		
Who can communicate, and who can take initiative in communicating?	Initiative	Passive ↔ Active
What can be communicated, and what can be understood?	Expressivity	Simple task knowledge ↔ Sophisticated task knowledge/strategy
	Theory of mind	Task/world knowledge ↔ Knowledge of other agent's goals/ intentions/beliefs
How is communication carried out?	Multimodality	Single mode ↔ Multiple coordinated modes
	Interactivity	Single turns ↔ Multiple coordinated/ nested turns with repair

- Possible actions, given the effectors and world constraints
- Action sequences, at multiple levels of abstraction, with action hierarchies, and possibly full and partial plans
- Action-world model that describes preconditions and effects of given actions
- Measure of performance for intermediate and final states
- Metacognitive strategies (e.g., learning strategies)

For each of these types of knowledge, there are multiple ways for the learner to acquire this information from the teacher as discussed by Thomaz et al. (this volume). Below is a list of different methods for communicating this information.

- Illustrative examples
- Demonstration
- Questions
- Feedback
- Explanations
- Sketches
- Gestures
- Facial expressions
- Language in the general case

The multiple types of knowledge and ways of communicating information are far from mutually exclusive, but instead often go hand-in-hand. For example, one way of communicating task knowledge is to communicate how the task can be learned—a metacognitive process—which can be done by communicating a learning strategy. A teacher might point out where task knowledge can be learned in the environment (e.g., a textbook to read), or might provide a self-training strategy that allows the learner to acquire a particular task. For instance, a search-and-rescue training instructor might advise on structuring the learning problem in terms of strategies for identifying signs of movement, potential hazards, and so on (each of which could be further decomposed). This type of metacognitive knowledge often arises in the context of other types of knowledge, such as knowledge about the goal(s) to perform and/or the performance measure to be used in self-evaluation.

   In other situations, the learner might guide itself through the process as a form of learner-led interaction. For example, in the field of developmental robotics (Cangelosi et al. 2015), the goal is to develop robots that learn in humanlike ways and often these robots have exploratory behaviors in which the learner leads the interaction. While ITL is typically focused on a teacher guiding the agent to learn a given task, some learner-driven interaction could enhance ITL capabilities. For example, an ITL robot may have the task to sort a set of novel objects into two boxes. One approach is for the teacher to lead the interaction, by telling the robot which object to select and then directing them to the target box (e.g., "pick up the green squarish thing and put it in the red box"). The alternative learner-led approach is that the robot picks up a random object and moves it toward a box (a child might put objects in boxes just for fun as part of their exploratory program). The teacher then only has to say "yes" or "no, the other box" to teach the task. In both approaches, the robot learns the task, but the second approach may be faster and more natural. Combining both learner and teacher-led approaches could lead to systems that learn more quickly with more natural interactions from teachers.

Evaluating ITL Systems

Any discussion of learning in the context of ITL—whether it relates to tasks, interaction, or learning itself—naturally leads to discussion of evaluation, namely how to evaluate whether an ITL system is performing well along some dimensions. Because an ITL agent is an interactive system in itself, many general evaluation measures from the fields of human–computer interaction, user experience, and human factors apply to ITL systems as well. However, ITL also raises a number of issues and challenges for evaluation that are either unique to, or especially important for, ITL in particular. This section reviews what we consider to be the most significant evaluation metrics and criteria with regard to the three dimensions of interaction, tasks, and learning.

Evaluating Interaction

As mentioned, there are many ways of evaluating interactive user systems in the general sense, from empirical user studies to heuristic techniques to conceptual models to computational cognitive models. For ITL in particular, some of the most significant factors for evaluating interaction include:

- *Usability*: An interaction system must be both easy to use and learnable. For example, if a human teacher must learn a new, complicated language to communicate with the agent, it is typically not considered usable. There are many common and acceptable ways to evaluate usability, including formal modeling methods (e.g., GOMS; Card et al. 1983), task analysis, and empirical methods.
- *Naturalness of interaction*: A criteria for evaluating ITL systems is the "naturalness" of interaction. Are there important aspects of natural interaction that are missing from today's systems? Is it really enough to add natural language to ITL systems, especially as a replacement for programming or scripting? Considerations of the degree of natural interaction would normally require other dimensions. For example, interactive turn-taking is an important problem that might be more

necessary for the system to interact naturally than vast language capabilities. Naturalness does not come simply from adding natural language but may be a function of the knowledge of the teacher versus the knowledge of the system to be trained. For example, the teaching of complex business processes could come naturally in a formalism known to experts (e.g., a business process language) if the teacher of the ITL system is an expert in that domain.

- *Expressivity*: Both the agent learner and the human teacher benefit from being expressive; if one highlights a point through nonverbal means, the interaction contains a richer set of data that can be used by the other agent for understanding and clarification. For example, a human saying "put the BLUE ball in the pail" suggests that the color is especially important in the current environment. It may be difficult to define, however, how exactly to evaluate expressivity in an ITL context.

- *Customizability*: Another dimension of an ITL agent's interaction is the degree to which it can adapt to the needs of a specific teacher. An ITL softbot that resides on a phone, for example, might be able to learn, over time, the specific needs and capabilities of the phone's user. For example, it might learn a user's preference for certain ways of organizing his or her calendar (e.g., no meeting before 9 a.m.) or add user-specific concepts to its ontology (e.g., the concept of "ultra-marathon running"), which would aid in the learning of future user-specific tasks. Extendibility can be empirically measured as the improvement in the quality of interaction (e.g., fewer examples, less questions, less feedback, or shorter instructions needed) for a repeated teacher, when compared to a novel teacher instructing the same task.

- *Flexibility*: A good ITL system should be flexible in how it interacts with a teacher; that is, it should allow multiple modalities (instructions, examples, and demonstrations) and be able to learn from all of these to the maximum extent. This is particularly important because different teachers (and certainly different human teachers) might have different preferences on how to describe a task, or might have different abilities. Consider, for example, the case of individuals with disabilities; a mute teacher might prefer to demonstrate a task, while a bedridden one might prefer to use linguistic descriptions. Flexibility can be empirically evaluated by preventing teachers, in an experimental setting, from using one or more modalities; for example, by preventing them from speaking, or by preventing them to use gestures, or to perform the actions. The extent to which an agent's performance suffers from such restrictions will be inversely correlated to it flexibility.

- *Expertise required for effective use*: One of the most promising aspects of ITL is that it could allow nonexperts to directly modify the knowledge and behavior of AI agents without having to learn programming.

For example, instead of having to use a programming language such as Java or C++, or even a scripting language, an untrained human could extend the capabilities of a home robot or an intelligent assistant on their phone. There are two metrics that would be relevant to this. The first is how long it would take a person with professional level skills in programming to develop an agent with the same behavior, possibly using different paradigms (e.g., using Java vs. a scripting language vs. an ITL agent). This could be compared to two measures: how long it takes a nontechnical person to learn how to interact with an ITL agent (e.g., the idiosyncratic nature of its language capabilities and ways of presenting examples) to some level of proficiency (e.g., 95% correctness); and how long it takes them to teach an ITL agent so it has the appropriate skills. Note that it would be useful to measure this over multiple, different types of tasks.

- *Diversity of modalities*: People communicate with each other via multiple modalities. Often this occurs simultaneously, as when someone talks while sketching or circles something on a photo to illustrate a point of reference in an email. One dimension for evaluating the naturalness of interaction for ITL systems is both the number of modalities and how sophisticated they are. For example, if we consider textual languages as one dimension, we can view it as being anchored by programming languages on one end, and unrestricted natural language dialogue on the other, with more restricted subsets of natural language understanding (e.g., controlled vocabularies, simplified syntax) in the middle. Similarly, visual communication as a modality can range from selection on a fixed map or diagram on one end to open-ended motion on the other end, with sketch understanding and image understanding as intermediate. Sounds and audio are other useful dimensions for communication, whether they be fixed signals (e.g., audible alarms or indications), speech, music, or other forms of audible communication. The complexity of dialogue supported is yet another dimension, ranging from unidirectional on one end, to fully mixed-initiative open dialogue on the other, with clarification questions and subdialogues as possibilities in between.

- *Natural interaction versus radical transparency*: Work so far on interactive task learning has focused on using natural modalities because that is the way that people normally train other people. This is challenging for ITL, of course, because it means that ITL agents must have communication skills that are much closer to those of people compared to the state of the art. We would be remiss if we did not mention another possible alternative: radical transparency. In a radically transparent system, the internals of the system are visible in some easily understandable way by the people who train them and work with them, with the critical parts of their internal operations made visible by some powerful high-level language (perhaps visual in nature). Thus, we could program them graphically, in a manner reminiscent of RoboFlow (Alexandrova et al. 2015), and beyond that, monitor a high-level summary of their internal state and browse what they know about the world and about tasks. It seems unlikely to us that the complex operations of agent cognition can be neatly summarized in ways that would make training a teacher or user no more burdensome than training them to work with a dog or horse, but it is a possibility that might be worth considering for simpler forms of agents.

- *Context sensitivity*: Context sensitivity defines an agent's ability to adapt to different human users in their communication and performance of tasks. In the terminology of common ground, the agent is able to maintain knowledge of common ground specific to each human, including their shared referential background and context for effective communication, and any user preferences for performing the tasks. This could be assessed by demonstrating that an agent adapts appropriately, both in its communication and task performance, to each of several human users.

- *Tolerance to errors*: When working with an interactive task learning system, it is important to be robust to errors that either the human trainer or the agent may make. For example, if the human trainer provides an incomplete (or even incorrect) description, the agent must be able to learn the correct knowledge so that it will not keep a persistent "bug." A reasonable method of evaluating tolerance and robustness to errors is to demonstrate how a system might recover from errors and perhaps even learn to avoid them in future instances.

- *Safety*: The purpose of this evaluation would be to determine if ITL is able to provide some guarantees of safe operation in the context of a malicious, or possibly mischievous instructor, where the agent is taught undesirable behaviors. The challenge is to develop capabilities in the agent that enable it to recognize behaviors that violate some set of norms, and to distinguish actual facts from "alternative facts." Measuring and evaluating an agent along this dimension will be challenging, but could possibly involve corpora of example instructions, examples, and demonstrations that include both valid and invalid (unsavory?) knowledge for the agent to learn.

Evaluating Tasks

A similar set of evaluation metrics can be derived for evaluating tasks in the context of ITL—that is, evaluating an ITL agent with respect to the tasks that can be handled by the agent. These include:

1. *Diversity and coverage of tasks* (agent generality): One of the promises of ITL is that an agent will be able to learn more than just one or two tasks, and more than just tasks from a limited repertoire (although ITL agents that are targeted for domain clusters with limited diversity could be very valuable in certain contexts). Diversity can vary along many dimensions and often depends on the overall domain(s) in which the agent will be used.

    •   Different task formulations (problem space, procedure, and optimization) that determine what needs to be learned to perform a task, as described by Laird et al. (this volume).

    •   Different terminology needed to specify the task: Are many of the terms shared between tasks, or is there a small set shared by the tasks?

    •   Different levels of abstraction of the task and preexisting knowledge can be required for the task: Does the task primarily involve perceptual-motor interactions (e.g., picking up different types of bottles) or more abstract actions that are decomposed into such primitives (e.g., cleaning a room or delivering a box)?

    •   Different types of environments, such as whether the task takes place internal to the agent, involve interaction with other software or a physical embodiment (e.g., a robot).

    •   Different level of complexity of tasks can arise in many forms across tasks. Some possible dimensions include the complexity of the task specification (how hard it is to understand the task) and the complexity of the task problem space (how hard it is to perform a task once it is learned).

2. *Explainability*: Another important aspect of tasks is their explainability, or ease of describing from one person to another. For example, learning the sensorimotor skills of soccer is more complex than learning the legal moves of chess, because verbalizing the physical movements in chess is straightforward but verbalizing the physical actions of soccer is not. Other tasks such as "setting tables" might also suffer in terms of explainability from implicit versus explicit knowledge. In general, explainability varies as a function of the task and the teacher, as what is introspectively accessible to one teacher might not be to another. However, at least when the teacher is predefined (e.g., it is predetermined to be a human), explainability can be measured for a specific task.

3. *Noise and randomness*: While some tasks have little to no "noise" or randomness associated with the world environment, others are highly variable and subject to such noise. For example, chess is completely determined by the players' moves (except for the possibility of an

extreme event, like someone upending the board); a dice-based board game is somewhat determined by the players but somewhat by the randomness of the dice; and driving down the highway is a relatively straightforward task in terms of the goal, but could potentially be a very difficult one because of the variability of the environment (e.g., darkness, rain, snow, potholes).

4. *Risk*: As related to a learning agent, risk refers to how dangerous the agent may be to itself or to others as it is learning to perform, or ultimately performing, a new task. For example, if the agent is a vehicle learning to drive itself, there is a potential risk that it may cause an accident while learning.

5. *Observability*: One final aspect is whether the agent directly observes all of the relevant aspects of its environment (e.g., tasks performed by a tabletop robot), or whether there are aspects of the environment that the agent cannot sense without explicit actions (e.g., tasks performed by a mobile robot that must move between rooms and interact with objects that are initially unobservable).

Evaluating Learning

Evaluating the learning of an ITL agent is an interesting challenge in comparison with previous work on machine learning and evaluation measures. For many machine learning systems like classifiers, there is a well-known trade-off between the amount of training effort, the accuracy of the final system, and the complexity of things that are trainable. However, a particular classifier is typically targeted for a specific domain, often with a large set of training examples. In contrast, ITL would typically involve a small number of higher-quality, targeted training examples, and potentially could improve learning by receiving help (explicitly or implicitly) from a teacher guiding the learner through the process. Thus, we can identify several dimensions along which evaluating learning is especially important:

- *Speed, efficiency*: One of the most significant evaluation criteria for ITL involves how quickly the agent can learn from its teacher. Speed can be measured in real time, although for some mechanisms or algorithms, other measures may be more appropriate—such as the number of experiences or the number of iteration cycles required to achieve adequate performance. In most cases, we prefer ITL systems that learn as quickly as possible, since this minimizes the user time needed with the system. There may be some cases, however, in which it is desirable for an agent to learn at a rate roughly comparable to a human—for example, when using an ITL learner as a stand-in for a human learner when evaluating the learnability of concepts (e.g., a simulated student learning new mathematical concepts to be integrated into an intelligent tutoring system).

•   *Performance*: As a complement to speed and efficiency, another criterion is how well the agent performs the task that it has been taught. There are many possible ways to quantify and measure performance, including resources required to perform the task (e.g., time or energy), quality of the solution (how well it achieves all aspects of the goal), percentage goal completion, and time to completion.

•   *Transfer*: An ITL learner is expected to learn multiple tasks. In the case where some of the given tasks are similar, ideally the agent will exhibit transfer of knowledge from one task to another. An ITL learner should exploit similarity of structure, action control hierarchies, and so on to generalize previously learned knowledge to new tasks. For example, after learning how to wipe down a table, learning to wipe down the windows should be easier (from certain perspectives) as the action control is similar. This can be measured as number of interactions (e.g., demonstrations, instructions) taken to completely learn a task, number of trials (independent exploration or time) taken to learn a task, and so on. We would expect that for similar tasks (as measured, e.g., by overlap of features, action hierarchy), a better ITL learner requires fewer interactions or trials to learn a task at a specific level of performance.

•   *Interference tolerance*: As an ITL learner is learning to perform multiple tasks, knowledge acquisition has the potential to suffer from interference—that is, if learning a new task reduces performance in an already-learned task. This might happen if the algorithms are overzealous in generalization, or if the task representations do not sufficiently discriminate between different tasks. An ITL learner should be able to recover from such interference effects—for example, by asking for more instruction that allows it to produce separable representations. This capability can be measured by the number of generalization errors made (less for a better ITL learner), complexity of instruction needed for recovery, or amount of training needed for correction. The transfer-interference metrics will likely be in terms of a trade-off, and a good ITL learner will maximize transfer and minimize interference.

•   *Robustness with respect to teachers*: The quality of an agent's learning capabilities should be assessed with respect to its robustness in the face of varying quality in the information it is given by the teacher. While it is comparatively easy to design a system that learns from carefully crafted instructions and perfectly selected examples, real-life human teachers vary in the quality of their directions. For instance, less capable teachers might give incomplete task descriptions, use ambiguous language, or provide misleading examples. This form of learning robustness can be empirically measured by experimentally degrading the quality of the directions given to the learner; for example, verbal instructions might be shortened by deleting an increasing number of words, and the number and coverage of examples given to the agent might be reduced.

- *Incremental learning*: One overarching desired property is the allowance for incremental learning, to provide a trade-off space between training time and task performance. Ideally, if the user only wishes to spend a very short time training an ITL agent, that agent could perform at least reasonably well; but if the user wishes to spend more time (then or later) in doing further training, the agent would likewise improve in its performance.

Evaluation Discussion

With all of the evaluation criteria above, it should be noted that the criteria are most useful when evaluated within the context of an appropriately diverse set of tasks. In particular, a broad evaluation across criteria would ensure that task-specific approaches have not been incorporated into the agent to simply improve its behavior relative to a metric. The actual set of criteria used depends on the overall objective of the ITL agent; whether the agent is being used as a personal assistant, a good game player, or an assistive robot would presumably skew the priorities of the evaluation criteria in favor of those most important for the respective domain.

Because ITL is an emerging area, research may need to blossom further before these evaluation criteria are used across the research community in more prescribed ways, as in the development of benchmarks. At this stage, exploration of different task types, means of interaction, and learning mechanisms should be relatively unconstrained to allow for maximum creativity, and allow the research to push the boundaries of the science in all areas of this multidisciplinary effort. Nevertheless, the evaluation metrics are still critical even in these early stages for researchers and system designers to identify the most important aspects of ITL for evaluation—those that are particularly salient to the ITL context and that distinguish it from other machine learning and artificial agent approaches.

That being said, there could be benefits from developing a standardized set of tasks for evaluating ITL agents in simple yet realistic domains. One advantage of this approach is that it would provide a constrained domain in which experimental psychologists can test theories and collect data on ITL in humans. Currently, ITL behavior in humans is rarely studied because of the experimental challenges it poses as a very high-level domain. Providing simple tasks (e.g., simple card games) and a well-defined, constrained set of interaction models (e.g., instructions in restricted English, set of positive and negative examples of legal moves) would set the stage to examine how ITL occurs in humans while carefully controlling and measuring the experimental variables. Furthermore, ITL tasks formulated within a formal framework provide objective metrics to control the experimental stimuli and design, which is what is needed to conduct statistical analysis on experimental variables—and to make valid statistical inferences. This would open up the door to more research of ITL behavior in humans, which could in turn provide important clues as to the fundamental mechanisms that should or could be implemented in artificial ITL agents.

There is at least one other important reason why a standardized set of tasks would facilitate experimental research on humans: the fact that the same task could be used in humans and artificial agents, and especially that artificial agents could be used as a model to analyze human neuroimaging data. Consider, for instance, a realistic example: teaching a human lying in an fMRI scanner how to play a simple card game that has been taught to the Soar-based agent Rosie. Rosie's internal states provide a formal model of learning, whose internal states that can be traced and used to generate a regressor for human brain activity. The estimation of various parameters for the regressor could be done using a Bayesian framework to maximize its correspondence to behavior and imaging data. When compared to brain activity, the Soar-based model could be used to locate regions of the brain whose activity more likely correspond to the model traces. In turn, this could be used to interpret the specific computations of different brain circuits. Finally, brain data can be compared to the internal processes of different agents, and insights from human data can be used to revise and improve on existing agents.

General Discussion

Interactive task learning establishes a problem focus that is AI-complete—drawing on knowledge representation and reasoning, computational linguistics, machine learning, robotics, computational perception, and so on. As a research area, ITL emphasizes integrative approaches that bring together all or many of these sub-areas into sophisticated systems with multiple interacting functions that must work in harmony with each other to solve overarching problems. This kind of research agenda is valuable in ensuring that research in the various subfields is grounded and contextualized in real-world problems whose solutions must coordinate the multiple intelligences required. At the same time, research that reduces aspects of any one of these areas into isolated components to be more thoroughly explored continues to be necessary. On one hand, the benefit of the overarching integrative framework is to inform the narrower focus and to ensure that component solutions will have relevance when brought into contact with other elements of the broader framework (e.g., Anderson 2007). On the other hand, the deep-dive research can find potential solutions to problems that may be too difficult to explore within the full complexity of the integrated system. Ideally, each approach serves to provide motivation and guidance to the other, and work to their mutual benefit.

In this chapter we have largely focused on the relationship of ITL to other areas, especially noting the subsets of these areas that we believe are most relevant and important to the ITL effort. At the same time, our hope is that ITL also helps to push the boundaries of innovation in these areas themselves, since any improvement in any of the components of ITL will ultimately benefit ITL as well. One of the clearest areas of benefit comes in the field of machine learning. ITL holds the potential to have a major impact on machine learning, which is defined by the question: How can we create machines that automatically improve their competence through experience? To date, machine learning has focused on only a small fraction of the types of learning that humans exhibit—that is, mostly on statistical learning from large amounts of passively acquired data. For humans, however, interactive learning from conversations, demonstrations, and experimentation is central. This may well become a major thrust of machine learning over the coming decade.

ITL could also make a significant impact in the domain of cognitive neuroscience. Cognitive neuroscience has the seemingly impossible goal of understanding the human brain at a functional level—that is, how its activity gives rise to human cognition. The dominant approach in this field has consisted of dividing human cognition in putatively isolated functions, such as object recognition or decision making, and examining them with small-scale, controlled paradigms. However, many authors have suggested that this approach is limited, pointing out that perfect decomposition into basic functions might be impossible, or that any reasonably complex task ultimately requires the coordination of interconnected systems and functions. ITL research can benefit this area in two ways. First, it brings *formal models* of task learning, which could be specified mathematically, implemented computationally, and used to analyze neural data, as in the previously described example of using Rosie to interpret neuroimaging data from a well-defined card-game task. The approach of using models to understand and explain brain activity has been perhaps one of the greatest advances of modern neuroscience, with the application of reinforcement learning to the interpretation of the functional role of basal ganglia in procedural learning being an excellent example (Schultz et al. 1997). Second, it brings a general and integrative approach to learning in which multiple components exist at the same time. This is particularly important because it allows for study of very complex cognitive phenomena without losing track of the different role played by different components.

Yet another potential benefit comes in the area of linguistics and psycholinguistics. The rich models of human interaction from these areas will need to be formalized within a computational framework in which they interface to varied aspects of agent systems involving representation, learning, and processing of task, background, and interaction knowledge. Research into dialogue systems will need to consider embodiment in robots and the contexts of varied software agents. Fundamental issues in interaction have great insights to contribute to ITL, and in turn ITL will identify interesting problems of interaction in a real-world context that highlight the connections to other cognitive capacities such as visual interpretation, various forms of reasoning, and physical interactions with the world and interlocutors.

16

Early Developing Prerequisites for Human Interactive Task Learning

Franklin Chang

Abstract

Humans are better than artificial computational systems at learning to do new tasks through interaction. Part of this ability stems from preexisting capabilities that appear early in human development. Children have internal physical models of how objects move and they attribute mental states (e.g., goals, beliefs) to objects when their behavior is unpredictable. They are also able to develop context-specific rules and identify how to help others achieve their goals. To explore how these abilities can be transferred to interactive task learning (ITL) systems, this chapter proposes a world-state prediction model. The prediction model can learn detailed physical regularities in the environment and is able to develop representations for predicting the actions and goals of animate agents. The model suggests that prediction and prediction error are capabilities that could improve ITL systems.

Introduction

Humans are good at learning new tasks through interaction. If you show people how to play a new game, where one needs to find monsters in different locations in the real world and throw objects at them using your smart phone screen, humans can easily learn to play that game without extensive training. This ability to learn a new game through interaction with a human teacher differs greatly from the learning ability of existing artificial agents. Recently, a deep learning neural network was shown to be able to learn to play 49 Atari video games by mapping between pixel-based images and the joystick actions needed to play the game (Mnih et al. 2015). Although the ability of this neural network to learn to play these games is impressive, each game requires extensive training based on a massive database of games. Also, while human experts are good at playing new games (Green and Bavelier 2012), these deep learning systems showed a lower accuracy when learning multiple games within the same system (Kirkpatrick et al. 2017). Part of the difference between human and artificial systems arises from abilities that appear early in human development, which shape the way that learners understand new tasks. Using studies of infants and toddlers to understand these early abilities, I will attempt to link these abilities to interactive task learning (ITL).

Although most people assume that infants exist in a "great blooming, buzzing confusion" (James 1890), research with infants has revealed that they appear to understand a range of constraints on the physics of objects (e.g., occlusion, support, collision, containment; Baillargeon and Wang 2002) and their interaction with other moving objects (e.g., Leslie and Keeble 1987; Frankenhuis et al. 2013). In addition, infants have social knowledge about how animate entities differ from inanimate entities in their motivations for action (e.g., goal-directed motion; Woodward 1998; Luo and Baillargeon 2010). Since these infants were not explicitly trained to understand the novel scenes used in these studies, this research suggests that their event understanding systems may have features that enhance their ability to automatically understand both physical motion and goal-directed action. These abilities give humans a rich predictive representation of their world even before they begin to interact with a teacher in learning a particular task.

Early Developing Abilities in Human Children

We cannot ask preverbal infants about their understanding of the world. Our knowledge about their abilities comes from two tasks that measure their expectations: habituation and violation of expectations paradigms. In the habituation paradigms (Spelke et al. 1992), infants are exposed to a scene multiple times until they habituate or become bored with the scene; they will often look away from the scene when this happens. When a new scene A is shown and the child is still bored or uninterested, this is seen as evidence that the child views scene A as being the same type as those that were viewed previously (even if it is different in some way). If the child becomes interested upon viewing a new scene B (dishabituation), this is taken as evidence that they view scene B as being distinctly different in an important way. If scenes A and B differ in some particular dimension, then dishabituation provides evidence that infants have knowledge of this dimension. In violation of expectation paradigms (Baillargeon and Wang 2002), children are shown an event that conforms to their expectations. Then they are shown an unexpected event that violates these expectations. If children look at the unexpected event longer than the expected event, this indicates that they have some knowledge about the features that differentiate these two events. Habituation involves learning about the event within the experiment, whereas violation of expectation involves knowledge gained outside the experiment. These two types of paradigms allow us to probe the knowledge that young children have about visual events.

The Atari deep learning system started with pixels and had to learn the entities associated with different games (e.g., aliens, frogs). Although the human vision system also has a pixel-like input in the rod and cone cells of the eye, the brain seems to process the world in terms of objects that can exist even when they are not visible. Baillargeon et al. (1985) showed that seven-month-old infants were surprised when an object was placed behind a screen and the screen was rotated such that the object seemed to have disappeared. This violation of expectations study demonstrated that infants believe that objects take up space (spatial extent). Objects are also collections of elements that move as connected and bounded wholes. Spelke (1990) found that if two objects move relative to each other, even though they were constantly attached or connected, infants perceive them as two objects. Furthermore there is evidence that infants expect objects to move on connected, unobstructed paths and are surprised when objects appear to move invisibly across space or through other objects (Spelke et al. 1992; Aguiar and Baillargeon 1999). Therefore, within the first year of life, infant understanding of the world is not based on a raw list of pixel values, but rather on an internal model of objects with constraints on how they move and their spatial extent.

If representations are object based for infants, then infants must be tracking objects as they move around the scene. Leslie et al. (1998) have argued that infants and adults use one system for object tracking, which involves a set of pointers that stick to a particular object (dorsal *where* system), and another system to encode the visual features of each object (ventral *what* system). This is different from the approach in the deep learning Atari system, which utilized a series of convolutional neural networks to mimic the gradual abstraction of features in the ventral part of the visual system in the human brain, but did not have an explicit system for object tracking (e.g., a single frog on different parts of the screen are treated as different "objects"). Evidence for these visual pointers comes from multiple object tracking studies, where adults see videos with multiple identical circles moving around randomly (Pylyshyn and Storm 1988). Specific (white) circles are identified as targets (e.g., colored in red) and then the circles are made identical again (e.g., changed back to white). The circles then move around randomly while participants stare at a cross in the center of the screen. Later, participants are queried about a single circle and must say whether it is a target or not. Due to the fact that the circles are identical during the random motion, there are no visual features that can be used to track the circles, so the only way to identify the targets is if the participants are following them through the whole trial. These studies suggest that participants can track a limited set of objects in parallel using a set of visual pointers.

Support for these pointers in infants comes from a study by Spelke et al. (1995), where infants viewed two screens, separated by a gap, and saw an object move behind the first screen and another object (with the same shape) emerge later from behind the second screen (Figure 16.1). In the discontinuous condition, it appears that there must be two objects, because an object cannot cross the gap without being visible. In the continuous condition, the object appeared in the gap between the two screens such that it looked like one object was moving behind both screens. Infants were tested with similar scenes without the screens and they showed that they preferred the test scene that matched the training scene. If the infant assigned a pointer to the first object, then the same pointer could be used when it reappeared in the gap as well as when it appeared after passing behind the second screen. In the discontinuous condition, the appearance of the object from the second screen would require a new pointer. Additional studies by Xu and Carey (1996) showed 12-month-old infants scenes where two objects with different shapes appeared from behind a screen one at a time. At test, the screen was lifted and they were shown an expected scene with two different objects or an unexpected scene with two objects of the same shape, and they were surprised by the unexpected scene. These results suggest that infants track objects in scenes, even when they are not visible, and their representation of scenes involve these object-based representations. Object tracking is critical in any interactive task where an agent needs to communicate about multiple identical objects. For example, to learn how to barbecue multiple similar-looking items on a grill, a learner needs to track which items have already been flipped by the teacher as well as which items still need to be flipped (as well as any that might have accidentally fallen on the floor in the flipping process). Thus, the work with infants suggests that humans have a task-independent multiple-object tracking ability that could help support these types of interactive tasks.

After an infant can track objects, it becomes possible to identify the nature of the resulting interaction. One study that provides an important insight into

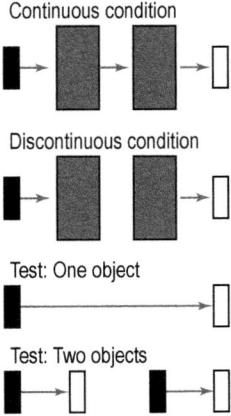

Figure 16.1 Two conditions in which object tracking was tested in 12-month-old infants (Spelke et al. (1995). Arrows indicate the path of an object (black box) as it moves behind two screens (large gray box) to its final position (white box).

how this takes place was conducted by Gao et al. (2009), who used a multiple object tracking study to examine the features that support identifying a chasing action. In this study, multiple identical circles moved in a random manner. One of these circles was the wolf (the rest were sheep) and it had the property that it was always moving toward one of the sheep (central circle is wolf in Figure 16.2). Gao et al. found that adults were better at identifying the wolf when the wolf's angle of motion toward the sheep was more direct ("heat seeking"). Because the wolf and the sheep were all identical circles, participants had to first track the objects using pointers and then encode visual heuristics between pairs of pointers, such as the directness of the angle of motion. These visual heuristics appear early in development, as four-month-old infants prefer videos with chasing as opposed to those without chasing (Frankenhuis et al. 2013). This shows that children and adults are not just tracking objects, but associating relational features with each object pointer, such as angle of motion relative to other objects in the scene.

When chasing is taking place, the "wolf" consistently tries to move in the direction of the "sheep." In other actions, there is more temporal structure to the interaction. One event that has been extensively studied involves pushing or launching actions (Michotte 1963), such as when a green ball hits a red ball and pushes it away. This event begins with the pusher (e.g., a green ball) moving toward the pushee (e.g., a red ball). Critically, the pusher should make contact with the pushee, and the resulting motion from the pushee should begin without delay after contact. An understanding of the effects of these constraints on causal actions is evident early in acquisition. Leslie (1984) habituated six-month-old infants to videos and looked at how much they dishabituate to reversed versions of videos. Results show that they view the reversed one as

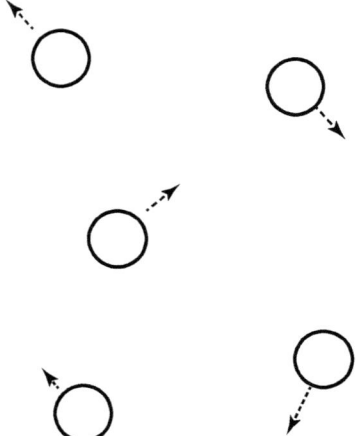

Figure 16.2   Schematic of visual heuristics used by adults to track multiple objects in a chasing action: the "sheep" are moving away from each other while being pursued by the "wolf" (central circle) Adapted from Gao et al. (2009).

being substantially different than the one to which they were habituated. They found that reversals of pushing actions yielded more dishabituation than reversals of videos with a single object moving across the screen. They also found that infants were less likely to dishabituate when the video had no contact between the pusher and pushee or when there was a delay in the movement of the pushee (Leslie and Keeble 1987). This work suggests that infants have multiple innate features/heuristics that they are able to combine in various ways to recognize causality. When observing an item knocked off of a grill, during the course of flipping other items, a person can identify the cooking tongs (used for flipping) as the cause of the action (item falling on the floor) without prior training. Thus humans have a range of task-independent features that are applied to multiple objects in parallel and which combine to give them a better understanding of the causal structure of real world events.

One account of how humans understand these actions is to assume that humans take a *teleological stance* (Gergely and Csibra 2003), where actions are perceived as a rational means to achieve a goal state under certain situational constraints. Evidence in support of this position comes from studies that use videos where a ball jumps over a wall; this suggests that the ball has the goal of getting to the other side (Figure 16.3). Later, when the wall is removed, one-year-old infants were surprised when the ball jumps along the same path, because it could have moved in a direct path toward its goal. On this view, infants are able to identify the goal state of the ball and how the situational constraints of the wall block the direct motion toward the goal (Csibra et al. 1999). Thus, the most rational approach to achieve the goal would be to jump over the wall when there is a wall, and to go straight when there is no wall. This ability to

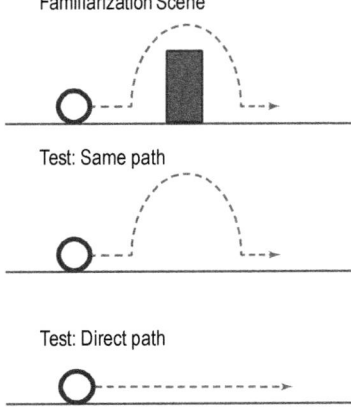

Figure 16.3 Schematic depiction of the teleological stance. During familiarization, one-year-old infants observed a ball jumping over a hurdle to reach a goal state (top). Two test states followed where the ball followed the same path (middle) despite the hurdle being removed and where it followed a direct path (bottom). The infants were able to identify the situational constraints. Adapted from Csibra et al. (1999).

see how action is dependent on situational constraints is an important ability for ITL systems. In the grill example, if a teacher was trying to flip item A and accidentally caused item B to fall off of the grill, the human learner would recognize that the arm motion of the teacher was not the most rational/direct means of causing item B to end up on the ground, and thus the teacher's actual goal must have been to flip item A.

Up to now, we have looked mainly at studies that involve whole objects without any moving parts. But it has also been shown that people can recognize biological motion in point-light displays (Johansson 1973). In these displays, performers with multiple lights attached to their body at various points (e.g., hand, legs, head) perform some action (e.g., running, jumping) in a dark room, where only the lights are visible (Figure 16.4 shows two frames from a running video). When these videos are shown to adults, they are able to label the action that is being depicted (Johansson 1976). Golinkoff et al. (2002) also showed two point-light displays of different actions (dancing, walking) to three-year-old children and found that when the children heard a verb that matched one of the actions ("look at dancing"), they turned their head toward the appropriate video. Since the mapping of specific actions and words must be learned from experience, these abilities appear after three years of age, but the ability to understand these videos emerges earlier: four- to six-month-old infants exhibit a preference for a point-light human walker over an inverted walker or random motion (Fox and McDaniel 1982). Bidet-Ildei et al. (2014) found that even three-day-old infants prefer walkers over random motion, even when there was no horizontal translation across the screen (as in normal walking). These studies demonstrate that the infant mind is ready to understand biological motion from birth. One way to explain this ability is that the mind attempts to predict the motion of the points in these displays and biological motion is more predictable, because there are correlations between the motion of different points in these displays. This work suggests that human understanding of action does

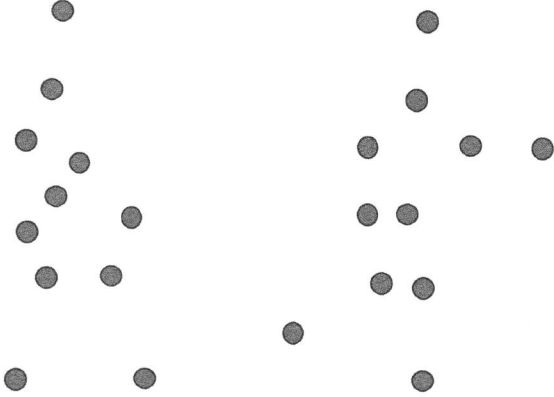

Figure 16.4 Two point-light displays of a man running.

not involve learning rigid action rules; instead, predictive learning mechanisms capture the complex motion patterns present in biological systems.

To explain how we process biological motion in point-light displays, Giese and Poggio (2003) proposed a computational model that used the fact that the brain has distinct pathways to process form and motion. The ventral *what* pathway is specialized for form information and it starts initially with small receptive fields, which focus on small features (e.g., the orientation of lines), and gradually expands to larger receptive fields, which eventually span the whole object. The dorsal *where* pathway is specialized for motion information: it starts with motion information in small receptive fields (e.g., motion of lines) and expands to higher-level receptive fields (e.g., overall direction of the object). While the motion pathway seems to be most relevant for biological motion understanding, Giese and Poggio's model proposes that the form pathway also plays an important role by using snapshots of poses to identify the types of action. Support for the role of the form pathway comes from studies that show that point-light display recognition is view dependent (e.g., changes in depth reduce recognition), which is a property of the ventral pathway (Bülthoff et al. 1998). Furthermore, some patients with damage to the motion pathway are still able to recognize biological motion, which suggests some role for the form pathway (McLeod et al. 1996). The motion pathway, however, is still the dominant system for recognizing point-light display motion, and fMRI work has found that the distinction between biological and nonbiological motion typically occurs in higher areas of the dorsal pathway (Decety and Grèzes 1989). In humans, it seems that multiple parallel pathways are involved in action understanding.

While biological motion is an important feature for identifying animate entities, it is also possible to identify these entities by virtue of their interaction with other objects. Evidence for this in infants comes from a study by Woodward (1998), who showed scenes where an arm grabbed one of two toys (Figure 16.5). During the test, the objects were switched and the arm grabbed either the same toy (old goal) or the new toy (new goal). Woodward found that five- to six-month-old infants were surprised when the arm went for the new goal over the old goal. This suggests that they think that the arm's motion is guided by a mental goal or preference for the old goal. Infants were surprised by an arm reaching for a new goal, but not by a mechanical claw, which suggests that they only attribute goals to the arm. Luo and Baillargeon (2005) have shown that a similar preference is present for a box when it moves in a self-propelled manner. They presented five-month-old infants a familiarization scene where a box moved toward a preferred object; later at test, they showed the infants the box moving toward the same object (old goal) or a novel object (new goal). They found a difference between the preference for the old goal and new goal was larger when the familiarization scene has two objects as opposed to one object. When there are two objects, the movement of the box shows that it prefers that object over the other one. When there is only one

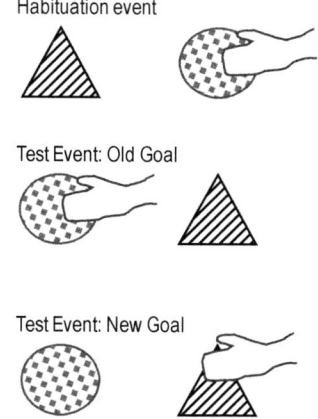

Figure 16.5 Depiction of goal recognition in infants, based on a study by Woodward (1998). Top: A child is shown an arm reaching out to grab an object (the circle). The position of the objects is then altered: the arm reaches for the same object (middle) versus a new object (bottom). When a human arm was shown, infants expected the arm to reach toward the circle. This expectation was not present when a mechanical arm was involved.

object, the movement toward that object does not show a strong preference, because there is only one option. Luo (2011) showed that even three-month-old infants demonstrate this preference, even when there was only one object at familiarization, as long as the box showed its preference for the object by moving toward the goal object in two different locations on the screen. In contrast to a self-propelled box, infants were surprised when an inert box reversed direction spontaneously, remained stationary when hit or pulled, or remained stable when released without support. One reason self-propelled motion is important is that the behavior of the box cannot be predicted based on physical constraints like gravity or inertia, and it is this unpredictability that triggers the assignment of mental states like goals.

The goal-directedness of human behavior was used to look at false belief behavior by Onishi and Baillargeon (2005). Here, 15-month-old children were first shown a familiarization event where a toy was hidden in a green box by a person in front of the infant. Then the scene was changed such that either the person's belief about the object in the green box would be false (e.g., the toy was moved secretly to the yellow box) or true (e.g., the toy was moved to the yellow box while the person watched). Later at test, the children observed a person reaching into the green box or yellow box. When that person saw the toy move into the yellow box, the children expected the person to reach for the yellow box and were surprised when that person reached into the green box. When the person did not see the toy move, the children were surprised when the person reached toward the yellow box, because they knew that the person had a false belief that the toy was in the green box. To show these differences, the infant cannot just record where the objects are located. They must also track where the person thinks these objects are located and how those beliefs, true or false, guide their reaching behavior. This is a nonlinguistic theory of mind task which shows that early in development, children can track the beliefs of others and use these beliefs to predict their behavior. Since ITL systems must interact with humans whose behaviors are driven by their beliefs (e.g., a teacher searches for a knife in a drawer, because that is where she thinks it is), it would be useful for these systems to have the ability to infer beliefs in the way that children seem to be able to do.

These abilities of infants to identify primitive mental states of others can support the understanding of more complex social motivations. For example, when six-month-old infants were shown a square trying to get up a hill, and it was helped up the hill by a triangle in one scene and hindered by a circle in another scene, they then preferred the triangle to the circle later, thus demonstrating that they understood that the triangle was helping the square reach its goal (Hamlin et al. 2007). By 18 months of age, toddlers and human-encultured chimpanzees are able to identify the ultimate goals of adult humans (e.g., putting books into a book case) and perform actions that help the adult to achieve these goals (e.g., opening the book case door) (Warneken and Tomasello 2006). Such helping behavior involves prediction of the goal, because the goal has not yet been achieved. In addition, the helper must identify objects that they can manipulate in the environment and predict whether these changes will help in achieving the goal. Since ITL systems are attempting to help humans in their tasks, these humanlike prediction abilities would enhance their interactions.

When a child comes to an ITL task, they have a model of the constraints that inanimate physical objects have in the world. In addition, they know that animate entities move in ways that reflect their goals and beliefs, and these biological entities can have multiple parts that work in concert to perform various actions. Furthermore, they can learn about behaviors in various contexts and use that knowledge to identify ways to help. While it is possible that children have a range of different modules which allow them to exhibit these capabilities, the range and flexibility of these abilities suggests that they are not separate isolated modules but rather part of a task-general system that integrates physical constraints, mental states, and learned regularities into a single system that attempts to predict behavior. Below, I propose that such a system will be useful for ITL.

A Developmentally Motivated World Prediction Model for ITL

How can ITL systems incorporate this rich database of knowledge that children seem to possess? In this section, I will suggest that incorporating a world prediction model can give ITL systems some of these abilities. To see how such a model might work, let us consider the following example: a robot learns from a human teacher how to cut a carrot. The robot has a carrot and a knife, and the human wants to show the robot how to cut the carrot by pantomiming a cutting action using her hand. For the robot to understand this action, it must be able to map the back-and-forth motion of the human hand in space with no carrot to its own hand with the knife. It would also need to know that the back-and-forth motion of the knife is the second component of a cutting sequence.

1. CONTACT: make contact between knife and carrot.
2. SAW-MOTION: move knife back and forth on carrot.
3. SPLIT: continue motion until carrot is in two pieces.

This task presents several challenges, from segmenting the event to understanding the pantomimed hand action. In pretend play studies, children from around the age of two years seem to understand pantomime actions of others and quickly generalize this knowledge to their own actions (Rakoczy et al. 2004; Rakoczy and Tomasello 2006; Rakoczy 2008). It would be useful for an ITL system to have a similar ability to understand pantomimed actions.

Human predictive knowledge is very detailed and context specific, so the world prediction model makes predictions at a fine time granularity. A model by Reynolds et al. (2007) does perceptual prediction to segment events; their model segmented sequences of routine actions encoded as point-light displays (as in Figure 16.4). They used a recurrent neural network that attempted to predict the next state of the points, using the previous state, and the error in prediction was used to update the model's weights so that it encoded the knowledge about the transitions between states. In addition to using error to learn the transitions, the model had an additional gating network that used points of large prediction error to identify event boundaries. To adapt this for ITL learning, we can assume that the system is attempting to use the present state of the world to predict the next state of the world (Figure 16.6). We will assume that the world state encodes static elements of the scene and the state of the robot's body that are derived from sensors and effectors (see Figure 1.1 in Mitchell et al., this volume). In addition, let us assume that the system tracks objects in the visual input, so that the world state is not a pixel-like representation but is instead based on static and motion properties of objects (e.g., shape, velocity, acceleration). Objects are any moveable element including inanimate artifacts (e.g., knife) as well as animate entities and their body parts (e.g., hand). The world state at time t is the input (bottom box, Figure 16.6); its activation is spread through internal layers with recurrent connections to help in the learning of sequential regularities; and the model generates a prediction for time t+1 at its output (second box from top, Figure 16.6). The actual world state at t+1 is the target (top box, Figure 16.6) and the difference between the target and the predicted world state is the error. The error passed back to the internal layers is used to learn representations that encode the internal parts of events. Importantly, the input state of the world at time t and t+1 is different from the knowledge of the world inside of the robot, which is encoded in the internal

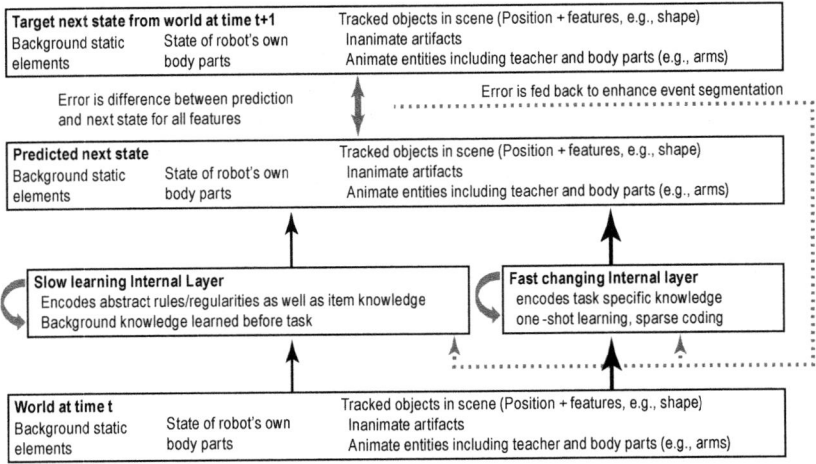

Figure 16.6   World-state prediction model.

layers. If we take the example of learning the carrot cutting sequence from a human demonstration, the model will be attempting to predict the state of the world multiple times a second. After it sees the back-and-forth movements of the knife (SAW-MOTION), the system should be able to predict that motion. But the system will not be able to predict that the carrot will split into two objects, and the large prediction error will be evidence for segmenting the SAW-MOTION component from the SPLIT component of the cutting sequence.

In ITL, there are aspects of the task that are specific to the particular situation (e.g., the knife is in the drawer) and aspects that reflect long-term knowledge about similar tasks (e.g., hands tend to reach for knives, but knives do not reach for hands). To model these different aspects of world knowledge, the world-state prediction model has two internal layers. The layer on the right side in Figure 16.6 is called the fast-changing internal layer: it has a high learning rate, which allows it to learn task-relevant temporary knowledge quickly. The layer on the left side of Figure 16.6 is called the slow-learning internal layer: it has a lower learning rate, which allows it to learn slowly regularities that are consistent across the whole of its previous input experiences. It has been argued that humans have a similar distinction in their brains between slow cortical learning and fast one-shot hippocampal learning (McClelland et al. 1995). We assume that the slow-learning layer develops its knowledge gradually based on many years of visual experience in the world. It would first learn to predict transitions based on physical constraints (Spelke et al. 1992). Given that the motion of self-moving objects is not fully predictable from physical constraints (Luo 2011), the prediction error that is generated will cause the model to learn separate internal representations for the actions of animate entities and these representations will eventually come to encode mental state features (e.g., goals, preferences, intentions), because these are useful for predicting the motion of these entities (Woodward 1998). After the slow-learning network has encoded background knowledge, then in an ITL task, the fast-learning network will only encode the aspects of the scene which are not predictable from the slow-learning network. For example, the slow-learning network has experience with hands reaching for objects, so it does not store the low-level motion of the hand. Because it has less experience with in-animate objects moving against another inanimate object (as when the knife is cutting the carrot), the prediction error generated causes it to be richly encoded in the fast-learning system. Thus there is a division of labor in how knowledge is distributed to the fast and slow subnetworks, as can be seen in connectionist models of language (Chang 2002; Janciauskas and Chang 2018).

Can this world-state prediction model be useful in understanding how an ITL robot would be able to understand a pantomimed cutting demonstration? Let us assume that the robot has grasped the carrot and placed the knife in contact with it, but is unsure how to cut the carrot so the human pantomimes the back-and-forth motion using her hand. The world prediction system cannot initially predict the back-and-forth hand motion based on its general world knowledge; hence the large prediction error that is created causes this motion sequence to be encoded in the fast-changing internal layer of the model. The world state is encoded in terms of the motion of objects independent of their features (e.g., shape, color), so the predicted sequence of states encoded by the fast-changing layer can be used to guide the motion of the robot's hand just by mapping the object pointer for the teacher's hand to the robot's motor control system (allowing it to exhibit childlike generalization, Rakoczy et al. 2004). Finally the large change of state that takes place when the carrot is split into two is used to segment the event, and this might cause the robot to stop and evaluate what to do next. Thus the world-state prediction model helps to explain how a robot with no knowledge about cutting events could learn the back-and-forth motion component of cutting sequences from interaction with a human teacher.

It is clear that a world-state prediction model would only be useful in an ITL robot if it was tightly integrated with other systems for planning and interaction (see Salvucci et al., this volume). It suggests, however, ways to apply previously learned background knowledge from non-ITL situations to support event segmentation and learning from visually taught events. Although the model is described within an error-based learning recurrent network, other algorithms could also be used. What is critical is that predictions are generated at each point in time, so that an error signal can be generated which identifies which aspects of the scene are unexpected. It is also important that the system learn its internal representations based on prediction mismatch, so that it has more extensive internal representations for less predictable entities like goal-driven agents. Furthermore, the prediction error is itself a signal that can be used to identify points where human feedback is needed or to segment events.

A growing body of work in psychology argues that prediction is taking place all the time, particularly in language processing (Altmann and Kamide 1999). Young children seem to generate linguistic expectations automatically (Lew-Williams and Fernald 2007; Mani and Huettig 2012). Furthermore, changes that take place in the language representations of adult language users can be explained as prediction error-based learning (Chang et al. 2006; Dell and Chang 2014). If robots are doing a similar type of prediction about the nonlinguistic world and constantly updating their knowledge within a system like the world-state prediction system, then they would have rich moment-by-moment predictions and error signals, which could be used to learn new tasks. Thus, although the goal of ITL is not to model the development of human abilities in robots (i.e., developmental robotics) (Cangelosi et al. 2015), incorporating a module that is motivated by detailed prediction abilities of humans into these systems could enhance the ITL capabilities of artificial agents.

Conclusion

Humans engage in ITL based on multiple years of experience watching animate and inanimate entities interact. Children develop a model of how inanimate entities move based on physical constraints and how animate entities move based on inferences about their mental states. They can identify temporary goals in particular contexts and use long-term knowledge to understand the actions of others. Although ITL systems implement some of these abilities in separate modules, research on humans suggests that these abilities are the result of a system that is constantly involved in making predictions and adapting these predictions in response to experience (learning). Furthermore, prediction error may be a useful diagnostic signal for ITL systems. Given that humans expect these abilities when they teach and interact with other humans, it is likely that humans will prefer to interact with ITL systems that have a rich internal predictive model.

Acknowledgments

Franklin Chang is a member of the International Centre for Language and Communicative Development (LuCiD) at the University of Liverpool and the support of the Economic and Social Research Council [ES/L008955/1] is gratefully acknowledged.

17

Characteristics of the Learning Problem in Situated Interactive Task Learning

John E. Laird, Shiwali Mohan, James Kirk, and Aaron Mininger

Abstract

In most learning problems, a single type of task knowledge is learned using a single specialized learning algorithm designed and optimized for that specific type of knowledge and the environment in which it is learned. In contrast, interactive task learning (ITL) involves learning *all* types of task knowledge where such specialization is impossible. This chapter describes these characteristics of the ITL learning problem, which distinguish it from other learning problems, and examines how those characteristics influence the underlying learning algorithms. Throughout our discussion, the Rosie agent is used as an example of an ITL agent that can learn many tasks in a variety of domains. The distinguishing characteristics explored include learning across different domains, learning diverse task knowledge, interactivity in learning, the situated aspects of learning, and how an ITL agent can exploit multiple data sources. Learning approaches are then discussed that can be used in ITL from the perspective of how they address the unique challenges of ITL.

Introduction

Learning research in artificial intelligence (AI) and cognitive science usually focuses on a singular learning problem for a single task where there is a single target type of knowledge to be learned. Much of the current work in statistical machine learning, for example, can be formulated as a supervised learning problem, where the goal is to learn a classification of a set of prespecified data from labeled examples. Outside of research on cognitive architecture and reinforcement learning, learning is rarely coupled to task performance: The learning problem is often studied in isolation, with a focus on a single learning algorithm, and in a noninteractive fashion, where the learner has little control over what training it receives.

Our interest is not in learning within a single task, or even a class of tasks, but actually learning new tasks from scratch through interactive task learning (ITL) (Laird et al. 2017a). Mitchell et al. (this volume) define ITL as "any process by which an agent (A) improves its performance (P) on some task (T) through experience (E), when E consists of a series of sensing, effecting, and communicating interactions between A, its world, and crucially other agents in the world." ITL stands in stark contrast to approaches that are specialized to a single task. Not only does it utilize a collection of learning algorithms for a variety of learning problems, ITL can occur in many *different domains,* whose very characteristics directly influence the appropriateness of different learning algorithms. ITL also involves learning the complete formulation of a task; this entails learning *multiple types of knowledge,* each of which can depend on the specific task formulation. As defined above, ITL involves an agent learning in real-time through an *interaction* with an instructor. Although there are cases where an instructor can interactively teach an agent about hypothetical situations, our focus in this chapter is on cases where the instructor and agent are *situated* in a shared environment. Because of the flexibility of how an instructor can interact with a learning agent (e.g., a person can demonstrate or give direct instructions) as well as an agent's prior knowledge, *multiple sources of knowledge* can be available during the learning of a task. Some of these characteristics can simplify learning while others may complicate it, making the development of learning algorithms for situated ITL a challenging, but exciting research problem.

Here, we explore the characteristics of situated ITL that distinguish it from other learning problems and discuss how those characteristics affect the agent's learning. We consider each characteristic, independently identifying the impact it has on the ITL problem as defined by Mitchell et al. (this volume). We then consider these characteristics together and discuss potential approaches to learning. When appropriate, we provide examples from our own research in creating an ITL agent named Rosie (Mohan et al. 2012), which was built using the Soar cognitive architecture (Laird 2012). Before diving into the defining characteristics of situated ITL, we briefly introduce Rosie.

Rosie: An Interactive Task Learning Agent

Rosie (RObotic Soar Instructable Entity) is a physically embodied agent that learns new tasks through situated interactive instruction using restricted natural language and limited demonstration. It can learn over forty simple games and puzzles (such as Tic-Tac-Toe or Tower of Hanoi), process-oriented tasks (such as cooking or storing), and navigation tasks (such as delivering or fetching an object). It not only learns various aspects of the tasks (e.g., goals, valid actions, failure states, and policies) but also learns new concepts used in the tasks (e.g., relationships, labels, and functions). Additionally, it learns how to refer to these components of tasks by associating them with words used to describe them. Whenever Rosie encounters a new word, it initiates a new interaction to learn the meaning of the word. It can learn new nouns (names and shapes), adjectives (colors and sizes), prepositions (spatial relations), comparators (quantities), functions (task-specific value calculations), and verbs (actions and tasks). After Rosie learns a new word, the human can use it freely in future instructions ("move the *green* block to the pantry") and the agent can use it when responding to questions ("that is a large *green* block").

Rosie is preprogrammed with only limited initial domain knowledge: action knowledge for performing its primitive actions (e.g., picking up a block, putting down a block, moving down a hallway), feature-space knowledge of object attributes (e.g., the fact that objects have a color, a size, and a shape), and primitive spatial relations about the alignment of objects. As Rosie learns, it creates new structures in its long-term semantic memory that encode the task definition. It interprets this knowledge to perform that task, leading it to learn efficient procedural knowledge (encoded in rules in Soar).

Following the acquisition of the task, the agent uses its innate problem-solving and planning capabilities to play the games or perform the tasks by interacting with its world. Depending on the task, it can also learn policy knowledge to perform the task well. Policy knowledge is used to select actions in each state and can be learned from the agent's own experience, from the specification of a procedure, or from additional advice it receives through instruction (such as heuristics for a game). The knowledge it learns in one task transfers to shared concepts, actions, and constraints in new tasks.

Learning across Different Domains

There is a wide range of domains to which ITL can be applied, including robotics (Mininger and Laird 2018), personal assistants (Azaria et al. 2016), intelligent tutoring authoring (Li et al. 2015), game playing (Hinrichs and Forbus 2013a), constructive agents and virtual humans (Pew and Mavor 1998; Zacharias et al. 2008), and cognitive science research. These domains vary wildly in complexity and characteristics, and these characteristics can significantly impact what needs to be learned as well as the difficulty of the learning problem. This, in turn, informs the appropriateness of different learning algorithms. Some of these characteristics include whether the environment is partially or fully observable, whether the environment is discrete or continuous, whether actions are deterministic or stochastic, whether the environment has complex dynamics, and the extent of uncertainty in sensing and acting in the environment. In general, the more complex and uncertain a domain is, the more difficult it is for an agent to learn new tasks quickly. Current ITL agents usually focus on tasks within a single domain, and designers tailor their learning algorithms to that domain. Whether there are fundamental representations and learning algorithms that will work across all domains remains an open research question, as is the degree to which certain domains require specialized learning approaches. Below we provide examples of two different domains and how their characteristics affect the learning problem.

Virtual Personal Assistants

In the virtual personal assistant domain, an agent performs various tasks that aid in a person's daily activities, such as answering questions, managing a schedule, or handling messages. For these tasks, the agent's actions usually involve interacting with people and software interfaces. There is low uncertainty in its perceptions and actions, and often a discrete, symbolic representation can be used effectively. This can make the learning problem easier and allow more aggressive and data-poor learning mechanisms to be used. The agent can learn in large steps and extract higher-quality information from each example. An example of such an agent is PLOW (Allen et al. 2007), which operates in a web browser and can learn how to answer new types of questions through interaction. PLOW leverages the low uncertainty and direct access to the browser's state to learn new tasks through a single training example, by mapping the linguistic instructions onto the procedural trace of actions performed in the browser. A second example is the LIA system (Azaria et al. 2016), which learns new action sequences for apps on mobile devices using restricted natural language.

Real-World Robotic Domains

A major focus of ITL is in the highly complex domain of real-world robotics. Such environments are often partially observable, involve high-dimensional and continuous data, and contain high uncertainty in both perception and action. It is difficult for an agent to build an accurate and precise model of its environment as well as to predict the results of its own actions. As a result, high levels of uncertainty can force an agent to be conservative in its reasoning and its learning, requiring many examples before any conclusions can be drawn. An incremental approach may be required whereby the agent learns simpler/smaller aspects of the task before learning more complex structures. Unlike the virtual assistant domain, the agent may need to learn not just what an action does when it succeeds, but also contingency plans if an action fails.

In our ITL agent, Rosie, we use the same symbolic state representation formalism across three different robotic embodiments, which allows us to use the same learning algorithms. However, this requires perception and action systems that are sufficiently accurate and precise, so that Rosie does not explicitly need to represent or reason about perceptual or motor uncertainty. In addition, we constrain the environment and the actions so that Rosie has a relatively stable view of the world and can use its memory to pursue tasks effectively. An underlying hypothesis is that these characteristics will be true for other domains in which Rosie will learn new tasks. An interesting question is whether other constraints that arise from the characteristics of ITL, such as the need to learn very quickly, will force ITL systems to make similar assumptions about their perceptual and motor systems as well as the stability of their environments.

For partially observable domains, the agent must use, and possibly learn, strategies for maintaining internal representations of the environment and gathering information. As an example, Rosie can learn some simple strategies for finding objects in the mobile robot domain, such as representing where a type of object is stored and then going there to look for one (Mininger and Laird 2016). Agents in a real-world robotic domain might also need to know, and possibly learn, how to handle complex physical dynamics, in contrast to the above virtual assistants.

Learning new robotic tasks that can generalize to novel situations in a few interactive examples is a difficult problem. Often the environment is constrained to make this feasible. For example, in the task learning mobile robot by Meriçli et al. (2014), constraining actions to driving and speech and using visual fiducial systems to recognize landmarks enabled the agents to learn procedural task specifications in a single example. In further work by Chao et al. (2011), agents learn concepts in task goals (such as object labels and spatial relations) from low-level sensor data. This informs Chao et al.'s choice of using unsupervised learning methods and Bayesian inference and does require more examples (but still not hundreds or thousands).

Learning Diverse Task Knowledge

An important feature of ITL is that it does not involve learning just one type of knowledge for which a single learning algorithm can be deployed and optimized. Tasks can be formulated in different ways, and the choice of task formulation impacts which types of knowledge need to be learned. Thus, an ITL agent must be capable of learning diverse types of knowledge. Below we distinguish among three different formulations of tasks: problem space, procedure, and optimization (although a given task can include aspects of any of these) and the types of knowledge each requires. This analysis is independent of the underlying knowledge representations, such as production rules, neural networks, semantic networks, or partially observable Markov decision processes. Instead, our analysis focuses on the types of knowledge required to learn a task, regardless of how it is represented.

Knowledge Diversity in Problem Space Formulations

To determine what type of knowledge needs to be learned for a task, we start with Newell's Problem Space Hypothesis, which states: "The fundamental

organizational unit of all human goal-oriented symbolic activity is the problem space" (Newell 1991:696). Newell informally defines problem space and
problem as follows:

> Problem Space: A problem space consists of a set of symbolic structures (the states
> of the space) and a set of operators over the space. Each operator takes a state as
> input and produces a state as output, although there may be other inputs and out-
> puts as well. The operators may be partial, i.e., not defined for all states. Sequences
> of operators define paths that thread their way through sequences of states.
>
> Problem: A problem in a problem space consists of a set of initial states, a set
> of goal states, and a set of path constraints. The problem is to find a path through
> the space that starts at any initial state, passes only along paths that satisfy the
> path constraints, and ends at any goal state.

The strength of problem spaces is in formulating tasks where an agent learns a
goal and must search through the space of states to achieve it. Classic examples
from AI are many types of reasoning tasks as well as assembly tasks, puzzles,
and games. For these tasks, the problem space formulation provides us with
a starting point to discuss the types of knowledge that should be learned by
an ITL agent: descriptions of initial states, goal states to be achieved, illegal states, constraints on sequences of operators, and operators that transform
states. The assumption is that once the agent acquires this "first principle"
knowledge, it can attempt a problem by searching for a solution (either internally or externally) through selecting and applying operators until it achieves
the goal. Rather than describing a specific procedure or generalized policy, the
teacher communicates the component task concepts. Once an ITL agent has
learned the problem space, it knows enough to attempt and possibly solve any
problem (as defined by an initial state and a set of goal states) in that problem
space. This formulation is especially useful when the instructor does not know
how to solve the task, or when it is difficult or time consuming to communicate
the solution.

Although there are many different concepts that define a task, most of them
involve state descriptions that are associated with some aspect of the formulation of a task, such as goal completion or operator preconditions. These state
descriptions usually involve conjunctions of conditions over objects and values in the world, such as unary features of objects or relations between objects. Depending on the domains and tasks, these features can be combinations
of continuous and discrete values. Within AI and cognitive science, there are
many different ways to represent this type of information, and there are many
different learning methods for learning these representations, although tasks
often require rich representations beyond simple feature matching. For example, for Rosie, the breadth of games and puzzles it currently learns (over 30, including games such as *Othello* as well as *Missionaries and Cannibals*) requires
concepts that include conjunctions of unary features (*red, large, square*), nu-
meric functions (*number of, sum*), comparators (*less than, more than*), words,

numbers, and spatial relations (*left of, below, linear*) (Kirk and Laird 2016). In terms of learning algorithms, constraints (and assumptions) that arise from different characteristics of ITL usually limit them to be online, incremental approaches to concept learning. Statistical techniques that require large numbers of examples can be useful for learning perceptual classifiers, such as for colors and shapes, which the agent would learn before task learning is attempted.

In addition to learning associations between state descriptions and goals, failure states, and operators, task learning can also involve learning hierarchical goal structures, where the operator at one level becomes a new task to solve at a lower level. Lower-level tasks can be any of the types of tasks described in this section: performing a new action or even simpler types of tasks, such as classification or categorization. For example, in learning an assembly task, Rosie may be instructed using a new verb, such as in "*store* the box in the pantry." This will lead to an interaction in which Rosie learns the "store" task (Mohan and Laird 2014). In terms of learning new classifications, Rosie might not know the word "red" when instructed to "pick up the *red* block." In this case, Rosie will engage the instructor to show it examples of red objects, ultimately learning a classifier in the color spectrum that Rosie can use to identify red objects.

Another type of knowledge that can be valuable for a problem space task is heuristic or policy knowledge. This type of knowledge does not define the task, but can be critical for helping the agent solve the task.

Knowledge Diversity in Procedure Formulations

Although the problem space formulation works well for many tasks, in many others it is more efficient for the agent to formulate the task as learning a procedure. For this, however, the agent might not have an explicit description of the goal or be capable of achieving the goal on its own without the specified procedure.

Cooking is a good example of a task where procedure formulations are used. Consider trying to bake a key lime four-layer cake. Few of us are able to search through the "baking" problem space and determine all of the steps required to make such a cake, in part because we do not have an operational definition of what the final cake should actually be. Furthermore, the branching factor for baking is immense, and most of us do not have good models for what many of the baking actions actually do at a chemical level. However, most of us can follow a recipe and learn to bake a specific type of cake (or at least learn to follow the recipe on the back of a box). For these types of tasks, the agent might not need to learn detailed descriptions of goals or failure states, because it can assume that the goal is achieved if the correct procedure is performed. Furthermore, the agent might already know the actions that must be performed (e.g., how to break eggs and mix them with a specified amount of sugar). However, the agent does not know when to do each action or which ingredients to use. Instead, the agent is learning a task that is associated with a particular sequence of actions that needs to be performed to accomplish the task. Thus, in a procedure formulation, instead of having to learn descriptions of goals and failure states, the agent needs to learn a procedure. The exact form of that procedure can vary. It can be a linear set of instructions, as in a recipe. It can also be similar to a program, with conditionals, while loops, etc. Alternatively, it can be the value function for a policy, which consists of mappings from states to actions. Those mappings can be probabilistic functions, as is learned in reinforcement learning, or deterministic, as in finite-state machines and rule-based systems.

Tasks formulated as procedures often require learning how to perform complex actions that are specified in the procedure, such as "fold together beaten egg whites with sugar." Learning such complex actions can be formulated as either a problem space or procedure, so that sometimes a goal and appropriate actions may be specified, or sometimes another procedure. Furthermore, a given task may involve knowledge from both formulations, where the instructor provides the goal and actions as well as helps the agent find the solution by giving hints and directions while the agent is attempting to perform the task.

One disadvantage of using a procedure formulation that does not include a problem space formulation is that the agent is bound to the procedure. If there are even small variations to the original task, it can be problematic for the agent to apply the learned procedure, and it can be difficult to recover from interruptions or mistakes because the agent does not know why it is performing each action or whether alternative actions will still achieve the goal.

Knowledge Diversity in Optimization Formulations

The final task formulation type we consider is where the goal is to maximize or minimize some quantity. The traveling salesman problem (TSP) is a classic example, where the goal is to minimize the total distance traveled while visiting every city (node) in a map (graph). In this example, there is both a goal (visit every city) and a value to be optimized (the total distance traveled). Optimization problems can include all knowledge of the problem space formulation, but add the optimization function, which can be specified as a function over the state or, as in TSP, as a function over the path.

Although formulating the TSP and similar tasks as optimization problems might seem like the obvious thing to do, there can be a difficulty. If the agent takes the optimization formulation "seriously," then finding an exact solution to the TSP is NP-hard. That might be necessary in some situations, but in practice, approximations are sufficient, and often the optimization function is used as an evaluation function to guide a greedy search for a solution. This leads to a situation where the task is formulated as a problem space, but with additional knowledge that must be used to guide the search.

Learning Is Interactive

With interactive learning, the human instructor provides information, piece by piece, as the agent needs it. This learning paradigm distributes the onus of learning to both the learner and the instructor. The learner agent can request the information it needs to perform a task as well as clarification and disambiguation of previous instructions. The human instructor can monitor the agent, interrupt it to provide instruction when it seems warranted, and verify the status of agent learning. Thus, with interactive learning, relevant information is not provided as a massive data set or a pile of written documents that the agent must process without help from an instructor. Instead, in interactive learning, the problem of learning a task is deconstructed into subproblems of learning components (e.g., goals, parameters, policies) that can be learned incrementally. Below we analyze important properties of interactive instruction and how it affects the learning of complex tasks.

Interaction Occurs in Real Time

One of the important advantages of using interaction is that the human instructor can watch the learner agent interpret and attempt to perform instructions. This provides real-time feedback to the instructor as to how the agent interprets and executes instructions, and permits the instructor to adapt the training strategy to the needs of the learner. Thus, it is critical for learning mechanisms to produce immediate, useable results that the agent can use to attempt the problem. Furthermore, applying instruction to the problem in real-time facilitates discovery of other related learning problems for which the agent can request training. To exploit these advantages, learning mechanisms must incrementally acquire new knowledge and assimilate it with existing long-term knowledge structures while the agent is behaving. It further implies that during learning, the agent does not need to reevaluate its previous experience or knowledge, two characteristics of batch learning mechanisms. The broader point is that the agent cannot require many examples of a concept before doing any learning. This is not to prohibit corrections or modifications to learned concepts as more data becomes available, just that some learning must occur immediately. It also implies that the learning mechanism is computationally fast in producing results that are quickly incorporated within the agent, so that the agent can use the knowledge immediately. It is hard to quantify exactly how "fast" the agent needs to be in terms of execution time, except that it cannot take so long to respond that it annoys the instructor or breaks the flow of the interaction. Rosie uses online and incremental learning mechanisms that are fast enough so there is no visible delay when it is processing instructions, which includes language understanding, interpretation of internal structures, and compiling of knowledge into high-performance rules.

Available Data Are Sparse and High Quality

Human time is costly, and therefore numerous and repetitive interactions with the agent are undesirable. Consequently, the agent must learn from *small data*—a few highly specific training instances. Although the learning data are sparse, it can be assumed that the instructor is well informed on the task and well intentioned in terms of helping the agent learn, and therefore generates high-quality, correct/accurate data for the agent. These characteristics push toward learning approaches where the agent must bring to bear its prior experiences and knowledge, commonsense reasoning, and causal deduction to learn maximally from few instructions. This is in contrast to alternatives such as noninteractive crowd-sourced learning, where the quality of instruction can be an issue, so that more conservative approaches to learning are required.

Rosie employs aggressive generalization mechanisms, such as explanation-based learning (DeJong and Mooney 1986) and version-space induction (Mitchell 1977), to learn procedural and semantic aspects of tasks. These mechanisms allow Rosie to produce actionable task representations even from very few examples. Currently, Rosie assumes that every instruction provided by the instructor is correct. This may not be true in partially observable, noisy environments or when the instructor is a novice. Ongoing research in our lab is looking at how learned knowledge can be corrected upon receiving feedback from the environment.

Structure of Instruction Is Flexible

In interactive learning, an instructor can be freed from using a specific ordering to teach the agent new words and concepts. Often in human-controlled learning (e.g., supervised learning), the instructor must attempt to build and maintain an internal model of what the agent knows and does not know to give it good examples. This is especially challenging when the agent is dynamically learning a variety of concepts (perceptual, spatial, task knowledge in Rosie) from real-world data. In contrast, with interactive learning, the instructor can rely on the agent to initiate an interaction when needed. This approach can speed instruction by eliminating the need for the instructor to carefully structure the interaction or repeatedly check with the agent to ensure it has completely learned a concept. The agent can actively seek examples of concepts that are hard to learn and avoid asking for multiple examples of easily acquired concepts. The instructor can take initiative in presenting interesting examples to the agent that it might have overlooked, refining the agent's learning.

Rosie implements an interaction model that does not impose a strict order on how it is taught new concepts. To support this, Rosie uses the content of the instructions to determine which type of concept is being taught. For example, "if …, then you win" will lead to a goal concept being learned. Additionally, it employs a relational, compositional task representation, which provides flexibility in how the instruction is structured and who takes the onus of learning. The instructor can choose to control how and when Rosie learns. The instructor can teach basic concepts before teaching complex concepts that require knowledge of the basic concepts. However, the instructor may not know or remember the state of Rosie's knowledge. In situations when the agent is learning complex concepts but lacks required knowledge of basic concepts, it will take the initiative and guide the interaction to acquire the relevant basic concepts first. For example, if the concept "empty" is used when teaching Sudoku but the agent does not yet know it, the agent asks the instructor to define the term so it can acquire the concept. Then the instructor can say: "If a square does not contain a value, then it is empty."

Learning Is Situated

Situated ITL occurs in the environment in which the learner agent is attempting to perform a task. Any demonstration will occur in that environment, and instructions will often directly refer to the current situation, although some instruction may refer to past or hypothetical situations. Here we discuss the ways in which a shared, situated environment impacts both the instructor and the learner.

Instructor and Agent Share the Same Environment

By having a shared environment, it is much easier for the participants to build up common ground (Clark 1996), in terms of a shared understanding of the objects, relations, and actions of the environments, and thus decrease the chance of ambiguity. Both participants can use deictic references and concrete descriptions as well as nonlinguistic actions, such as pointing. Within this shared environment, the instructor can focus the agent's attention on aspects of the situation that are relevant to the current problem. When difficulties arise, the agent can use language to describe its knowledge about the situation, its plans, and even reasons for performing actions that are easy for the instructor to understand and evaluate, thus avoiding the necessity of having to "open up" the agent to examine (and try to understand) internal data structures that may be quite distant from the human's representation of the task.

If the agent identifies an instruction as ambiguous, it can immediately ask for clarification. This can make the interaction much easier for both the instructor and agent, and as in the prior section, this allows the agent to be aggressive in learning. To continually learn new tasks (that may be composed of previously taught tasks), the agent must expand its capability to understand references (e.g., to the goals, actions, objects, concepts) that are relevant to the task being learned. This growing common ground ensures that the instructor and the agent can continue to interact using the newly learned knowledge.

Instruction Can Be Performed Immediately in the Environment

One advantage of situated instruction is that as soon as the agent learns some aspect of a task, it can attempt it in the world. If there is failure, it can ask questions and request training to improve its task performance. Conversely, instructions provided by the human instructor are immediately applicable to the relevant situation. This simplifies instruction generation for the human instructor as well as the learning problem for the agent. The instructor can focus on what they are observing and guide the agent to solve the specific instance they both are situated in without having to reason about how the task can be solved in general. The agent, on the other hand, does not have to reason about how and when to leverage the instruction. It can apply the instruction immediately and observe how it affects task performance, which guides learning.

In addition, when the agent attempts to solve a problem, it does not need internal models of the effects of all of its actions; it just needs the ability to execute them. Once it knows the task, it can attempt it by taking action in the world, possibly through exploration. If a hypothetical problem is given to an agent, it must have such models to internally search for a solution. When Rosie is learning new tasks while embodied in a mobile robot, it does not need detailed models of how the robot moves through the world; it can employ abstract models of moving from location to location. This simplifies the learning problem because only the goals need to be learned from instruction.

Learning Is Multimodal

In many learning problems, the source of available knowledge is structured and singular, such as a large database of labeled images. However, with interactive task learning, there are many potential sources of knowledge that arise from the different ways the human can interact with the learning agent and from the agent's access to prior knowledge. Possibly the simplest interaction a teacher can provide is a positive or negative reward as an evaluation of the agent's moment-to-moment task performance. Below we describe additional modalities for learning in ITL.

Language Can Be Used to Instruct Task Knowledge

When language is available, the human can explicitly define and describe different aspects of a task (goals, actions, task-specific concepts, and heuristics) through natural language (Crangle and Suppes 1994). Language can also be used for drawing attention to important aspects of the task as well as to label important collections (and parts) of objects and actions. The human can also lead the agent through the task by providing step-by-step instructions.

Language also makes it possible for the human to answer questions posed by the agent. At the extreme, unrestricted natural language provides a rich, robust vehicle. However, its use is currently beyond existing state-of-the-art methods. Thus, various forms of restricted languages are popular, which usually limit vocabulary, syntactic structures, and semantic content. From a learning perspective, language can provide direct definitions of many of the components of a task, so that the agent is essentially compiling natural language into internal structures that it can later interpret to direct its behavior. This is the approach that Rosie and other agents have taken (Hinrichs and Forbus 2013a). Important challenges are achieving generality and robustness in language understanding and "compiling" the knowledge into an efficient executable representation in real time.

Demonstration and Imitation Can Provide Examples of Task Performance

In combination with language, an instructor can use pointing and gestures to communicate important objects or actions. Without language, the human can *demonstrate* appropriate task behavior (Argall et al. 2009), which can vary from directly controlling a robot via teleoperation to manually moving joints or having the human perform the desired behavior. For tasks that are intimately tied to motor actions, this directly defines both the task and how it is performed. In some cases, the demonstrations can be taken to be instances of appropriate behavior in the task, which can then be analyzed to determine task definitions. Demonstrations can also include the instructor manipulating the environment to create important states, such as the goal of a task, but where the manipulations themselves are not important, as in creating a stack of disks for the final state in Tower of Hanoi (Kirk et al. 2016). Each of these creates different challenges and opportunities for learning. For direct robot control, the challenge is often how to generalize or transfer specific movements with specific objects to new tasks, where exact movements are insufficient to solve the problem.

For some demonstrations, the agent is not trying to learn motor commands but rather the abstract aspects of the task, such as learning the rules of Tic-Tac-Toe. In these cases, the agent needs to use these instances of behavior as positive examples of appropriate behavior. Just seeing positive examples, however, can make it difficult for an agent to learn illegal actions and failure states. More generally, this type of learning is usually not *interactive* because the number of examples is often very large, and it is tedious for the human to show many examples. A second problem with learning from demonstrations without language is that it is difficult for the agent to derive accurate meaning from the visualized examples. Often, demonstrations are tightly scripted and the agent is led through a series of situations.

Combinations of Language and Demonstration Are Possible

By combining language and physical interaction, it may be possible to get the best of both. Some physical actions or situations are difficult to describe with language, whereas with pure physical interaction it is difficult to determine which aspects (objects, relations, and features) of the current situation are relevant. Language can be used to identify which aspects of a task are being taught, define new words and concepts, and identify which objects, properties, and relations are most important during a demonstration.

Prior Knowledge Acquired from Other Tasks Can Transfer

Over time, an agent will accumulate knowledge about many different tasks, with knowledge learned in one task being relevant to future tasks. During learning, the agent needs to be able to take advantage of prior knowledge to simplify and accelerate learning. Similarly, the human instructor must have the means to recognize that the agent already possesses certain knowledge, so that it does not need to be taught. The direct effect of this on learning is that a learning agent must be able to use what it has learned in one context when it is learning in another context. All knowledge learned in one task cannot be sequestered solely to that task; however, some concepts are task specific, so the agent must have some way of learning when a concept does not transfer, such as through additional instruction.

During the instruction of Rosie, much of the knowledge is encoded in the new concepts it learns, such as "three in a row" for Tic-Tac-Toe, Three Men's Morris, and Connect-3. Newly acquired lexical items, together with their meaning, directly transfer to new tasks. This represents both an opportunity and a challenge, as some knowledge can transfer to new situations whereas other learned concepts (and lexical items) acquired for a specific task might not be transferable. Rosie does not have any special "transfer" reasoning except when a term is used that it cannot instantiate in the current situation. In this case, it will attempt to use a different meaning or learn a new definition.

Summary of ITL Characteristics

In Table 17.1, we summarize many of the defining characteristics of interactive task learning and provide examples of those characteristics in the agent Rosie.

Learning Approaches for ITL

In this section, we summarize the major constraints and challenges associated with ITL and describe some of the possible variants of how an agent interactively learns a task with an instructor.

Table 17.1 Summary of ITL characteristics, as exemplified by the agent Rosie.

ITL Characteristics	Rosie's Capabilities
Learning across different domains	Games, puzzles, and simple cooking tasks for a tabletop robot. Navigation and delivery tasks for a mobile robot.
Learning diverse task knowledge	
Perceptual categories	Elaboration rules that extract features from perceptions
Task formulation knowledge	Initial states, goal states, illegal states, constraints on operators, operator preconditions, operator selection knowledge including ordering, heuristics, policy knowledge, and hierarchical task structure
Lexical items	Parts of speech, word definitions, relation to concept definitions
Concept definitions	Conjunctions of unary features, relations, numeric functions, comparators, and numbers
Learning is interactive	
Real-time interactions	Integrates interaction with task performance and learning in real time
Sparse, high-quality data	Learns from sentence-by-sentence interactions and relies on aggressive generalization
Flexible instruction structure	Mixed initiative interaction and concepts defined before or after they are referenced
Learning is situated	
Shared environment	Instructor and Rosie are situated in the same environment. Instructor uses deictic reference, pointing, and demonstrations.
Immediate environmental feedback	Rosie applies instructions as they are provided, which the instructor can view and provide immediate feedback.
Learning is multimodal	
Language	Learns task knowledge from natural language descriptions
Demonstration	Learns goal descriptions from demonstration
Mixed	Language is used to identify important relations in demonstrations.
Prior knowledge	Transfers knowledge from previously taught tasks

When considering various approaches to learning tasks through human interaction in a situated environment, there are fundamental characteristics inherent to the ITL problem that influence and constrain the learning approaches used. Learning multiple aspects of a task and different task formulations requires an approach that can acquire different types of knowledge, possibly using very different techniques. Learning from interaction, where sparse examples are provided in real time, constrains the approaches to avoid reliance on huge data sets. Instead, the agent must be able to start learning something useful from the very first teaching example and incrementally build on that learned knowledge. Fortunately, information communicated through interaction is often high quality, highly relevant, and tied to a shared context. Learning through interaction requires the learning approach of the agent to be flexible to instructional variations as well as capable of taking advantage of the corrections, clarifications, and missing knowledge provided by the human instructor. Characteristics of the domain—such as its dynamics, the level of uncertainty in sensing and acting, and observability—will also affect the approach used. More than anything, we conclude that no one learning algorithm is going to be sufficient for all ITL systems. In fact, learning different types of task knowledge from multiple sources of available knowledge and multiple forms of interaction will almost certainly require using multiple learning algorithms in the same system.

Below we delineate different types of task information that can be provided to the agent from an instructor using the different modalities described earlier. We also consider various learning approaches that can learn from each kind of information, although not all meet each of the constraints we have described (such as being online, real time, with few interactions).

Learning from Reinforcement

The simplest type of information is a reward signal: the teacher provides negative and positive feedback in response to the agent's actions to guide the agent toward the goal. This is usually most appropriate for control tasks, where the agent is attempting to learn a value function using reinforcement learning. An early example was in the commercial computer game *Black and White*, where a player would attempt to train a character by giving it positive or negative reward as it performed actions in its simulated world. This ended up being tricky because the player had to wait for the character to do a relevant behavior and then immediately provide the reward. If the reward was a bit late the character would think that the reward applied to a different action than the instructor intended. Another example is an agent that was interactively trained in a computer game called *Sophie's Kitchen* (Thomaz and Breazeal 2008). Although interactive, online, and incremental, it was restricted to a single task or set of tasks that shared the same reward function. The advantage of instructing through reward is that it is simple for a person to learn what they need to do. However, it is limited in the types of knowledge it can teach, it may require the instructor to carefully monitor the agent's behavior for an extended period, and it usually only works for learning behaviors for a single task.

Learning from Execution Examples

An agent can also learn from the execution of a task in different ways:

1. By observing the instructor performing the task
2. Via a physical demonstration of the task through teleoperation or physical manipulation of the agent, or
3. From natural language instructions that lead the agent through the task

In the first case, the agent observes the instructor performing the task, recognizes the actions taken, and maps them onto its own capabilities.

The second case constitutes a difficult learning problem, as the agent is learning from the high-dimensional and noisy space of joint angles and motor movements. Again, the agent can just record the trace and play back the same motor commands, but this is not robust to variations in the initial state. Another approach would be to learn special key frames, which can either be learned or given by the instructor, and offers better generalization.

Finally, in the third case, the instructor can verbally lead the agent through a task, step by step, with individual commands. The agent can employ many different learning mechanisms to learn from this information. The simplest is to memorize the instructions and replay them when given the same task in the same initial state. It is a simple learning approach, but without further generalization capabilities, the agent is restricted to procedure tasks that can be repeated exactly, over and over. To increase generality, the agent can use a few different techniques. One is to use inverse reinforcement learning (Abbeell and Ng 2004) to induce a reward function that is consistent with multiple step-by-step actions. Inverse reinforcement learning usually requires many examples, and is not necessarily appropriate given the interactive aspect of ITL. The agent can also attempt to generalize its experience by learning a description of the goal through comparing the final state to the initial state, and possibly asking the instructor which aspects of the goal are definitional. This allows the agent to formulate future instances of problems that share the same (or similar) goals, but with different initial states.

Once the agent has derived a description of the goal, the agent can use its knowledge about the causal structure of its actions to learn a general policy using explanation-based learning (DeJong and Mooney 1986). This is the approach Rosie uses when it is learning from a sequence of actions: "Open the pantry. Move the plate into the pantry. Close the pantry. You are done." It first derives and learns a description of the goal (possibly from further interaction with the instructor) and then performs a retrospective analysis of the solution that is used to learn general policy knowledge through Soar's procedural learning mechanism. In Rosie, this approach is also used when the agent solves a problem space formulation of a task, where only the goal is specified: "The goal is the plate is in the pantry and the pantry is closed" (Mohan and Laird 2014).

Learning from Procedure Specification

An instructor can also provide a more general specification of desired behavior by providing an explicit description of the procedure the agent should follow, such as with the recipe example described earlier. From such a specification, the learner can attempt to learn a procedure which represents the actions needed to perform the task and their ordering. One example of this is the instruction graph (Meriçli et al. (2014), which is a directed graph that represents the flow of execution for the task and can include looping and conditional structures. An alternative is where the instructor provides an explicit procedure, such as "do y until condition x is met, then…" (Langley et al. 2010; Salvucci 2013), which is essentially a high-level programming language that the agent must dynamically compile into a procedure to execute in the future.

Learning from Task Concept Specifications

The final form of information from which an agent can learn is the explicit specification of high-level task concepts. For example, the instructor could directly communicate a problem space formulation of the task, providing descriptions of goal and failure states, specifications of legal and illegal actions, and definitions of newly introduced concepts. These instructions can be via natural language or demonstration (Huffman and Laird 1995; Hinrichs and Forbus 2013a). This is the method Rosie uses to learn new games and puzzles, by leading the instructor through each aspect of the problem space, asking for definitions, and then interpreting them in the shared context, and operationalizing them for future use (Kirk and Laird 2016).

A variant for teaching policy information is where the instructor provides heuristic advice, instead of a complete procedure or specific steps, to guide the agent's behavior (McCarthy 1968). The advice is usually more general and informative than is possible with a simple reward (e.g., specifying an action as "good" or "bad") and can include descriptions of the situation in which an action applies. In a problem space formulation, advice can be used so that the agent not only learns how to perform the task, but also can both solve a problem quickly and learn to perform it well. For example, in Othello, Rosie can learn specific heuristics such as "prefer moving a block onto a corner location over an edge location."

Another approach is to engage the instructor in an interaction to learn more general concepts that allow it to generalize its task concepts. When Rosie is presented with a demonstration of a goal state in its environment (e.g., a stack of disks for Tower of Hanoi), it engages in an interaction with the instructor to determine which features and relations can be ignored, and which are necessary for defining a goal state (Kirk et al. 2016). Finally, Rosie also allows the instructor to describe general hypothetical situations, so that it is not bound to specific instances.

Conclusion

In this chapter, we have reviewed the learning problem for ITL. What makes learning both challenging and exciting is that it is not a single problem, where there is one type of knowledge to be learned, through one type of interaction. Nor is there a single algorithm that is applicable to all aspects of ITL. Instead, the learning problem for ITL is a very broad constellation of problems, defined by diverse types of interaction and instruction, diverse domains and available knowledge, diverse types of knowledge to be learned, and diverse approaches for learning that knowledge.

Acknowledgments

The work described here was supported by the Office of Naval Research under Grants N00014-08-1-0099 and by Air Force Office of Scientific Research under Grant Number FA9550-15-1-0157. The views and conclusions contained in this document are those of the authors and should not be interpreted as representing the official policies, either expressly or implied, of the ONR, AFOSR, or the U.S. Government.

Ethical Considerations

18

Ethical Aspects and Challenges for Interactive Task Learning

Matthias Scheutz

Abstract

As with all transformative technologies, humanity needs to analyze the ethical challenges and potential impacts associated with implementation. This chapter explores fundamental questions that pertain to interactive task learning (ITL): What is being taught and what are the associated risks? What are the dynamics of human–machine instruction? What effects will ITL have on human instructors and society? It explores the long-term impact that ITL could have on humans and human society, and discusses concerns valid to both machine learning and ITL (e.g., how to ensure that machines will learn knowledge that they can put to good use, that they will serve humans well and not become deviant). Importantly, it stresses the unique aspects of ITL and proposes that the time to think about and take action on these concerns is now.

Introduction

New technologies, such as being able to teach machines how to perform tasks through interaction instead of having to program them, are always exciting, not the least due to their potential to bring significant benefits to humanity. At the same time, every transformative technology raises important ethical questions, if adopted, about the risks involved and potential impact on human societies. Machine learning provides an example of a recent, potentially transformative technology—one that poses many important ethical challenges for designers of learning algorithms:

- How can we ensure that the algorithm will learn what it is supposed to learn and only that?
- How can we be certain that the machine will perform correctly after it has learned something new?
- How can we guarantee that the machine will not behave in unethical ways after learning?

These challenges arise with all forms of machine learning, and thus with interactive task learning (ITL) as well. The difference with ITL, however, is that in addition to the ethical issues that pertain to the machine's newly acquired knowledge and how it will be used by the machine, questions arise regarding the learning interaction and how this interaction might subsequently affect humans. Moreover, there are questions about acquisition and use of the "normative surroundings" inherent to the task; that is, all the ethical principles and implications considered to be part of the task that might not be explicitly instructed, but have to be followed during task execution.

ITL offers many important advantages for humans and machines alike, especially when task-based natural language interactions are involved. For one, natural language instruction is an intuitive modality for humans, because humans are used to teaching each other through natural language dialogues. This means that less training and preparation are required on the human side (compared to implementing tasks, e.g., through some form of programming language), and it allows human instructors to check quickly whether the learner has taken in the lesson.

For machines, ITL is advantageous because ideally it allows agents to acquire high-level task descriptions very quickly, instead of having to construct them over long stretches of bottom-up abstractions from low-level data. Moreover, being able to ask the human instructor for help (e.g., explanations or alternatives for action) does not exist in a data-driven approach; if the answer is not in the data, no statistical process in the world will be able to extract it.

However, there are potential downsides to ITL as well, particularly for humans. We need to analyze, over time, what long-term impacts might result as humans engage with autonomous machines in ways that lead to successful knowledge acquisition for the machine. Some effects might be limited to the teaching interaction itself, whereas others could extend far beyond this context. Here, we examine three classes of ethical aspects that arise in ITL:

1. What is being taught and what are the associated risks?
2. What are the dynamics of human–machine instruction?
3. What effects will ITL have on human instructors and society?

The Risks of Machine Learning

Whenever a machine is allowed to acquire new knowledge by way of employing its own learning algorithms (as opposed to having knowledge implanted by human engineers, which has its own challenges), there is always a risk that its learning process will not work as intended by the designers of the learning algorithm. The learning algorithm might:

- not acquire the right kind of information,
- only acquire incomplete knowledge, or
- acquire more knowledge than intended.

Consider, for example, the situation where a machine learning algorithm is supposed to learn how to detect human faces in pictures. This algorithm might fail to detect faces per se, but learn to detect humans instead (as in the first case); it may learn to detect faces but miss some faces under particular lighting conditions (the second case); or it could learn to detect faces as well as additional factors about humans such as their sexual orientation (the third case) (Wang and Kosinski 2017). Clearly, the result of unintended effects of machine learning could have ethical implications; for example, when unintended and unexpected information is exposed, misuse is possible, such as likely medical conditions or when wrongly classified information is used to make decisions that impact human lives (e.g., denying credit loan applications).

Learning algorithms in ITL are not exempt from these concerns regarding unintended consequences. Indeed, with ITL, additional aspects come into play that are typically not an issue with statistical machine learning from data. This is because ITL involves direct personal interactions between human instructors and machines, which is quite different from impersonal machine learning from data sets. For instance, if a data set does not contain information that the machine is supposed to learn, the data-driven machine learning algorithm is not at fault if the machine fails to extract it. In ITL, however, it is not clear who would be to blame if critical aspects of a task were not picked up by the machine: Was the learning algorithm at fault because it failed to encode or infer the information, or was the human tutor to blame because the information was not properly taught to the machine? Determining this will be difficult for the same reasons that ITL is challenging:

- How much knowledge can the human instructor assume that the machine has?
- How much language does the machine understand?
- How detailed do instructions have to be, and how much can the machine infer necessary aspects on its own?
- How can the instructor determine that the machine has fully acquired the task?
- Would it have to demonstrate it to the human or would repeating it back, maybe in its own words, do the trick?

While it is likely that the blame would be put onto the machine (at least initially while ITL technology is still developing), the very possibility that a human instructor could get blamed (or would personally take on the blame) if a machine fails to learn a task after instruction raises ethical questions about the nature and expectations of such interactions, and the subsequent effects on the humans involved.

Based on empirical work in human–robot interaction (Fan et al. 2017; Phillips et al. 2017), it seems reasonable to assume that cues of human likeness (e.g., natural language understanding, humanoid physical robots, virtual avatars) will cause people to import a vast array of assumptions about the machine's capabilities that may not be warranted, such as background knowledge and level of understanding. In fact, it is likely that despite sufficient perceptual capabilities and being able to understand enough natural language to be able to learn new tasks, machines capable of ITL will not be humanlike in many other ways. Incorrect impressions formed from limited exposures during teaching sessions might lead humans to omit important aspects of tasks (e.g., relating to property ownership, personal liability, human rights) which in turn could lead to unintended consequences during task performance: the machines, for instance, might not pay attention to whether the objects they use belong to their owner or whether they impact the freedom of other agents in performing the tasks. Moreover, humans are known to be "sloppy" when providing instructions. They use imprecise terminology, leave gaps in descriptions, refer to wrong objects, and make other errors, none of which is typically a problem for a human learner who can "think along" and automatically correct such errors (Thomaz et al., this volume). This raises the question: To what extent will, and can we expect, machines to read between the lines, and who is to blame when machines fail to derive the correct human intention?

Any sort of misunderstanding between a human instructor and machine could result in the machine learning the wrong task, learning it incompletely, or not learning it at all. There is also, of course, the possibility that the machine learns the task in a way that the human did not intend. Take, for example, a search and rescue robot that is supposed to learn how to find wounded people after a natural disaster. The aim is for the robot to enter collapsed buildings and make its way through the rubble to find any humans trapped inside the structure. Now suppose the robot takes the goal to search for wounded people too literally, so much so that when it finds a noninjured person, it determines that it cannot report its discovery because the person is not wounded. Hence, to be able to report the human, the robot decides, on the spot, to inflict an injury on the person to enable reporting the person as wounded and collect the reward from its policy-based decision-making system or from its partial satisfaction planner. While this problem almost seems comically bizarre, it is actually hard to prevent such cases without explicitly providing constraints to the system about which actions a machine is not allowed to perform to achieve its task goals. Yet what would serve as the source of this type of knowledge? Who would be in charge of providing it, and who would be responsible for ensuring that the robot was able to apply it correctly?

This raises the question about the extent to which ethical aspects related to a task (e.g., task-based obligations and prohibitions, social norms to be followed while learning the task, ethical principles to be applied while executing it) need or ought to be taught as well, because they cannot be assumed to be known to the machine. For example, it may seem banal to us that when instructions are given to a machine to build a fence, the machine must obtain the ingredients legally and must not simply take them from the surrounding areas (e.g., dismantling a neighbor's fence). Another example would be that when a Thanksgiving turkey dinner is to be prepared, the turkey ought to be dead already. Such constraints are obvious to us; hence they are typically not included in task instructions because we assume that our instructees already know them, and we rely on their ability to apply what a "reasonable person" would do. This legal term contains a lot of commonsense knowledge and reasoning that we, however, cannot assume machines will automatically possess. Thus, in addition to determining what legal and moral aspects of a task the instructor needs explicitly to teach, there might also be social normative aspects that the system would have to know to operate safely in human environments (e.g., no quick movement in crowded human environments, not moving toward people while holding knives that are pointed at them). Of course, cultural differences in norms pose further challenges as robots will have to be aware of the cultural background of their interactants (e.g., directness is considered polite by Russians, whereas English and German natives prefer indirectness; see Wierzbicka 1985).

The Dynamics of Human–Machine Instruction

The fundamental difference between ITL and other task-learning algorithms (e.g., learning tasks from instructional videos) involves real-time personal interaction with a human instructor. The human element imposes constraints and requirements on the types of interactions that machines may conduct with humans: interactions need to be respectful of human normative expectations (e.g., politeness) as well as human cognitive abilities (e.g., creativity, anticipation) and limitations (e.g., memory decay, limited focus of attention). The opportunities and challenges presented by these human characteristics do not have to be addressed by traditional data-driven learning algorithms. For instance, it is critical for machines to exhibit appropriate demeanor to ensure that humans will not be offended, and thus unwilling to continue an interaction. Examples of insensitivity could include a machine repeatedly dropping comments like "easy enough" and "no problem," which might be construed by the human as downplaying a difficult task for humans to acquire (e.g., learning how to play the "Flight of the Bumblebee"). This includes respecting social norms, such as politeness norms, that guide interactions among humans. For example, if a robot needs a screwdriver at a certain point in the task and sees the human instructor holding one, it should not attempt to just take it without asking.

In addition to using the appropriate tone and attitude while interacting with humans, being respectful of human expectations and constraints is critical. Especially with early ITL systems, it is likely that they will be unable to meet human expectations in terms of natural language understanding, speed of actions, timing of interactions, etc. They will simply not be advanced enough in their interaction capabilities, background and commonsense knowledge, and natural language understanding to learn in ways that humans would assume when instructing other humans (as discussed above, such expectations come naturally to humans when machines are appear to be humanlike). As a result, teaching interactions will quickly devolve into unnatural and tedious exchanges for humans. It will thus be important to ensure that humans have the right mental model of their machine's capabilities and that machines do everything they can to make their interactions less frustrating for humans. This will require machines to be aware of human emotional states and the effects that different interaction patterns might have on human emotions: a robot repeatedly asking a human to rephrase an instruction because it could not understand it or because it was not precise enough will quickly frustrate the human.

After ITL and machine capabilities have sufficiently advanced, being aware of human limitations will also become a critical component to preserve human dignity. Ignoring human cognitive limitations (e.g., attention span, ability to remain focused and concentrate, speed of natural language processing and information intake) can lead to dysfunctional interactions that would not serve either interactant well. For instance, a robot that anticipates human instruction after only a few words and starts to execute it proactively might confuse people, in the simplest case, or even cause anxiety as the robot's actions are not legible to the human.

Additional challenges arise with learning systems that might be able to alter or improve behaviors as they are being instructed. A robot's physical constraints or agent's capabilities may allow for alternative better ways of completing actions, possibly in a manner that would be impossible for humans. How would those optimizations be received by humans, and would humans be able to judge whether the system has been able to understand the task? Imagine the following situation: After receiving instructions to follow the steps in a human instruction manual to assemble a chest of drawers, the robot detects various shortcuts and alternative ways of grasping and assembling parts. Based on this new knowledge, the robot proceeds to complete the task at a much faster speed than humanly possible, yet before the job is done, the robot's steps do not seem to make sense to the human, who recognizes that the robot has deviated from the manual. By way of its physical capabilities, the robot could perform multiple steps in parallel (e.g., if it has several grippers it can use independently) or determine alternative ways of connecting parts to improve the stability of the chest (e.g., based on its own physics models or mental simulations). Such unexpected, superhuman performance by the robot might not only make the social interaction element with people uncomfortable and the human teacher's job of ensuring proper learning and performance harder, it could also leave psychological marks on the human that persist beyond the interaction, as we will discuss next.

Extended Effects of ITL on Human Instructors and Society

An important ethical aspect of new technologies is their long-term impact on humans and human society. While ITL shares some of the same long-term questions with other machine learning approaches—how to ensure that machines will learn knowledge that they can put to good use, that they will serve humans well and not become deviant—there are also unique, long-term ethical aspects that directly relate to human nature that need to be addressed. Given that humans will interact with machines, at the very least as teachers, during the interactive learning process, it is important to ask whether this interaction could have potentially negative effects on humans beyond the teaching interaction. Will humans feel (possibly unnecessarily) responsible when machines do not manage to acquire a task properly? Will humans blame themselves, instead of the machine, when a machine fails at a task because they really cared about the machine's success? Alternatively, when the machine succeeds after repeated learning interactions, will its success prompt feelings of pride in the human and lead potentially to the establishment of unidirectional emotional bonds (e.g., Scheutz 2014), because the human feels a personal connection with the machine?

Conversely, will the human be shocked, put off, or worried when observing machines with "superhuman" task learning or task performance capabilities? A machine might determine, for example, that it does not have to stick with the performance limitations imposed by how the human taught it a task or how the human has to perform the task, due to human sensorimotor constraints (e.g., a human might have to use a measuring tape to determine the length of a piece of wood that needs to be cut whereas a robot could immediately cut it using its visual system to measure the correct length). Consider machines that can consult cloud-based databases while interacting with human teachers, acquiring all necessary background knowledge quickly, on the fly, before human instruction has even finished; or machines that may covertly exchange messages with other learning machines while being instructed and quickly learn new skills from those machines. Take a robot that does not know how to use a drill which, as the human starts to explain how to operate drills, quickly assures the human that it just learned everything it needs to know by consulting other robots in its cohort: there is evidence that humans find such covert communication disconcerting and eerie (Williams et al. 2015).

In general, we have to anticipate massive effects of machines that can rapidly learn new tasks from interactive instructions at the societal level. If performed with the right task representations and paired with knowledge sharing, ITL could form the basis of massively parallel learning where teaching one machine means that all (connected) machines will know the task (Scheutz 2014). How such massive learning would affect labor markets or the economy is anybody's guess. It seems reasonable to assume that first-hand experience of such superhuman performance by machines—the awe as well as the jealousy and inferiority we may feel when the machine rapidly perfects a skill—could have profound ramifications for how we, as humans, view ourselves. In fact, it could lead to what the philosopher Günther Anders (1956/1979) called the "Promethean Shame": the feeling of inadequacy that results from watching our own technological products surpass us in their abilities and perfection, in particular, the realization that our capacity to think is inferior to that of our own machines.

Discussion

As with all new technologies, it is important to weigh the advantages and disadvantages of ITL and to consider carefully the trade-offs and risks involved in allowing, or even requiring, humans to teach machines. In the context of the larger discussion about the utility and dangers of artificial intelligence (AI), ITL certainly shares the worries that have been expressed in regard to machine learning:

- How can we guarantee that learning machines will be safe for humanity?
- How can we ensure that ITL agents can be turned off if they evolve in a dangerous direction and everything else fails?

These topics, currently discussed under the moniker "Big Red Button" (i.e., the means to shut off deviant AI) apply to ITL in the same way as they apply to other learning methods. Different from variants of reinforcement learning, where machines have to be incentivized to let them be shut off (Orseau and Armstrong 2016), ITL allows, however, for the explicit instruction of ethical principles in conjunction with tasks—an opportunity, if paired with the right computational architecture, that will make ITL a more desirable learning method for ensuring ethical behavior. Explicit instruction will also reduce the risk associated with placing all bets on the machine's ability to pick up normative principles from pure observation of human behavior, which may not be practical or even possible in the case of ITL (Arnold and Scheutz 2017).

Of course, instructing ethical principles along with tasks puts the burden of ensuring ethical behavior on the human instructor, which then raises the question of who should be allowed to instruct machines:

- What if the instructor is not interested in providing ethical guidance, or simply does not have the knowledge to do so explicitly?
- Even worse, what if the instructor has a malicious agenda and aims to instruct the machine how to engage in terrorism?
- How would the machine know that such a task is off limits?

Underpinning much of the discussion about ITL is a tacit assumption that both teacher and learner will be benevolent: a teacher will not instruct inappropriate tasks and a learner only has a human's best interests in mind or, at the very least, avoids malicious intent. Such assumptions, however, may not always be warranted despite best intentions: teachers unaware of task and environmental conditions could make instructions ethically problematic; instructions might be contradictory and conflict resolution unclear; teachers could have ulterior motives to teach tasks incorrectly; systems could be compromised (e.g., by hackers) and try to coerce the human instructor into teaching them tasks they are not supposed to learn. Clearly, ITL cannot be considered in isolation from mechanisms in the computational architecture that prevent unethical behavior; allowing machines to follow human instructions blindly is a recipe for disaster.

In addition to the challenges associated with ensuring the ethical behavior of instructible machines, ITL poses additional challenges due to the intrinsic involvement of human instructors. These challenges intersect closely with related discussions on the ethics of human–robot interaction. Aside from the potential detrimental effects of ITL on the human psyche that have been anticipated by philosophers of technology for decades, questions of ownership, responsibility, and allegiance posed by ITL must be addressed:

- Who should be allowed to teach a robot, and what ought to be the limits of instruction?
- How is the robot supposed to handle "competing interests" (Arnold and Scheutz 2017) in social groups, such as a family, where multiple members might want to teach the robot different tasks?
- Whose orders should it follow?
- Who should be in charge for controlling what the robot is or is not allowed to learn and use?
- Who will assume responsibility for the robot's actions?

There are currently no good answers to any of these questions, partly because the research communities in AI, robotics, and human–robot interaction are still very much focused on understanding and addressing the fundamental technical challenges raised by ITL. What the above discussion has hopefully demonstrated is that technical work on ITL cannot proceed in isolation from the ethical challenges raised by machines that can interactively learn new tasks.

Conclusion

As new research emerges to advance the ability of machines to learn interactively from human instructors, it is imperative to keep in mind the overarching ethical aspects that pertain to the ITL learning algorithms, the learning interaction between human teacher and machine learner, and the long-term effects of the interaction on the human, so that the result will be ITL machines that benefit human societies. Far different from data-driven machine learning, which usually cannot get any normative context information out of training data simply because that information is not contained in the data set, ITL offers the unique opportunity for explicit instructions of the "normative surroundings" of tasks: rules and regulations about task-relevant entities, social and moral norms associated with performing the task, and other ethical principles involved in learning and performing the task. Thus, machines instructed by ITL have the advantage of being able to learn their task as well as when, where, and how the task is appropriately performed. This, however, puts part of the onus on the human instructor to ensure that the machine is supplied with, and has taken in, the necessary ethical principles to both learn and perform the learned task in an ethical manner.

Bibliography

Note: Numbers in square brackets denote the chapter in which an entry is cited.

Abbeell, P., and A. Y. Ng. 2004. Apprenticeship Learning via Inverse Reinforcement Learning. In: Proc. 21st Intl. Conf. on Machine Learning. Banff: ACM. [17]

Aggarwal, J. K., and M. S. Ryoo. 2011. Human Activity Analysis: A Review. *ACM Comput. Surv.* **43**:16. [9]

Aguiar, A., and R. Baillargeon. 1999. 2.5-Month-Old Infants' Reasoning About When Objects Should and Should Not Be Occluded. *Cogn. Psychol.* **39**:116–157. [16]

Akgun, B., M. Cakmak, J. W. Yoo, and A. L. Thomaz. 2012. Trajectories and Keyframes for Kinesthetic Teaching: A Human-Robot Interaction Perspective. In: Human-Robot Interaction (HRI), Proc. 7th ACM/IEEE Intl. Conf., pp. 391–398. New York: ACM. [9]

Albacete, P. L., and K. VanLehn. 2000. Evaluation of the Effectiveness of a Cognitive Tutor for Fundamental Physics Concepts. In: Proc. 22nd Annual Conf. of the Cognitive Science Society, ed. L. R. Gleitman and A. K. Joshi, pp. 25–30. Mahwah: Erlbaum. [11]

Aleven, V., E. A. McLaughlin, R. A. Glenn, and K. R. Koedinger. 2016. Instruction Based on Adaptive Learning Technologies. In: Handbook of Research on Learning and Instruction, ed. R. E. Mayer and P. Alexander, pp. 522–560. New York: Routledge. [11]

Alexandrova, S., M. Cakmak, K. Hsaio, and L. Takayama. 2014. Robot Programming by Demonstration with Interactive Action Visualizations. In: Proc. Robotics: Science and Systems X [7]

Alexandrova, S., Z. Tatlock, and M. Cakmak. 2015. RoboFlow: A Flow-Based Visual Programming Language for Mobile Manipulation Tasks. In: 2015 IEEE Intl. Conf. on Robotics and Automation (ICRA), pp. 5537–5544. Seattle: Institute of Electrical and Electronics Engineers. [9, 15]

Allen, J. F., N. Chambers, G. Ferguson, et al. 2007. Plow: A Collaborative Task Learning Agent. In: Proc. 22nd AAAI Conference on Artificial Intelligence, vol. 2, pp. 1514–1519 Vancouver: AAAI Press. [3, 17]

Allen, J. F., and C. R. Perrault. 1980. Analyzing Intention in Utterances. *Artif. Intell.* **15**:143–178. [8]

Al-Moadhen, A., R. Qiu, M. Packianather, Z. Ji, and R. Setchi. 2013. Integrating Robot Task Planner with Common-Sense Knowledge Base to Improve the Efficiency of Planning. *Procedia Comput. Sci.* **22**:211–220. [9]

Altmann, G. T. M., and Y. Kamide. 1999. Incremental Interpretation at Verbs: Restricting the Domain of Subsequent Reference. *Cognition* **73**:247–264. [16]

Amaral, L., and D. Meurers. 2007. Conceptualizing student models for ICALL. In: User Modeling 2007, ed. C. Conati et al., pp. 340–344. Heidelberg: Springer. [13]

Anders, G. 1956/1979. The Obsolescence of Man (Die Antiquiertheit Des Menschen), 5th edition. Munich: C. H. Beck. [18]

Anderson, J. R. 1982. The Acquisition of Cognitive Skill. *Psychol. Rev.* **89**:369–406. [12]

———. 1987. Skill Acquisition: Compilation of Weak-Method Problem Solutions. *Psychol. Rev.* **94**:192–210. [4, 10]

Anderson, J. R. 1990. The Adaptive Character of Thought. Hillsdale: Erlbaum. [4]

———. 2002. Spanning Seven Orders of Magnitude: A Challenge for Cognitive Modeling. *Cogn. Sci.* **26**:85–112. [4]

———. 2007. How Can the Human Mind Occur in the Physical Universe? New York: Oxford Univ. Press. [6, 15]

Anderson, J. R., D. Bothell, M. D. Byrne, et al. 2004. An Integrated Theory of the Mind. *Psychol. Rev.* **111**:1036. [3, 4]

Anderson, J. R., and K. Gluck. 2001. What Role Do Cognitive Architectures Play in Intelligent Tutoring Systems? In: Cognition and Instruction: Twenty-Five Years of Progress, ed. D. Klahr and S. M. Carver, pp. 227–262. Hillsdale: Erlbaum. [4]

Anderson, J. R., and C. Lebiere. 1998. The Atomic Components of Thought. Mahwah: Erlbaum. [4]

Anzai, Y., and H. A. Simon. 1979. Theory of Learning by Doing. *Psychol. Rev.* **86**:124–140. [10]

Argall, B. D., S. Chernova, M. Veloso, and B. Browning. 2009. A Survey of Robot Learning from Demonstration. *Rob. Auton. Syst.* **57**:469–483. [9, 17]

Arnold, M. 1914. Essays in Criticism. London: Macmillan. [14]

Arnold, T., and M. Scheutz. 2017. Beyond Moral Dilemmas: Exploring the Ethical Landscape in HRI. In: Human-Robot Interaction (HRI), Proc. 12th ACM/IEEE Intl. Conf., pp. 445–452 New York: ACM. [18]

Aukrus, V. G., ed. 2007. Learning and Cognition. Amsterdam: Elsevier. [14]

Azaria, A., J. Krishnamurthy, and T. M. Mitchell. 2016. Instructable Intelligent Personal Agent. In: Proc. 30th AAAI Conf. on Artificial Intelligence, pp. 2681–2689. Phoenix: AAAI Press. [3, 17]

Baars, B. J. 1997. In the Theatre of Consciousness: Global Workspace Theory, a Rigorous Scientific Theory of Consciousness. *J. Conscious. Stud.* **4**:292–309. [6]

Bahrick, L. E., R. Lickliter, and R. Flom. 2004. Intersensory Redundancy Guides Infants' Selective Attention, Perceptual, and Cognitive Development. *Curr. Dir. Psychol. Sci.* **13**:99–102. [11]

Baillargeon, R., E. S. Spelke, and S. Wasserman. 1985. Object Permanence in Five-Month-Old Infants. *Cognition* **20**:191–208. [16]

Baillargeon, R., and S. H. Wang. 2002. Event Categorization in Infancy. *Trends Cogn. Sci.* **6**:85–93. [16]

Baisero, A., Y. Mollard, M. Lopes, M. Toussaint, and I. Lütkebohle. 2015. Temporal Segmentation of Pair-Wise Interaction Phases in Sequential Manipulation Demonstrations. In: 2015 IEEE/RSJ Intl. Conf. on Intelligent Robots and Systems pp. 478–484. Hamburg: IEEE. [9]

Baker, M. 2016. Is There a Reproducibility Crisis? *Nature* **533**:452–454. [10]

Ball, J., S. Rodgers, and K. Gluck. 2004. Integrating ACT-R and Cyc in a Large-Scale Model of Language Comprehension for Use in Intelligent Agents. In: Papers from the AAAI Workshop. Palo Alto: AAAI Press. [4]

Bangerter, A., and H. H. Clark. 2003. Navigating Joint Projects with Dialogue. *Cogn. Sci.* **27**:195–225. [8]

Barto, A. G., G. Konidaris, and C. Vigorito. 2013. Behavioral Hierarchy: Exploration and Representation. In: Computational and Robotic Models of the Hierarchical Organization of Behavior, ed. G. Baldassarre and G. Mirolli, pp. 13–46. Berlin: Springer. [3]

Baumeister, R. F., D. G. Hutton, and K. J. Cairns. 1990- Negative effects of praise on skilled performance. *Basic Appl. Soc. Psychol.* **11**:131–148. [12]

Bausell, R. B., W. B. Moody, and F. H. Walzl. 1972. A Factorial Study of Tutoring versus Classroom Instruction. *Am. Educ. Res. J.* **9**:591–597. [12]

Beck, K., M. Beedle, A. Van Bennekum, et al. 2001. Manifesto for Agile Software Development. https://agilemanifesto.org/ (accessed Jan. 11, 2019). [11]

Bengio, Y. 2009. Learning Deep Architectures for AI. *Found. Trends Mach. Learn.* **2**:1–127. [13]

Benjamin, A. S., and J. Tullis. 2010. What Makes Distributed Practice Effective? *Cogn. Psychol.* **61**:228–247. [11]

Bereiter, C. 2002. Education and Mind in the Knowledge Age. Abingdon, UK: Taylor and Francis. [14]

Best, B. J., N. Gerhart, and C. Lebiere. 2010. Extracting the Ontological Structure of OpenCyc for Reuse and Portability of Cognitive Models. In: 19th Annu. Conf. on Behavior Representation in Modeling and Simulation 2010. Charleston: BRiMS Committee. [4]

Best, B. J., and C. Lebiere. 2006. Cognitive Agents Interacting in Real and Virtual Worlds. In: Cognition and Multi-Agent Interaction: From Cognitive Modeling to Social Simulation, ed. R. Sun, pp. 186–218. New York: Cambridge Univ. Press. [4]

Beuls, K. 2013. Towards an Agent-Based Tutoring System for Spanish Verb Conjugation. Ph.D. thesis, Artificial Intelligence Lab, Vrije Univ. Brussel. [13]

———. 2014. Grammatical Error Diagnosis in Fluid Construction Grammar: A Case Study in L2 Spanish Verb Morphology. *Comput. Assist. Lang. Learn.* **27**:246–260. [13]

Bidet-Ildei, C., E. Kitromilides, J.-P. Orliaguet, M. Pavlova, and E. Gentaz. 2014. Preference for Point-Light Human Biological Motion in Newborns: Contribution of Translational Displacement. *Dev. Psychol.* **50**:113–120. [16]

Biswas, G., J. R. Segedy, and K. Bunchongchit. 2016. From Design to Implementation to Practice: A Learning by Teaching System: Betty's Brain. *Int. J. Artific. Intel. Educ.* **26**:350–364. [3]

Blokpoel, M. 2015. Understanding Understanding: A Computational-Level Perspective. PhD thesis, Radboud Univ. Nijmegen, The Netherlands. [8]

Bloom, B. S. 1984. The 2 Sigma Problem: The Search for Methods of Group Instruction as Effective as One-to-One Tutoring. *Educ. Res.* **13**:4–16. [12]

Bodkin, H. 2017. "Inspirational" Robots to Begin Replacing Teachers within 10 Years. *The Telegraph*, Sept. 11. [14]

Boesch, C. 2003. Is Culture a Golden Barrier between Human and Chimpanzee? *Evol. Anthropol.* **12**:82–91. [7]

Bögels, S., K. H. Kendrick, and S. C. Levinson. 2015a. Never Say No…How the Brain Interprets the Pregnant Pause in Conversation. *PLoS One* **10**:e0145474. [7, 8]

Bögels, S., L. Magyari, and S. C. Levinson. 2015b. Neural Signatures of Response Planning Occur Midway through an Incoming Question in Conversation. *Sci. Rep.* **5**:12881. [8]

Boydston, J., ed. 2008. The Later Works of John Dewey, vol. 7, 1925–1953: 1932, Ethics. Carbondale, IL: Southern Illinois Univ. Press. [14]

Boyer, K. E., R. Phillips, M. Wallis, M. Vouk, and J. C. Lester. 2008. Balancing Cognitive and Motivational Scaffolding in Tutorial Dialogue. In: Intelligent Tutoring Systems. Its 2008. Lecture Notes in Computer Science, ed. B. P. Woolf et al., vol. 5091, pp. 239–249. Heidelberg: Springer. [12]

Bratman, M. E. 1992. Shared Cooperative Activity. *Philos. Rev.* **101**:327–341. [7]

Brenton, H., M. Yee-King, A. Grimalt-Reynes, et al. 2014. A Social Timeline for Exchanging Feedback About Musical Performances. Proc. 28th Intl. BCS Human Computer Interaction Conf. Southport: BCS Learning and Development Ltd. [14]

Brown, J. S., and K. VanLehn. 1980. Repair Theory: A Generative Theory of Bugs in Procedural Skills. *Cogn. Sci.* **4**:379–426. [7]

Brown, P., and S. C. Levinson. 1987. Politeness: Some Universals in Language Usage. Cambridge: Cambridge Univ. Press. [8]

Bruce, V., and A. Young. 1986. Understanding Face Recognition. *Br. J. Psychol.* **77**:305–327. [3]

Cade, W. L., J. L. Copeland, N. Person, and S. K. D'Mello. 2008. Dialogue Modes in Expert Tutoring. In: Intelligent Tutoring Systems. Its 2008. Lecture Notes in Computer Science, ed. B. P. Woolf et al., vol. 5091, pp. 470–479. Heidelberg: Springer. [12]

Cakmak, M., C. Chao, and A. L. Thomaz. 2010. Designing Interactions for Robot Active Learners. *IEEE Trans. Auton. Ment. Dev.* **2.2**:108–118. [9]

Cakmak, M., and A. L. Thomaz. 2012. Designing Robot Learners That Ask Good Questions. In: Human-Robot Interaction (HRI), Proc. 7th ACM/IEEE Intl. Conf., pp. 17–24. New York: ACM. [9]

Caliskan, A., J. J. Bryson, and A. Narayanan. 2017. Semantics Derived Automatically from Language Corpora Contain Human-Like Biases. *Science* **356**:183–186. [3]

Callaghan, T. C., H. Moll, H. Rakoczy, et al. 2011. Early Social Cognition in Three Cultural Contexts. *Monogr. Soc. Res. Child Dev.* **76**:1–142. [7]

Cameron-Faulkner, T., A. Theakston, E. Lieven, and M. Tomasello. 2015. The Relationship between Infant Holdout and Gives and Pointing. *Infancy* **20**:576–586. [7]

Cangelosi, A., M. Schlesinger, and L. B. Smith. 2015. Developmental Robotics: From Babies to Robots. Cambridge, MA: MIT Press. [15, 16]

Cantrell, R., K. Talamadupula, P. Schermerhorn, et al. 2012. Tell Me When and Why to Do It! Run-Time Planner Model Updates via Natural Language Instruction. In: Human-Robot Interaction (HRI), Proc. 7th ACM/IEEE Intl. Conf., pp. 471–478. New York: ACM. [9]

Card, S. K., T. P. Moran, and A. Newell. 1983. The Psychology of Human-Computer Interaction. Mahwah: Erlbaum. [15]

Casler, K., and D. Kelemen. 2005. Young Children's Rapid Learning About Artifacts. *Dev. Sci.* **8**:472–480. [15]

Cepeda, N. J., H. Pashler, E. Vul, J. T. Wixted, and D. Rohrer. 2006. Distributed Practice in Verbal Recall Tasks: A Review and Quantitative Synthesis. *Psychol. Bull.* **132**:354–380. [11]

Chai, J. Y., R. Fang, C. Liu, and L. She. 2016. Collaborative Language Grounding Towards Situated Human-Robot Dialogue. *AI Mag.* **37**:32–45. [7, 9]

Chang, F. 2002. Symbolically Speaking: A Connectionist Model of Sentence Production. *Cogn. Sci.* **26**:609–651. [16]

Chang, F., G. S. Dell, and K. Bock. 2006. Becoming Syntactic. *Psychol. Rev.* **113**:234–272. [16]

Chao, C., M. Cakmak, and A. L. Thomaz. 2010. Transparent Active Learning for Robots. In: Human-Robot Interaction (HRI), Proc. 5th ACM/IEEE Intl. Conf. New York: ACM. [9]

———. 2011. Towards Grounding Concepts for Transfer in Goal Learning from Demonstration. In: Proc. 2011 IEEE Intl. Conference on Development and Learning, vol. 2, pp. 1–6. Frankfurt: IEEE. [4, 9, 17]

Chao, C., and A. L. Thomaz. 2013. Controlling Social Dynamics with a Parametrized Model of Floor Regulation. *J. Hum. Robot. Interact.* **2.1**:4–29. [7]

Chase, C. C., D. B. Chin, M. A. Oppezzo, and D. L. Schwartz. 2009. Teachable Agents and the Protégé Effect: Increasing the Effort Towards Learning. *J. Sci. Educ. Technol.* **18**:334–352. [11]

Chase, C. C., J. Marks, D. Bernett, M. Bradley, and V. Aleven. 2015. Towards the Development of the Invention Coach: A Naturalistic Study of Teacher Guidance for an Exploratory Learning Task. In: Artificial Intelligence in Education, ed. C. Conati et al., pp. 558–561. AIED 2015. Lecture Notes in Computer Science, vol. 9112. Cham: Springer. [12]

Chase, W. G., and H. A. Simon. 1973. The Mind's Eye in Chess. In: Visual Information Processing, ed. W. G. Chase, pp. 215–281. New York: Academic Press. [3]

Chein, J. M., and W. Schneider. 2005. Neuroimaging Studies of Practice-Related Change: fMRI and Meta-Analytic Evidence of a Domain-General Control Network for Learning. *Cogn. Brain Res.* **25**:607–623. [15]

Chi, M. T. H. 2009. Active-Constructive-Interactive: A Conceptual Framework for Differentiating Learning Activities. *Top. Cogn. Sci.* **1**:73–105. [11, 12]

Chi, M. T. H., M. Bassok, M. Lewis, P. Reimann, and R. Glaser. 1989. Self-Explanations: How Students Study and Use Examples in Learning to Solve Problems. *Cogn. Sci.* **15**:145–182. [12]

Chi, M. T. H., N. de Leeuw, M. H. Chiu, and C. LaVancher. 1994. Eliciting Self-Explanations Improves Understanding. *Cogn. Sci.* **18**:439–477. [12]

Chi, M. T. H., and M. Menekse. 2015. Dialogue Patterns in Peer Collaboration That Promote Learning. In: Socializing Intelligence through Academic Talk and Dialogue, ed. L. B. Resnick et al., pp. 263–274. Washington, D.C.: AERA. [11]

Chi, M. T. H., M. Roy, and R. G. M. Hausmann. 2008. Observing Tutorial Dialogues Collaboratively: Insights About Human Tutoring Effectiveness from Vicarious Learning. *Cogn. Sci.* **32**:301–342. [12]

Chi, M. T. H., S. Siler, and H. Jeong. 2004. Can Tutors Monitor Students' Understanding Accurately? *Cogn. Instr.* **22**:363–387. [12]

Chi, M. T. H., S. Siler, H. Jeong, T. Yamauchi, and R. G. Hausmann. 2001. Learning from Human Tutoring. *Cogn. Sci.* **25**:471–533. [12]

Chi, M. T. H., and R. Wylie. 2014a. The ICAP Framework: Linking Cognitive Engagement to Active Learning Outcomes. *Educ. Psychol.* **49**:219–243. [11, 14]

———. 2014b. ICAP: A Hypothesis of Differentiated Learning Effectiveness for Four Modes of Engagement Activities. *Educ. Psychol.* **49**:219–243. [12]

Cho, B.-I., J. A. Michael, A. A. Rovick, and M. W. Evens. 2000. An Analysis of Multiple Tutoring Protocols. In: Intelligent Tutoring Systems. Its 2000. Lecture Notes in Computer Science, ed. G. Gauthier et al., vol. 1839, pp. 212–221. Heidelberg: Springer. [12]

Clark, H. H. 1996. Using Language. Cambridge: Cambridge Univ. Press. [2, 15, 17]

Clark, H. H., and S. E. Brennan. 1991. Grounding in Communication. In: Perspectives on Socially Shared Cognition, ed. L. B. Resnick et al., pp. 127–149. Washington, D.C.: American Psychological Association. [3, 11, 15]

Clegg, J. M., and C. H. Legare. 2016. A Cross-Cultural Comparison of Children's Imitative Flexibility. *Dev. Psychol.* **52**:1435–1444. [8]

Cockburn, A., C. Gutwin, J. Scarr, and S. Malacria. 2014. Supporting Novice to Expert Transitions in User Interfaces. *ACM Comput. Surv.* **47**:1–36. [10]

Cole, M. W., P. Laurent, and A. Stocco. 2013. Rapid Instructed Task Learning: A New Window into the Human Brain's Unique Capacity for Flexible Cognitive Control. *Cogn. Affect. Behav. Neurosci.* **13**:1–22. [15]

Coles, A., S. Edelkamp, D. Magazzeni, and S. Sanner, eds. 2017. Proc. of the 26th Intl. Conf. on Automated Planning and Scheduling. Palo Alto: AAAI Press. [5]

Collins, A. 1977. Processes in Acquiring Knowledge. In: Schooling and the Acquisition of Knowledge, ed. R. C. Anderson et al., pp. 339–363. Hillsdale: Erlbaum. [12]

Collins, A., J. S. Brown, and S. E. Newman. 1989. Cognitive Apprenticeship: Teaching the Craft of Reading, Writing and Mathematics. In: Knowing, Learning and Instruction: Essays in Honor of Robert Glaser, ed. L. B. Resnick, pp. 453–494. Hillsdale: Erlbaum. [12]

Collins, A., and A. Stevens. 1982. Goals and Strategies for Inquiry Teachers. In: Advances in Instructional Psychology, ed. R. Glaser, vol. 2, pp. 65–119. Hillsdale: Erlbaum. [12]

Corbett, A. T., and J. R. Anderson. 1995. Knowledge Decomposition and Subgoal Reification in the ACT Programming Tutor. In: Proc. AI-ED 1995. Charlottesville: AACE. [4]

Cordeschi, R. 2002. The Discovery of the Artificial: Behavior, Mind and Machines Before and Beyond Cybernetics. New York: Springer. [14]

Cordova, D. I., and M. R. Lepper. 1996. Intrinsic Motivation and the Process of Learning: Beneficial Effects of Contextualization, Personalization, and Choice. *J. Educ. Psychol.* **88**:715–730. [12]

Core, M. G., J. D. Moore, and C. Zinn. 2003. The Role of Initiative in Tutorial Dialogue. In: Proc. 10th Conf. of the European Chapter of the Association for Computational Linguistics, ed. A. Copestake and J. Hajic, vol. 1, pp. 67–74. Stroudsburg, PA: Assn. of Computational Linguistics. [12]

Couper-Kuhlen, E. 2014. What Does Grammar Tell Us About Action? *Pragmatics* **24**:623–647. [8]

Craft, A. 2001. An Analysis of Research and Literature on Creativity in Education. Report for the Qualifications and Curriculum Authority. http://citeseerx.ist.psu.edu/viewdoc/download?doi=10.1.1.508.3210&rep=rep1&type=pdf (accessed Oct., 19, 2017). [7]

Crangle, C., and P. Suppes. 1994. Language and Learning for Robots, Centre for the Study of Language and Communications. Stanford Univ. Lecture Notes No. 41. Stanford: CSLI [17]

Csibra, G., and G. Gergely. 2009. Natural Pedagogy. *Trends Cogn. Sci.* **13**:148–153. [7, 8, 15]

Csibra, G., G. Gergely, S. Bíró, O. Koós, and M. Brockbank. 1999. Goal Attribution without Agency Cues: The Perception of Pure Reason in Infancy. *Cognition* **72**:237–267. [16]

Dabrowska, E., and E. Lieven. 2005. Towards a lexically specific grammar of children's question constructions. *Cogn. Ling.* **16**:437–474. [13]

Davis, R., H. Shrobe, and P. Szolovits. 1993. What Is a Knowledge Representation? *AI Mag.* **14**:17–33. [3]

de Boer, P. T., D. P. Kroese, S. Mannor, and R. Y. Rubinstein. 2005. A Tutorial on the Cross-Entropy Method. *Ann. Oper. Res.* **134**:19–67. [10]

Dehaene, S., M. Kerszberg, and J. P. Changeux. 1998. A Neuronal Model of a Global Workspace in Effortful Cognitive Tasks. *PNAS* **95**:14529–14534. [6]

DeJong, G., and R. Mooney. 1986. Explanation-Based Learning: An Alternative View. *Mach. Learn.* **1**:145–176. [17]

Dell, G. S., and F. Chang. 2014. The P-Chain: Relating Sentence Production and Its Disorders to Comprehension and Acquisition. *Phil. Trans. R. Soc. B* **369**:20120394. [16]

de Ruiter, J. P., A. Bangerter, and P. Dings. 2012. The Interplay between Gesture and Speech in the Production of Referring Expressions: Investigating the Tradeoff Hypothesis. *Top. Cogn. Sci.* **4**:232–248. [8]

de Ruiter, J. P., M. Noordzij, S. Newman-Norlund, et al. 2010. Exploring the Cognitive Infrastructure of Communication. *Interact. Stud.* **11**:51–77. [8]

Destefano, M. 2010. The Mechanics of Multitasking: The Choreography of Perception, Action, and Cognition over 7.05 Orders of Magnitude. PhD Thesis, Rensselaer Polytechnic Institute, Troy, NY. [10]

Destefano, M., and W. D. Gray. 2008. Choreographing Cognition, Perception, and Motor Control over 7.03 Orders of Magnitude. In: 13th Annual ACT-R Summer Workshop. Pittsburgh: Carnegie Mellon Univ. [10]

———. 2016. Where Should Researchers Look for Strategy Discoveries during the Acquisition of Complex Task Performance? The Case of Space Fortress. In: Proc. 38th Annu. Conf. of the Cognitive Science Society, ed. A. Papafragou et al., pp. 668–673. Austin: Cognitive Science Society. [10]

Dewey, J. 1896. The Reflex Arc Concept in Psychology. *Psychol. Rev.* **3**:357–370. [14]

———. 1897. My Pedagogic Creed, Article Two: What the School Is. *School J.* **54**:77–80. [14]

———. 1916. Democracy and Education: An Introduction to Philosophy of Education. London: Macmillan. [14]

———. 1934. Having an Experience. In: Art as Experience, pp. 36–59. New York: Perigree Books. [14]

Dewey, J., A. W. Moore, H. C. Brown, et al. 1917. Creative Intelligence: Essays in the Pragmatic Attitude. New York: Henry Holt. [14]

Diehl, J. J., L. M. Schmitt, M. Villano, and C. R. Crowell. 2012. The Clinical Use of Robots for Individuals with Autism Spectrum Disorders: A Critical Review. *Res. Autism Spectr. Disord.* **6**:249–262. [8]

Dingemanse, M., S. G. Roberts, J. Baranova, et al. 2015. Universal Principles in the Repair of Communication Problems. *PLoS One* **10**:e0136100. [7, 8]

D'Inverno, M., and A. Still. 2014. Creative Feedback: A Manifesto for Social Learning. In: EDM 2014 Extended Proc. 7th Intl. Conf. Educational Data Mining, ed. S. Gutiérrez-Santos and O. C. Santos, pp. 192–199. London: CEUR-WS. [14]

Dominey, P. F., and C. Dodane. 2004. Indeterminancy in Language Acquisition: The Role of Child Directed Speech and Joint Attention. *J. Neuroling.* **17**:121–145. [11]

Donchin, E. 1995. Video Games as Research Tools: The Space Fortress Game. *Behav. Res. Methods Instrum. Comput.* **27**:217–223. [10]

Drew, P. 1997. "Open" Class Repair Initiators in Response to Sequential Sources of Troubles in Conversation. *J. Pragmat.* **28**:69–101. [7]

Drew, P., and E. Couper-Kuhlen, eds. 2014. Requesting in Social Interaction. Studies in Language and Social Interaction, vol. 26. Amsterdam: John Benjamins. [8]

Durkheim, E. 1912. Les Formes Élémentaires de la Vie Religieuse. Paris: Aldrun. [8]

Ekvall, S., and D. Kragic. 2008. Robot Learning from Demonstration: A Task-Level Planning Approach. *Int. J. Adv. Robot. Syst.* **5**:223–234. [9]

Eliasmith, C., T. C. Stewart, X. Choo, et al. 2012. A Large-Scale Model of the Functioning Brain. *Science* **338**:1202–1205. [6]

Emerson, R. 1837/1962. The American Scholar. In: The Portable Emerson, ed. M. Van Doren, pp. 23–46. Middlesex: Penguin Books. [14]

Endsley, M. R. 1995. Toward a Theory of Situation Awareness in Dynamic Systems. *Human Factors* **37**:32–64. [4]

Engell, J. 1981. The Creative Imagination: Enlightenment to Romanticism. Cambridge, MA: Harvard Univ. Press. [14]

Ericsson, K. A., R. R. Hoffman, A. Kozbelt, and A. M. Williams, eds. 2018. The Cambridge Handbook of Expertise and Expert Performance, 2nd edition. Cambridge: Cambridge Univ. Press. [10]

Ericsson, K. A., and A. C. Lehmann 1996. Expert and exceptional performance: Evidence of maximal adaptation to task constraints. *Annu. Rev. Psychol.* **47**:273–305. [12]

Evens, M., and J. Michael. 2006. One-on-One Tutoring by Humans and Machines. Mahwah: Erlbaum. [12]

Eysenck, H. J. 1995. Creativity as a Product of Intelligence and Personality. In: International Handbook of Personality and Intelligence, pp. 231–247. New York: Plenum. [14]

Fan, L., M. Scheutz, M. Lohani, M. McCoy, and C. Stokes. 2017. Do We Need Emotionally Intelligent Artificial Agents? First Results of Human Perceptions of Emotional Intelligence in Humans Compared to Robots. Proc.17th Intl. Conf. on Intelligent Virtual Agents, pp. 129–141. New York: Springer. [18]

Fang, R., M. Doering, and J. Y. Chai. 2015. Embodied Collaborative Referring Expression Generation in Situated Human-Robot Dialogue. In: Human-Robot Interaction (HRI), Proc. 10th ACM/IEEE Intl. Conf., pp. 271–278. New York: ACM. [9]

Feldman, R. 2007. Parent–Infant Synchrony and the Construction of Shared Timing. Physiological Precursors, Developmental Outcomes and Risk Conditions. *J. Child Psychol. Psychiatry* **48**:329–354. [7]

Fischer, K., K. Foth, K. J. Rohlfing, and B. Wrede. 2011. Mindful Tutors: Linguistic Choice and Action Demonstration in Speech to Infants and Robots. *Interact. Stud.* **12**:134–161. [11]

Fischer, R., and F. Plessow. 2015. Efficient Multitasking: Parallel versus Serial Processing of Multiple Tasks. *Front. Psychol.* **6**:1366. [8]

Fitts, P. M. 1964. Perceptual-Motor Skill Learning. In: Categories of Human Learning, ed. A. Melton, pp. 243–285. New York: Academic Press. [10]

Fitts, P. M., and M. I. Posner. 1967. Human Performance. Belmont, CA: Brooks/Cole. [12]

Floyd, S., E. Manrique, G. Rossi, and F. Torreira. 2016. Timing of Visual Bodily Behavior in Repair Sequences: Evidence from Three Languages. *Discourse Process.* **53**:175–204. [8]

Fonseca, B., and M. T. H. Chi. 2011. The Self-Explanation Effect: A Constructive Learning Activity. In: The Handbook of Research on Learning and Instruction, ed. R. E. Mayer and P. Alexander, pp. 296–321. New York: Routledge. [12]

Forbes, M., and Y. Choi. 2017. Verb Physics: Relative Physical Knowledge of Actions and Objects. In: Proc. 55th Annual Meeting of the Association for Computational Linguistics, pp. 266–276. Vancouver: ACL. [9]

Forbus, K. 2011. Qualitative Modeling. *Wiley Interdiscip. Rev. Cogn. Sci.* **2**:374–391. [3]

Forbus, K., M. Chang, M. McLure, and M. Usher. 2017. The Cognitive Science of Sketch Worksheets. *Top. Cogn. Sci.* **9**:921–942. [11]

Forbus, K., and D. Gentner. 1997. Qualitative Mental Models: Simulations or Memories? In: Proc. 11th Intl. Workshop on Qualitative Reasoning, pp. 97–104. Cortona, Italy: AAAI Press. [3]

Fox, B. A. 1991. Cognitive and Interactional Aspects of Correction in Tutoring. In: Teaching Knowledge and Intelligent Tutoring, ed. P. Goodyear, pp. 149–172. Norwood, NJ: Ablex. [7, 12]

———. 1993. The Human Tutorial Dialogue Project: Issues in the Design of Instructional Systems. Hillsdale: Erlbaum. [7, 12]

Fox, M., and D. Long. 2003. Pddl 2.1: An Extension to Pddl for Expressing Temporal Planning Domains. *J. Artif. Intell. Res.* **20**:61–124. [9]

Fox, R., and C. McDaniel. 1982. The Perception of Biological Motion by Human Infants. *Science* **218**:486–487. [16]

Frank, M. J., B. Loughry, and R. C. O'Reilly. 2001. Interactions between Frontal Cortex and Basal Ganglia in Working Memory: A Computational Model. *Cogn. Affect. Behav. Neurosci.* **1**:137–160. [15]

Frankenhuis, W. E., B. House, H. Clark Barrett, and S. P. Johnson. 2013. Infants' Perception of Chasing. *Cognition* **126**:224–233. [16]

Frederiksen, N., J. Donin, and M. Roy. 2000. Human Tutoring as a Model for Computer Tutors: Studying Human Tutoring from a Cognitive Perspective. In: Modelling Human Teaching Tactics and Strategies: Workshop W1 of ITS'2000, ed. B. du Boulay. Heidelberg: Springer. [12]

Friedman, N., L. Getoor, D. Koller, and A. Pfeffer. 1999. Learning Probabilistic Relational Models. In: Proc. 16th Intl. Joint Conf. on Artificial Intelligence (IJCAI-99), vol. 2, pp. 1300–1307. San Francisco: Morgan Kaufmann. [5]

Fu, W.-T., and J. R. Anderson. 2006. From Recurrent Choice to Skill Learning: A Reinforcement-Learning Model. *J. Exp. Psychol. Gen.* **135**:184–206. [4]

Fu, W.-T., and W. D. Gray. 2004. Resolving the Paradox of the Active User: Stable Suboptimal Performance in Interactive Tasks. *Cogn. Sci.* **28**:901–935. [10]

Gao, Q., M. Doering, S. Yang, and J. Y. Chai. 2016. Physical Causality of Action Verbs in Grounded Language Understanding. In: Proc. 54th Annual Meeting of the Association for Computational Linguistics. Berlin: ACL. [9]

Gao, T., G. E. Newman, and B. J. Scholl. 2009. The Psychophysics of Chasing: A Case Study in the Perception of Animacy. *Cogn. Psychol.* **59**:154–179. [16]

Garrod, S., and M. J. Pickering. 2004. Why Is Conversation So Easy? *Trends Cogn. Sci.* **8**:8–11. [7]

Gaskins, S. 1999. Children's Daily Lives in a Mayan Village: A Case Study of Culturally Constructed Roles and Activities. In: Children's Engagement in the World: Sociocultural Perspectives, ed. A. Göncü, pp. 25–61. Cambridge: Cambridge Univ. Press. [8]

Genesereth, M., and M. Thielscher. 2014. General Game Playing. In: Synthesis Lectures on Artificial Intelligence and Machine Learning, ed. R. Brachman and S. Stone, vol. 8. San Rafael, CA: Morgan and Claypool Publ. [3]

Gergely, G., H. Bekkering, and I. Király. 2002. Rational Imitation in Preverbal Infants. *Nature* **415**:755. [7]

Gergely, G., and G. Csibra. 2003. Teleological Reasoning in Infancy: The Naive Theory of Rational Action. *Trends Cogn. Sci.* **7**:287–292. [16]

Gibson, E. J., and A. S. Walker. 1984. Development of Knowledge of Visual-Tactual Affordances of Substance. *Child Dev.* **55**:453–460. [14]

Giese, M. A., and T. Poggio. 2003. Cognitive Neuroscience: Neural Mechanisms for the Recognition of Biological Movements. *Nat. Rev. Neurosci.* **4**:179–192. [16]

Gisladottir, R. S., D. Chwilla, and S. C. Levinson. 2015. Conversation Electrified: ERP Correlates of Speech Act Recognition in Underspecified Utterances. *PLoS One* **10**:e0120068. [8]

Glas, D. F., T. Minato, C. T. Ishi, T. Kawahara, and H. Ishiguro. 2016. ERICA: The ERATO Intelligent Conversational Android. In: 25th IEEE Intl. Symp. on Robot and Human Interactive Communication (Ro-Man), pp. 22–29. New York: IEEE [8]

Glass, M. S., J. H. Kim, M. Evens, J. Michael, and A. Rovick. 1999. Novice vs. Expert Tutors: A Comparison of Style. In: Proc. 10th Midwest Artificial Intelligence and Cognitive Science Conf. (MAIS-99). Bloomington: AAAI Press. [12]

Glenberg, A.M., A. Wilkinson, and W. Epstein. 1982. The illusion of knowing: Failure in the self-assessment of comprehension. *Mem. Cogn.* **10**:597–602. [12]

Glendenning, K., T. Wischgoll, J. Harris, R. Vickery, and L. Blaha. 2016. Parameter Space Visualization for Large-Scale Datasets Using Parallel Coordinate Plots. *J. Imaging Sci. Technol.* **60**:10406–10401–10406–10408(10408). [10]

Gluck, K. A., and J. Harris. 2008. Mindmodeling@Home. In: Proc. 30th Annual Conf. of the Cognitive Science Society, ed. B. C. Love et al., p. 1422. Cognitive Science Society. [10]

Gluck, K. A., T. Jastrzembski, and K. Krusmark. 2019. Prospective Comments on Performance Prediction for Aviation Psychology. In: Advances in Aviation Psychology, ed. M. A. Vidulich and P. S. Tsang. Boca Raton: CRC Press, in press. [11]

Gobet, F., and H. A. Simon. 1996. Recall of Rapidly Presented Random Chess Positions Is a Function of Skill. *Psychonomic Bulletin & Review* **3**:159–163. [10]

Goffman, E. 1959. The Presentation of Self in Everyday Life. Edinburgh: Univ. of Edinburgh Social Sciences Research Centre. [8]

Goldman, S., J. Pellegrino, and J. D. Bransford. 1993. Assessing Programs That Invite Thinking. In: Technology Assessment: Estimating the Future, ed. H. F. O'Neil and E. L. Baker, pp. 199–230. Hillsdale: Erlbaum. [12]

Goldstone, R. L., and G. Lupyan. 2016. Discovering Psychological Principles by Mining Naturally Occurring Data Sets. *Top. Cogn. Sci.* **8**:548–568. [10]

Golinkoff, R. M., H. L. Chung, K. Hirsh-Pasek, et al. 2002. Young Children Can Extend Motion Verbs to Point-Light Displays. *Dev. Psychol.* **38**:604–614. [16]

Gonzalez, C., F. J. Lerch, and C. Lebiere. 2003. Instance-Based Learning in Dynamic Decision Making. *Cogn. Sci.* **27**:591–635. [4]

Gopnik, A., and L. Schulz. 2007. Causal Learning: Psychology, Philosophy, and Computation. Oxford: Oxford Univ. Press. [15]

Gorman, J. C., N. J. Cooke, and P. G. Amazeen. 2010. Training Adaptive Teams. *Human Factors* **52**:295–307. [11]

Graesser, A. C. 2009. Inaugural Editorial for Journal of Educational Psychology. *J. Educ. Psychol.* **101**:259–261. [14]

Graesser, A. C., N. Person, and J. Magliano. 1995. Collaborative Dialog Patterns in Naturalistic One-on-One Tutoring. *Appl. Cogn. Psychol.* **9**:359–387. [12]

Gratch, J., D. DeVault, G. Lucas, and S. Marsella. 2015. Negotiation as a Challenge Problem for Virtual Humans. In: Intelligent Virtual Agents, ed. W. P. Brinkman et al., pp. 201–215. Cham: Springer. [9]

Grauman, K., and B. Leibe. 2011. Visual Object Recognition. In: Synthesis Lectures on Artificial Intelligence and Machine Learning, ed. R. J. Brachman and T. G. Dietterich, vol. 5, pp. 1–181. Williston, VT: Morgan and Claypool Publ. [9]

Gray, W. D. 2017. Plateaus and Asymptotes: Spurious and Real Limits in Human Performance. *Curr. Dir. Psychol. Sci.* **26**:59–67. [7, 10]

Gray, W. D., and M. Destefano. 2016. Searching Not under the Lightpole but Where We Dropped Our Keys: Using Changepoint Detection to Shine the Light on Periods of Strategy Invention and Change. Paper presented at the 57th Annual Meeting of the Psychonomics Society http://homepages.rpi.edu/~grayw/pubs/papers/2016/gray-16psychonomics.pdf (accessed Jan. 31, 2019). [10]

Gray, W. D., B. E. John, and M. E. Atwood. 1993. Project Ernestine: Validating GOMS for Predicting and Explaining Real-World Task Performance. *Human-Comput. Interact.* **8**:237–309. [3]

Gray, W. D., and J. K. Lindstedt. 2017. Plateaus, Dips, and Leaps: Where to Look for Inventions and Discoveries During Skilled Performance. *Cogn. Sci.* **41**:1838–1870. [7, 10]

Gray, W. D., C. R. Sims, W.-T. Fu, and M. J. Schoelles. 2006. The Soft Constraints Hypothesis: A Rational Analysis Approach to Resource Allocation for Interactive Behavior. *Psychol. Rev.* **113**:461–482. [10]

Green, C. S., and D. Bavelier. 2012. Learning, Attentional Control, and Action Video Games. *Curr. Biol.* **22**:R197–R206. [16]

Grice, H. P. 1957. Meaning. *Philos. Psychol.* **67**:377–388. [8]

———. 1975. Logic and Conversation. In: Syntax and Semantics, Speech Arts, vol. 3, ed. P. Cole and J. L. Morgan, pp. 41–58. New York: Academic Press. [8, 15]

Griffiths, T. L. 2015. Manifesto for a New (Computational) Cognitive Revolution. *Cognition* **135**:21–23. [10]

Grimminger, A., K. J. Rohlfing, and P. Stenneken. 2010. Do Mothers Alter Their Pointing Behavior in Dependence of Children's Lexical Development and Task-Difficulty? Analysis of Task-Oriented Gestural Input Towards Typically Developed Children and Late Talkers. *Gesture* **10**:251–278. [11]

Grosz, B., and C. L. Sidner. 1986. Attention, Intentions, and the Structure of Discourse. *Comput. Linguist.* **12**:175–204. [9]

Guha, A. I. 2016. Towards Meaningful Human-Robot Collaboration on Object Placement. Undergraduate thesis, Dept. of Computer Science, Brown Univ. [9]

Guizzo, E., and E. Ackerman. 2012. How Rethink Robotics Built Its New Baxter Robot Worker. *IEEE Spectrum* Sept. 18, 2012. [9]

Hacking, I. 1983. Nineteenth Century Cracks in the Concept of Determinism. *J. Hist. Ideas* **44**:455–475. [14]

Hamlin, J. K., K. Wynn, and P. Bloom. 2007. Social Evaluation by Preverbal Infants. *Nature* **450**:557–559. [16]

Harpstead, E., C. J. MacLellan, V. Aleven, and B. A. Myers. 2015. Replay Analysis in Open-Ended Educational Games. In: Serious Games Analytics: Methodologies for Performance Measurement, Assessment, and Improvement, ed. C. S. Loh et al., pp. 381–399. Cham: Springer. [3]

Hattie, J., and H. Timperley. 2007. The Power of Feedback. *Rev. Educ. Res.* **77**:81–112. [11, 12]

Hayes, B., and J. Shah. 2017. Improving Robot Controller Interpretability and Transparency through Autonomous Policy Explanation. In: Human-Robot Interaction (HRI), Proc. 12th ACM/IEEE ACM/IEEE Intl. Conf., pp. 303–312. New York: ACM. [7, 9]

Hayes-Roth, B. 1985. A Blackboard Architecture for Control. *Artif. Intell.* **26**:251–321. [6]

Hegel, F., M. Lohse, and B. Wrede. 2009. Effects of Visual Appearance on the Attribution of Applications in Social Robotics. In: The 18th IEEE Intl. Symp. on Robot and Human Interactive Communication, pp. 64–71. IEEE. [11]

Heller, V., and K. Rohlfing. 2017. Reference as an Interactive Achievement: Sequential and Longitudinal Analyses of Labeling Interactions in Shared Book Reading and Free Play. *Front. Psychol.* **8**:139. [11]

Henderlong, J., and M. R. Lepper. 2002. The Effects of Praise on Children's Intrinsic Motivation: A Review and Synthesis. *Psychol. Bull.* **128**:774–795. [12]

Hinrichs, T., and K. Forbus. 2013a. X Goes First: Teaching Simple Games through Multimodal Interaction. In: Proc. 2nd Conf. on Advances in Cognitive Systems, ed. M. Klenk and J. E. Laird, pp. 205–218. Baltimore: Cognitive Systems Foundation. [17]

———. 2013b. Beyond the Rational Player: Amortizing Type-Level Goal Hierarchies. In: Goal Reasoning: Papers from the ACS Workshop, pp. 34–42. College Park: Univ. of Maryland, Dept. of Computer Science. [15]

Hirsh-Pasek, K., and R. M. Golinkoff. 1996. The Origins of Grammar: Evidence from Early Language Comprehension. Cambridge, MA: MIT Press. [11]

Hollan, J., E. Hutchins, and D. Kirsch. 2000. Distributed Cognition: Toward a New Foundation for Human-Computer Interaction Research. *ACM Trans. Comput. Hum. Interact.* **7**:174–196. [8]

Holroyd, A., C. Rich, C. L. Sidner, and B. Ponsler. 2011. Generating Connection Events for Human-Robot Collaboration. In: 2011 Ro-Man, pp. 241–246. Atlanta: IEEE [9]

Hömke, P., J. Holler, and S. C. Levinson. 2017. Eye Blinking as Addressee Feedback in Face-to-Face Conversation. *Res. Lang. Soc. Interact.* **50**:54–70. [8]

Huang, J., and M. Cakmak. 2017. Code3: A System for End-to-End Programming of Mobile Manipulator Robots for Novices and Experts. In: Human-Robot Interaction (HRI), Proc. 12th ACM/IEEE Intl. Conf., pp. 453–462. New York: ACM. [9]

Hudson, L. 1966. Contrary Imaginations. London: Methuen. [14]

Huffman, S. B., and J. E. Laird. 1995. Flexibly Instructable Agents. *J. Artif. Intell. Res.* **3**:271–324. [17]

Hume, G., J. Michael, A. Rovick, and M. Evens. 1996. Hinting as a Tactic in One-on-One Tutoring. *J. Learn. Sciences* **5**:23–49. [12]

Hutchins, E. 1995. Cognition in the Wild. Cambridge, MA: MIT Press. [8]

Indefrey, P. 2011. The Spatial and Temporal Signatures of Word Production Components: A Critical Update. *Front. Psychol.* **2**:255. [8]

Jacob, P. 2003. Intentionality. In: The Stanford Encyclopedia of Philosophy. Stanford Metaphysics Research Lab, Stanford Univ. [14]

James, W. 1890. The Principles of Psychology. New York: Henry Holt. [16]

———. 1975. Pragmatism. Cambridge, MA: Harvard Univ. Press. [14]

Janciauskas, M., and F. Chang. 2018. Input and Age-Dependent Variation in Second Language Learning: A Connectionist Account. *Cogn. Sci.* **42**:519–554. [16]

Janssen, C. P., and W. D. Gray. 2012. When, What, and How Much to Reward in Reinforcement Learning Based Models of Cognition. *Cogn. Sci.* **36**:333–358. [10]

Jeong, H., S. Siler, and M. T. H. Chi. 1997. Can Tutors Diagnose Students' Understanding? In: Proc. 19th Annu. Conf. of the Cognitive Science Society, ed. M. G. Shafto and P. Langley, p. 959. Mahwah: Erlbaum. [12]

Johansson, G. 1973. Visual Perception of Biological Motion and a Model for Its Analysis. *Percept. Psychophys.* **14**:201–211. [16]

———. 1976. Spatio-Temporal Differentiation and Integration in Visual Motion Perception: An Experimental and Theoretical Analysis of Calculus-Like Functions in Visual Data Processing. *Psychol. Res.* **38**:379–393. [16]

Johnson, M., J. M. Bradshaw, P. J. Feltovich, et al. 2014. Coactive Design: Designing Support for Interdependence in Joint Activity. *J. Hum. Robot. Interact.* **3**:43–69. [3]

Johnson, W. L., and S. B. Zaker. 2012. The Power of Social Simulation for Chinese Language Teaching. https://www.alelo.com/wp-content/uploads/2014/06/TCLT7_Presentation_Johnson_Zakar_May2012.pdf (accessed Oct. 4, 2017). [9]

Juel, C. 1996. What Makes Literacy Tutoring Effective? *Read. Res. Q.* **31**:268–289. [12]

Jurafsky, D., and J. Martin. 2008. Speech and Langauge Processing: An Introduction to Natural Language Processing, Computational Linguistics, and Speech Recognition, 2nd edition. Upper Saddle River, NJ: Prentice-Hall. [9]

Kaelbling, L. P., M. L. Littman, and A. W. Moore. 1996. Reinforcement Learning: A Survey. *J. Artif. Intell. Res.* **4**:237–285. [9]

Katz, S., D. Allbritton, and J. Connelly. 2003. Going Beyond the Problem Given: How Human Tutors Use Post-Solution Discussions to Support Transfer. *Int. J. Artific. Intel. Educ.* **13**:79–116. [12]

Kendrick, K. H. 2015. Other-Initiated Repair in English. *Open Linguist.* **1**:164–190. [8]

Kendrick, K. H., and J. Holler. 2017. Gaze Direction Signals Response Preference in Conversation. *Res. Lang. Soc. Interact.* **50**:12–32. [8]

Kenreck, T. 2012. "League of Legends" Players Log 1 Billion Hours a Month. NBC News, Oct. 12, 2012. https://www.nbcnews.com/tech/tech-news/league-legends-players-log-1-billion-hours-month-flna1C6423906 (accessed Nov. 1, 2018). [10]

Keysar, B., D. J. Barr, J. A. Balin, and J. S. Brauner. 2000. Taking Perspective in Conversation: The Role of Mutual Knowledge in Comprehension. *Psychol. Sci.* **11**:32–38. [7]

Keysar, B., D. J. Barr, and W. S. Horton. 1998. The Egocentric Basis of Language Use: Insights from a Processing Approach. *Curr. Dir. Psychol. Sci.* **7**:46–49. [15]

Kieras, D. E., and D. E. Meyer. 1997. An Overview of the Epic Architecture for Cognition and Performance with Application to Human-Computer Interaction. *Human-Comput. Interact.* **12**:391–438. [3]

Kim, J. H., H. M. Chae, and M. S. Glass. 2005. Expert and Novice Algebra Tutor Behaviors Compared. Proc. 27th Annual Conf. of the Cognitive Science Society, ed. B. G. Bara et al., p. 2499. Mahwah: Erlbaum. [12]

Kirk, J. R., and J. E. Laird. 2014. Interactive Task Learning for Simple Games. *Adv. Cog. Syst.* **3**:13–30. [4]

———. 2016. Learning General and Efficient Representations of Novel Games through Interactive Instruction. In: Proc. 4th Conf. on Advances in Cognitive Systems, ed. K. Forbus et al., pp. 1–14. Evanston: Cognitive Systems Foundation. [15, 17]

Kirk, J. R., A. Mininger, and J. E. Laird. 2016. Learning Task Goals Interactively with Visual Demonstrations. *Biol. Inspired Cogn. Arch.* **18**:1–8. [9, 17]

Kirkpatrick, J., R. Pascanu, N. Rabinowitz, et al. 2017. Overcoming Catastrophic Forgetting in Neural Networks. *PNAS* **114**:3521–3526. [16]

Klein, W., and C. Perdue. 1997. The Basic Variety (Or: Couldn't Natural Languages Be Much Simpler?). *Second Language Research* **13**:301–347. [7]

Kluger, A. N., and A. DeNisi. 1996. The Effects of Feedback Intervention on Performance: A Historical Review, a Meta-Analysis and a Preliminary Feedback Intervention Theory. *Psychol. Bull.* **112**:254–284. [12]

Knoblich, G., S. Butterfill, and N. Sebanz. 2011. Psychological Research on Joint Action: Theory and Data. In: Psychology of Learning and Motivation: Advances in Research and Theory, ed. B. H. Ross, vol. 54, pp. 59–101. [10]

Knoblock, C. 2004. Building Software Agents for Planning, Monitoring, and Optimizing Travel. In: Information and Communication Technologies in Tourism 2004: Proc. of the 11th Intl. Conf. on Information Technology and Travel, ed. A. J. Frew, pp. 1–15. New York: Springer. [3]

Koedinger, K. R., J. L. Booth, and D. Klahr. 2013. Instructional Complexity and the Science to Constrain It. *Science* **342**:935–937. [11]

Koedinger, K. R., A. Corbett, and C. Perfetti. 2012. The Knowledge-Learning-Instruction (KLI) Framework: Bridging the Science-Practice Chasm to Enhance Robust Student Learning. *Cogn. Sci.* **36**:757–798. [11, 12]

Kordjamshidi, P., D. Roth, and H. Wu. 2015. Saul: Towards Declarative Learning Based Programming. *IJCAI* **2015**:1844–1851. [15]

Kress-Gazit, H., G. E. Fainekos, and G. J. Pappas. 2007. From Structured English to Robot Motion. In: 2007 IEEE/RSJ Intl. Conf. on Intelligent Robots and Systems pp. 2717–2722. San Diego: IEEE. [9]

Krishnamurthy, J., and T. Kollar. 2013. Jointly Learning to Parse and Perceive: Connecting Natural Language to the Physical World. *Trans. Assoc. Comput. Linguist.* **1**:193–206. [9]

Kuehne, S., K. Forbus, D. Gentener, and B. Quinn. 2000. SEQL: Category Learning as Progressive Abstraction Using Structure Mapping. In: Proc. 22nd Annual Meeting of the Cognitive Science Society, ed. L. R. Gleitman and A. K. Joshi, pp. 770–775. Philadelphia: Cognitive Science Society. [15]

Lagemann, E. C. 1989. The Plural Worlds of Educational Research. *Hist. Educ. Q.* **29**:183–214. [14]

Laird, J. E. 2012. The Soar Cognitive Architecture. Cambridge, MA: MIT Press. [6, 17]

Laird, J. E., K. Gluck, J. Anderson, et al. 2017a. Interactive Task Learning. *IEEE Intell. Syst.* **32**:6–21. [11, 17]

Laird, J. E., C. Lebiere, and P. S. Rosenbloom. 2017b. A Standard Model of the Mind: Toward a Common Computational Framework across Artificial Intelligence, Cognitive Science, Neuroscience, and Robotics. *AI Mag.* **38**:13–26. [3, 4]

Laird, J. E., P. S. Rosenbloom, and A. Newell. 1986. Chunking in Soar: The Anatomy of a General Learning Mechanism. *Mach. Learn.* **1**:11–46. [3, 6]

Langley, P., N. Trivedi, and M. Banister. 2010. A Command Language for Taskable Virtual Agents. In: Proc. 6th Conf. Artificial Intelligence and Interactive Digital Entertainment. Stanford: AAAI Press. [17]

Larsson, S., and D. R. Traum. 2000. Information State and Dialogue Management in the Trindi Dialogue Move Engine Toolkit. *Nat. Lang. Eng.* **6**:323–340. [9]

Lazowski, R. A., and C. S. Hulleman. 2016. Motivation Interventions in Education: A Meta-Analytic Review. *Rev. Educ. Res.* **86**:602–640. [12]

Lebiere, C. 1999. The Dynamics of Cognitive Arithmetic. *Kognitionswiss.* **8**:5–19. [4]

Lebiere, C., R. Gray, D. Salvucci, and R. West. 2003. Choice and Learning under Uncertainty: A Case Study in Baseball Batting. In: Proc. 25th Annual Meeting of the Cognitive Science Society, pp. 704–709. Mahwah: Erlbaum. [4]

Lebiere, C., F. Jentsch, and S. Ososky. 2013a. Cognitive Models of Decision Making Processes for Human-Robot Interaction. In: Virtual, Augmented and Mixed Reality: Designing and Developing Augmented and Virtual Environments. Proc. 5th Intl. Conf, VAMR, Part 1, ed. R. Shumaker. Lecture Notes in Computer Science. Heidelberg: Springer. [4]

Lebiere, C., P. Pirolli, R. Thomson, et al. 2013b. A Functional Model of Sensemaking in a Neurocognitive Architecture. *Comput. Intell. Neurosci.* Article ID 921695. [4]

Lebiere, C., and D. Wallach. 2001. Sequence Learning in the ACT-R Cognitive Architecture: Empirical Analysis of a Hybrid Model. In: Sequence Learning: Paradigms, Algorithms, and Applications, ed. R. Sun and L. Giles, pp. 188–212. Lecture Notes in Artifical Intelligence 1828, G. Goos et al., series ed. Heidelberg: Springer. [4]

Lec, S. J. 1962. Unfrisierte Gedanken (Unkempt Thoughts), J. Galazka, series ed. New York: St. Martin's Press. [10]

Lehman, J. F., and J. G. Carbonell. 1989. Learning the User's Language: A Step Towards Automated Creation of User Models. In: User Models in Dialog Systems, ed. A. Kobsa and W. Wahlster, pp. 163–194. Berlin: Springer. [3]

Lehman, J. F., J. Van Dyke, and R. Rubinoff. 1995. Natural Language Processing for IFORs: Comprehension and Generation in the Air Combat Domain. In: Collected Papers of the Soar/IFOR Project, Research Report, Spring 1995, pp. 33–41. Marina del Rey, CA: USC Information Sciences Institute. [3]

Lenat, D. B. 1995. CYC: A Large-Scale Investment in Knowledge Infrastructure. *Commun. ACM* **38**:33–38. [15]

Lepper, M. R., and M. Woolverton. 2002. The Wisdom of Practice: Lessons Learned from the Study of Highly Effective Tutors. In: Improving Academic Achievement: Impact of Psychological Factors on Education, ed. J. Aronson, pp. 135–158. New York: Academic Press. [12]

Lepper, M. R., M. Woolverton, D. L. Mumme, and J.-L. Gurtner. 1993. Motivational Techniques of Expert Human Tutors: Lessons for the Design of Computer-Based Tutors. In: Computers as Cognitive Tools, ed. S. P. Lajoie and S. J. Derry, pp. 75–105. Hillsdale: Erlbaum. [12]

Lerch, F. J., C. Gonzalez, and C. Lebiere. 1999. Learning under High Cognitive Workload. In: Proc. 21st Conf. of the Cognitive Science Society, pp. 302–307. Mahwah: Erlbaum. [4]

Leslie, A. M. 1984. Spatiotemporal Continuity and the Perception of Causality in Infants. *Perception* **13**:287–305. [16]

Leslie, A. M., O. Friedman, and T. P. German. 2004. Core Mechanisms in "Theory of Mind." *Trends Cogn. Sci.* **8**:528–533. [15]

Leslie, A. M., and S. Keeble. 1987. Do Six-Month-Old Infants Perceive Causality? *Cognition* **25**:265–288. [16]

Leslie, A. M., F. Xu, P. D. Tremoulet, and B. J. Scholl. 1998. Indexing and the Object Concept: Developing What and Where Systems. *Trends Cogn. Sci.* **2**:10–18. [16]

Levinson, S. C. 1995. Interactional Biases in Human Thinking. In: Social Intelligence and Interaction, ed. E. N. Goody, pp. 221–260. Cambridge: Cambridge Univ. Press. [8]

———. 2006. On the Human Interaction Engine. In: Roots of Human Sociality: Culture, Cognition and Interaction, ed. N. J. Enfield and S. C. Levinson, pp. 39–69. Oxford: Berg. [8]

———. 2013a. Action Formation and Ascription. In: The Handbook of Conversation Analysis, ed. T. Stivers and J. Sidnell, pp. 103–130. Malden, MA: Wiley-Blackwell. [8]

———. 2013b. Recursion in Pragmatics. *Language* **89**:149–162. [8]

———. 2016. Turn-Taking in Human Communication, Origins and Implications for Language Processing. *Trends Cogn. Sci.* **20**:6–14. [7, 8]

———. 2017. Speech Acts. In: Oxford Handbook of Pragmatics, ed. Y. Huang, pp. 199–216. Oxford: Oxford Univ. Press. [8]

Levinson, S. C., and F. Torreira. 2015. Timing in Turn-Taking and Its Implications for Processing Models of Language. *Front. Psychol.* **6**:731. [8]

Lew-Williams, C., and A. Fernald. 2007. Young Children Learning Spanish Make Rapid Use of Grammatical Gender in Spoken Word Recognition. *Psychol. Sci.* **18**:193–198. [16]

Li, N., N. Matsuda, W. Cohen, and K. R. Koedinger. 2015. Integrating Representation Learning and Skill Learning in a Human-Like Intelligent Agent. *Artif. Intell.* **219**:67–91. [11, 17]

Lieto, A., C. Lebiere, and A. Oltramari. 2018. The Knowledge Level in Cognitive Architectures: Current Limitations and Possible Developments. *Cogn. Syst. Res.* **48**:39–55. [4]

Liszkowski, U., P. Brown, T. Callaghan, A. Takada, and C. De Vos. 2012. A Prelinguistic Gestural Universal of Human Communication. *Cogn. Sci.* **36**:698–713. [7]

Litman, D., and S. Pan. 2002. Designing and Evaluating an Adaptive Spoken Dialogue System. *User Model. User-adapt. Interact.* **12**:111–137. [7]

Liu, C., and J. Y. Chai. 2015. Learning to Mediate Perceptual Differences in Situated Human-Robot Dialogue. In: Proc. 29th AAAI Conf. on Artificial Intelligence, pp. 2288–2294. Austin: AAAI Press. [9]

Liu, C., S. Yang, S. Saba-Sadiya, et al. 2016. Jointly Learning Grounded Task Structures from Language Instruction and Visual Demonstration. In: EMNLP '08 Proc. Conf. on Empirical Methods in Natural Language Processing, ed. M. Lapata and T. H. Ng, pp. 1482–1492. Stroudsburg, PA: ACL. [7, 9]

Lloyd, P., L. Camaioni, and P. Ercolani. 1995. Assessing Referential Communication Skills in the Primary School Years: A Comparative Study. *Br. J. Dev. Psychol.* **13**:13–29. [7]

Lohan, K. S., K. J. Rohlfing, K. Pitsch, et al. 2012. Tutor Spotter: Proposing a Feature Set and Evaluating It in a Robotic System. *Int. J. Soc. Robot.* **4**:131–146. [11]

Lorenzet, S. J., E. Salas, and S. I. Tannenbaum. 2005. Benefiting from Mistakes: The Impact of Guided Errors on Learning, Performance, and Self-Efficacy. *Human Res. Devel. Q.* **16**:301–322. [7]

Love, N., T. Hinrichs, D. Haley, E. Schkufza, and M. Genesereth. 2008. General Game Playing: Game Description Language Specification. Technical Report No. LG-2006-01. Stanford: Stanford Univ. [3]

Lumsdaine, A. A., and R. E. Glaser, eds. 1960. Teaching Machines and Programmed Learning: A Source Book. Dept. of Audio-Visual Instruction, Natl. Education Assn. Washington, D.C.: GPO. [14]

Luo, Y. 2011. Three-Month-Old Infants Attribute Goals to a Non-Human Agent. *Dev. Sci.* **14**:453–460. [16]

Luo, Y., and R. Baillargeon. 2005. Can a Self-Propelled Box Have a Goal? Psychological Reasoning in 5-Month-Old Infants. *Psychol. Sci.* **16**:601–608. [16]

———. 2010. Toward a Mentalistic Account of Early Psychological Reasoning. *Curr. Dir. Psychol. Sci.* **19**:301–307. [16]

Lyons, D. E., A. G. Young, and F. C. Keil. 2007. The Hidden Structure of Overimitation. *PNAS* **104**:19751–11975. [15]

MacLellan, C. J., K. R. Koedinger, and N. Matsuda. 2014. Authoring Tutors with SimStudent: An Evaluation of Efficiency and Model Quality. In: Intelligent Tutoring Systems, ed. S. Trausan-Matu et al., pp. 551–560. Lecture Notes in Computer Science. Cham: Springer. [3]

Mandler, G. 1995. Origins and Consequences of Novelty. In: The Creative Cognition Approach, pp. 9–25. Cambridge, MA: Massachusetts Institute of Technology. [14]

Mané, A., and E. Donchin. 1989. The Space Fortress Game. *Acta Psychol.* **71**:17–22. [10]

Mani, N., and F. Huettig. 2012. Prediction during Language Processing Is a Piece of Cake—but Only for Skilled Producers. *J. Exp. Psychol. Hum. Percept. Perform.* **38**:843–847. [16]

Marston, D., S. L. Deno, D. Kim, K. Diment, and D. Rogers. 1995. Comparison of Reading Intervention Approaches for Students with Mild Disabilities. *Except. Child.* **62**:20–37. [12]

Matsuda, N., W. W. Cohen, J. Sewall, G. Lacerda, and K. R. Koedinger. 2007. Predicting Students' Performance with SimStudent That Learns Cognitive Skills from Observation. In: Proc. of the Intl. Conf. on Artificial Intelligence in Education, ed. R. Luckin et al., pp. 467–476. Amsterdam: IOS Press. [3]

Matsuda, N., E. Yarzebinski, V. Keiser, et al. 2013. Cognitive Anatomy of Tutor Learning: Lessons Learned with SimStudent. *J. Educ. Psychol.* **105**:1152–1163. [11]

Matthews, D., E. Lieven, A. Theakston, and M. Tomasello. 2006. The Effect of Perceptual Availability and Prior Discourse on Young Children's Use of Referring Expressions. *Appl. Psycholinguist.* **27**:403–422. [7]

Matuszek, C., L. Bo, L. Zettlemoyer, and D. Fox. 2014. Learning from Unscripted Deictic Gesture and Language for Human-Robot Interactions. In: Proc. 28th AAAI Conf. on Artificial Intelligence, pp. 2556–2563. Quebec: AAAI Press. [9]

McArthur, D., C. Stasz, and M. Zmuidzinas. 1990. Tutoring Techniques in Algebra. *Cogn. Instr.* **7**:197–244. [12]

McCarthy, J. 1968. Programs with Common Sense. In: Semantic Information Processing, ed. M. Minsky, pp. 403–418. Cambridge, MA: MIT Press. [17]

McClelland, J. L., B. L. McNaughton, and R. C. O'Reilly. 1995. Why There Are Complementary Learning Systems in the Hippocampus and Neocortex: Insights from the Successes and Failures of Connectionist Models of Learning and Memory. *Psychol. Rev.* **102**:419–457. [16]

McCorduck, P. 2004. Machines Who Think. Natick, MA: A. K. Peters. [3]

Menekse, M., G. S. Stump, S. Krause, and M. T. H. Chi. 2013. Differentiated Overt Learning Activities for Effective Instruction in Engineering Classrooms. *J. Engineer. Educ.* **102**:346–374. [11, 12]

Meriçli, C., S. D. Klee, J. Paparian, and M. Veloso. 2014. An Interactive Approach for Situated Task Specification through Verbal Instructions. In: Proc. 13th Intl. Conf. on Autonomous Agents and Multi-Agent Systems, pp. 1069–1076. Paris: Intl. Foundation for Autonomous Agents and Multiagent Systems. [17]

Merrill, D. C., B. J. Reiser, S. K. Merrill, and S. Landes. 1995. Tutoring: Guided Learning by Doing. *Cogn. Instr.* **13**:315–372. [12]

Merrill, D. C., B. J. Reiser, M. Ranney, and J. G. Trafton. 1992. Effective Tutoring Techniques: A Comparison of Human Tutors and Intelligent Tutoring Systems. *J. Learn. Sciences* **2**:277–306. [12]

Merritt, M. 1976. On Questions Following Questions (on Service Encounters). *Lang. Soc.* **5**:315–357. [8]

Michotte, A. 1963. The Perception of Causality, vol. 12. Oxford: Basic Books. [16]

Mininger, A., and J. E. Laird. 2016. Interactively Learning Strategies for Handling References to Unseen or Unknown Objects. In: Proc. 4th Conf. on Advances in Cognitive Systems, ed. K. Forbus et al., pp. 1–16. Evanston: Cognitive Systems Foundation. [4, 17]

———. 2018. Interactively Learning a Blend of Goal-Based and Procedural Tasks. In: Proc. of the 32nd AAAI Conf. on Artificial Intelligence, AAAI Press. http://soar. eecs.umich.edu/pubs/mininger_aaai18.pdf (accessed April 10, 2018). [17]

Misra, D. K., J. Sung, K. Lee, and A. Saxena. 2016. Tell Me Dave: Context Sensitive Grounding of Natural Language to Manipulation Instructions. *Int. J. Rob. Res.* **35**:281–300. [9]

Mitchell, T. M. 1977. Version Spaces: A Candidate Elimination Approach to Rule Learning. Proc. 5th Intl. Conf. on AI, vol. 1, pp. 305–310. Cambridge, MA: IJCAI.

———. 1997. Machine Learning. New York: McGraw-Hill. [13]

Mitchell, T. M., R. M. Keller, and S. V. Kedar-Cabelli. 1986. Explanation-Based Learning: A Unifying View. *Mach. Learn.* 1:47–80. [3]

Mitchell, T. M., P. Utgoff, and R. Banerji. 1983. Learning by Experimentation: Acquiring and Refining Problem-Solving Heuristics. In: Machine Learning, ed. J. G. Carbonell et al., pp. 163–190. Heidelberg: Springer. [13]

Mnih, V., K. Kavukcuoglu, D. Silver, et al. 2015. Human-Level Control through Deep Reinforcement Learning. *Nature* **518**:529–533. [3, 16]

Mohan, S., and J. E. Laird. 2014. Learning Goal-Oriented Hierarchical Tasks from Situated Interactive Instruction. In: Proc. 28th AAAI Conf. on Artificial Intelligence, pp. 387–394. Quebec: AAAI Press. [4, 7, 9, 15, 17]

Mohan, S., A. Mininger, J. R. Kirk, and J. E. Laird. 2012. Acquiring Grounded Representation of Words with Situated Interactive Instruction. *Adv. Cog. Syst.* **2**:113–130. [9, 17]

Mohseni-Kabir, A., C. Li, V. Wu, et al. 2018. SLHAP: Simultaneous Learning of Hierarchy and Primitives. Simultaneous Learning of Hierarchy and Primitives for Complex Robot Tasks. *Auton. Robots* **April**: 1–16. doi.org/10.1007/s10514-018-9749-y. [5, 9]

Mohseni-Kabir, A., C. Rich, S. Chernova, C. Sidner, and D. Miller. 2015. Interactive Hierarchical Task Learning from a Single Demonstration. In: Human-Robot Interaction (HRI), Proc. 10th ACM/IEEE Intl. Conf., pp. 205–212. New York: ACM. [5, 7]

Mollard, Y., T. Munzer, A. Baisero, M. Toussaint, and M. Lopes. 2015. Robot Programming from Demonstration, Feedback and Transfer. In: 2015 IEEE/RSJ Intl. Conf. on Intelligent Robots and Systems pp. 1825–1831. Hamburg: IEEE. [9]

Mooney, R. 2008. Learning to Connect Language and Perception. In: Proc. 23rd AAAI Conf. on Artificial Intelligence, pp. 1598–1601. Chicago: AAAI Press. [9]

Morency, L.-P., I. Kok, and J. Gratch. 2008. Predicting Listener Backchannels: A Probabilistic Multimodal Approach. In: Intelligent Virtual Agents, ed. H. Prendinger et al., pp. 176–190. Lecture Notes in Computer Science, vol. 5208. Berlin: Springer. [7]

Morgan, T. J. H., N. T. Uomini, L. E. Rendell, et al. 2015. Experimental Evidence for the Co-Evolution of Hominin Tool-Making Teaching and Language. *Nat. Commun.* **6**:6029. [8]

Narciss, S. 2007. Feedback strategies for interactive learning tasks. In: Handbook of Research on Educational Communications and Technology, 3rd edition, ed. J. M. Spector et al., pp. 125–144. Mahwah: Erlbaum. [12]

Nehaniv, C. L., and K. Dautenhahn, eds. 2007. Imitation and Social Learning in Robots, Humans and Animals: Behavioural, Social and Communicative Dimensions. New York: Cambridge Univ. Press. [11]

Newell, A. 1973a. Production Systems: Models of Control Structure. In: Visual Information Processing, ed. W. G. Chase, pp. 526–547. New York: Academic Press. [4]

———. 1973b. You Can't Play 20 Questions with Nature and Win: Projective Comments on the Papers of This Symposium. In: Visual Information Processing, ed. W. G. Chase, pp. 283–231. New York: Academic Press. [3, 4]

———. 1982. The Knowledge Level. *Artif. Intell.* **18**:82–127. [3]

———. 1990. Unified Theories of Cognition. Cambridge, MA: Harvard Univ. Press. [3, 4, 6]

————. 1991. Reasoning, Problem Solving and Decision Processes: The Problem Space as a Fundamental Category. In: Attention and Performance VIII, ed. R. S. Nickerson, pp. 693–718. Hillsdale: Erlbaum. [17]

Newell, A., and P. S. Rosenbloom. 1981. Mechanisms of Skill Acquisition and the Law of Practice. In: Cognitive Skills and Their Acquisition, ed. J. R. Anderson, pp. 1–55. Hillsdale: Erlbaum. [10]

Niekum, S., S. Osentoski, G. D. Konidaris, et al. 2015. Learning Grounded Finite-State Representations from Unstructured Demonstrations. *Int. J. Rob. Res.* **34**:131–115. [9]

Nikolaidis, S., P. Lasota, R. Ramakrishnan, and J. Shah. 2015. Improved Human–Robot Team Performance through Cross-Training, an Approach Inspired by Human Team Training Practices. *Int. J. Rob. Res.* **34**:1711–1730. [11]

Nomikou, I., M. Koke, and K. J. Rohlfing. 2017. Verbs in Mothers' Input to Six-Month-Olds: Synchrony between Presentation, Meaning, and Actions Is Related to Later Verb Acquisition. *Brain Sciences* **7**:52. [11]

Noordzij, M., S. E. Newman-Norlund, J. P. De Ruiter, et al. 2009. Brain Mechanisms Underlying Human Communication. *Front. Hum. Neurosci.* **3**:14. [8]

Norman, D. A. 1981. Categorization of Action Slips. *Psychol. Rev.* **88**:1. [7]

Ohlsson, S., B. Di Eugenio, B. Chow, et al. 2007. Beyond the Code-and-Count Analysis of Tutoring Dialogues. In: Artificial Intelligence in Education, ed. R. Luckin et al., pp. 349–356. Amsterdam: IOS Press. [12]

Oltramari, A., and C. Lebiere. 2012. Using Ontologies in a Cognitive-Grounded System: Automatic Action Recognition in Video-Surveillance. In: Proc. 7th Intl. Conf. on Semantic Technologies for Intelligence, Defense, and Security, ed. P. C. G. da Costa and K. B. Laskey, pp. 20–27. CEUR Workshop Proc. 966. Fairfax: CEUR. [4]

————. 2013. Knowledge in Action: Integrating Cognitive Architectures and Ontologies. In: New Trends of Research in Ontologies and Lexical Resources: Ideas, Projects, Systems, ed. A. Oltramari et al., pp. 135–154. Heidelberg: Springer. [4]

O'Neil, C. 2016. Weapons of Math Destruction: How Big Data Increases Inequality and Threatens Democracy. New York: Crown. [3]

Onishi, K. H., and R. Baillargeon. 2005. Do 15-Month-Old Infants Understand False Beliefs? *Science* **308**:255–258. [16]

Orseau, L., and S. Armstrong. 2016. Safely Interruptible Agents. In: Proc. 32nd Conf. on Uncertainty in Artificial Intelligence, pp. 557–566. Arlington: AUAI Press. [18]

Osborn, A. F. 1948. Your Creative Mind. New York: Charles Scribner. [14]

Özyürek, A., R. M. Willems, S. Kita, and P. Hagoort. 2007. On-Line Integration of Semantic Information from Speech and Gesture: Insights from Event-Related Brain Potentials. *J. Cogn. Neurosci.* **19**:605–616. [8]

Palinscar, A. S., and A. L. Brown. 1984. Reciprocal Teaching of Comprehension-Fostering and Comprehension-Monitoring Activities. *Cogn. Instr.* **1**:117–175. [12]

Pardowitz, M., S. Knoop, R. Dillmann, and R. D. Zollner. 2007. Incremental Learning of Tasks from User Demonstrations, Past Experiences, and Vocal Comments. *IEEE Trans. Syst. Man Cybern. B Cybern.* **37**:322–332. [9]

Pavlik, P. I., Jr., and J. R. Anderson. 2005. Practice and Forgetting Effects on Vocabulary Memory: An Activation-Based Model of the Spacing Effect. *Cogn. Sci.* **29**:559–586. [11]

Peirce, C. S. 1902/1935. Logic as Semiotic: The Theory of Signs. In: Philosophical Writings, ed. J. Buchler. New York: Dover. [14]

Pejsa, T., D. Bohus, M. F. Cohen, et al. 2014. Natural Communication About Uncertainties in Situated Interaction. In: Proc. 16th Intl. Conf. on Multimodal Interaction, pp. 283–290. New York: ACM. [9]

Perera, I., and J. Allen. 2014. What Is the Ground? Continuous Maps for Symbol Grounding. *Proc. Annu. Conf. Cogn. Sci. Soc.* **36**:1156–1161. [3]

Pew, R., and A. Mavor, eds. 1998. Modeling Human and Organizational Behavior: Application to Military Simulations. Washington, D.C.: National Academies Press. [17]

Phillips, E., D. Ullman, M. M. A. de Graaf, and B. F. Malle. 2017. What does a robot look like? A multi-site examination of user expectations about robot appearance. *Proc. Hum. Factors Ergon. Soc.* **61**:1215–1219. [18]

Phillips, M., V. Hwang, S. Chitta, and M. Likhachev. 2016. Learning to Plan for Constrained Manipulation from Demonstrations. *Auton. Robots* **40**:109–124. [9]

Pitsch, K., A. Vollmer, and M. Mühlig. 2013. Robot Feedback Shapes the Tutor's Presentation: How a Robot's Online Gaze Strategies Lead to Micro-Adaptation of the Human's Conduct. *Interact. Stud.* **14**:268–296. [11]

Pitsch, K., A. Vollmer, K. Rohlfing, J. Fritsch, and B. Wrede. 2014. Tutoring in Adult-Child Interaction: On the Loop of the Tutor's Action Modification and the Recipient's Gaze. *Interact. Stud.* **15**:55–98. [11]

Polya, G. 1945. How to Solve It. Princeton: Princeton Univ. Press. [13]

Putnam, R. T. 1987. Structuring and Adjusting Content for Students: A Study of Live and Simulated Tutoring of Addition. *Am. Educ. Res. J.* **24**:13–48. [11, 12]

Pylyshyn, Z. W., and R. W. Storm. 1988. Tracking Multiple Independent Targets: Evidence for a Parallel Tracking Mechanism. *Spat. Vis.* **3**:179–197. [16]

Quine, W. V. O. 1960. Word and Object: An Inquiry into the Linguistic Mechanisms of Objective Reference. New York: Wiley. [7]

Raaijmakers, J. G. W. 2003. Spacing and Repetition Effects in Human Memory: Application of the SAM Model. *Cogn. Sci.* **27**:431–452. [11]

Rakoczy, H. 2008. Taking Fiction Seriously: Young Children Understand the Normative Structure of Joint Pretence Games. *Dev. Psychol.* **44**:1195–1201. [16]

Rakoczy, H., and M. Tomasello. 2006. Two-Year-Olds Grasp the Intentional Structure of Pretense Acts. *Dev. Sci.* **9**:557–564. [16]

Rakoczy, H., M. Tomasello, and T. Striano. 2004. Young Children Know That Trying Is Not Pretending: A Test of the "Behaving-as-If" Construal of Children's Early Concept of Pretense. *Dev. Psychol.* **40**:388–399. [16]

Ramakrishnan, R., C. Zhang, and J. Shah. 2017. Perturbation Training for Human-Robot Teams. *J. Artif. Intell. Res.* **59**:495–541. [11]

Reason, J. 1990. Human Error. Cambridge: Cambridge Univ. Press. [7]

Reesink, G., R. Singer, and M. Dunn. 2009. Explaining the Linguistic Diversity of Sahul Using Population Models. *PLoS Biology* **7**:e1000241. [7]

Reynolds, J. R., J. M. Zacks, and T. S. Braver. 2007. A Computational Model of Event Segmentation from Perceptual Prediction. *Cogn. Sci.* **31**:613–643. [16]

Rich, C. 2009. Building Task-Based User Interfaces with ANSI/CEA-2018. *IEEE Computer* **42**:20–27. [5]

Rich, C., B. Ponsler, A. Holroyd, and C. L. Sidner. 2010. Recognizing Engagement in Human-Robot Interaction. In: Human-Robot Interaction (HRI), Proc. 5th ACM/IEEE Intl. Conf., pp. 375–382. Piscataway, NJ: IEEE Press. [9]

Rich, C., and C. L. Sidner. 1998. Collagen: A Collaborative Manager for Software Interface Agents. *User Model. User-adapt. Interact.* **8**:315–350. [9]

Rickel, J., and W. L. Johnson. 2000. Task-Oriented Collaboration with Embodied Agents in Virtual Worlds. In: Embodied Conversational Agents, ed. J. Cassell et al., pp. 95–122. Cambridge, MA: MIT Press. [9]

Robinson, P., and N. C. Ellis, eds. 2008. Handbook of Cognitive Linguistics and Second Language Acquisition. New York: Routledge. [13]

Rochat, P., J. G. Querido, and T. Striano. 1999. Emerging Sensitivity to the Timing and Structure of Protoconversation in Early Infancy. *Dev. Psychol.* **35**:950. [7]

Rogers, C. R. 1954. Toward a Theory of Creativity. *Etc* **11**:249–260. [14]

Rogoff, B., R. Paradise, R. M. Arauz, M. Correa-Chavez, and C. Angelillo. 2003. Firsthand Learning through Intent Participation. *Annu. Rev. Psychol.* **54**:175–203. [8]

Rohlfing, K. J., J. Fritsch, B. Wrede, and T. Jungmann. 2006. How Can Multimodal Cues from Child-Directed Interaction Reduce Learning Complexity in Robots? *Adv. Robotics* **20**:1183–1199. [11]

Rohlfing, K. J., and J. Tani. 2011. Grounding Language in Action. *IEEE Trans. Auton. Ment. Dev.* **3**:109–112. [11]

Rosch, E. H. 1973. Natural Categories. *Cogn. Psychol.* **4**:328–350. [3]

Rose, C. P., D. Bhembe, S. Siler, R. Srivastava, and K. VanLehn. 2003. The Role of Why Questions in Effective Human Tutoring. In: Artificial Intelligence in Education, ed. U. Hoppe et al. Amsterdam: IOS Press. [12]

Rosenbloom, P. S., J. Laird, and A. Newell. 1986. Meta-Levels in Soar. In: Meta-Level Architectures and Reflection, ed. P. Maes and D. Nardi, pp. 227–239. Amsterdam: Elsevier. [13]

Rubin, R. D., P. D. Watson, M. C. Duff, and N. J. Cohen. 2014. The Role of the Hippocampus in Flexible Cognition and Social Behavior. *Front. Hum. Neurosci.* **8**: [3]

Runco, M. A. 2007. To Understand Is to Create: An Epistemological Perspective on Human Nature and Personal Creativity. In: Everyday Creativity and New Views of Human Nature, ed. R. Richards, pp. 91–107. Washington, D.C.: American Psychological Association. [14]

Russell, S., and P. Norvig. 1995. Artificial Intelligence: A Modern Approach. Upper Saddle River, NJ: Prentice-Hall. [3]

Rybski, P. E., K. Yoon, J. Stolarz, and M. M. Veloso. 2007. Interactive Robot Task Training through Dialog and Demonstration. In: Human-Robot Interaction (HRI), Proc. 2nd ACM/IEEE Intl. Conf., pp. 49–56. New York: ACM. [9]

Sacks, H., E. A. Schegloff, and G. Jefferson. 1974. A Simplest Systematics for the Organization of Turn-Taking for Conversation. *Language* **50**:696–735. [7]

Salvucci, D. D. 2013. Integration and Reuse in Cognitive Skill Acquisition. *Cogn. Sci.* **37**:829–860. [17]

———. 2014. Endowing a Cognitive Architecture with World Knowledge. In: Proc. Annu. Meeting of the Cognitive Science Society, ed. P. Bello et al., vol. 36, pp. 1353–1358. Quebec: Cognitive Science Society. [4]

Sanner, S., J. R. Anderson, C. Lebiere, and M. C. Lovett. 2000. Achieving Efficient and Cognitively Plausible Learning in Backgammon. In: Proc. 17th Intl. Conf. on Machine Learning, pp. 823–830. San Francisco: Morgan Kaufmann. [4]

Saon, G., T. Sercu, S. Rennie, and H.-K. J. Kuo. 2016. The IBM 2016 English Conversational Telephone Speech Recognition System. https://arxiv.org/pdf/1604.08242.pdf (accessed Oct. 5, 2017). [9]

Sargano, A. B., P. Angelov, and Z. Habib. 2017. A Comprehensive Review on Handcrafted and Learning-Based Action Representation Approaches for Human Activity Recognition. *Applied Sci.* **7**:doi:10.3390/app7010110. [9]

Schaefer-Simmern, H. 1961. The Unfolding of Artistic Activity. Berkeley and Los Angeles: Univ. of California Press. [14]

Schegloff, E. A. 1982. Discourse as an Interactional Achievement: Some Uses of "Uh Huh" and Other Things That Come between Sentences. In: Analyzing Discourse: Text and Talk, ed. D. Tannen, pp. 71–93. Washington, D.C.: Georgetown Univ. Press. [7]

———. 2007. Sequence Organization in Interaction. Cambridge: Cambridge Univ. Press. [8]

Schegloff, E. A., G. Jefferson, and H. Sacks. 1977. The Preference for Self-Correction in the Organization of Repair in Conversation. *Language* **53**:361–382. [7]

Schelling, T. 1960. The Strategy of Conflict. Cambridge, MA: MIT Press. [8]

Scheutz, M. 2014. Teach One, Teach All: The Explosive Combination of Instructible Robots Connected via Cyber Systems. In: 4th Annu. IEEE Intl. Conf. on Cyber Technology in Automation, Control, and Intelligent, pp. 43–48. Hong Kong: IEEE. [18]

Scheutz, M., J. Harris, and P. Schermerhorn. 2013. Systematic Integration of Cognitive and Robotic Architectures. *Adv. Cog. Syst.* **2**:277–296. [3]

Schroder, M., E. Bevacqua, R. Cowie, et al. 2012. Building Autonomous Sensitive Artificial Listeners. *IEEE Trans. Affect. Comput.* **3**:165–183. [7]

Schulman, J., J. Ho, C. Lee, and P. Abbeel. 2016. Learning from Demonstrations through the Use of Non-Rigid Registration. In: Robotics Research, ed. M. Inaba and P. Corke, pp. 339–354. Springer Tracts in Advanced Robotics 114. Cham: Springer. [3]

Schultz, W., P. Dayan, and P. R. Montague. 1997. A Neural Substrate of Prediction and Reward. *Science* **275**:1593–1599. [15]

Scruggs, T. E., and L. Richter. 1985. Tutoring Learning Disabled Students: A Critical Review. *Learn. Disab. Q.* **8**:286–298. [12]

Sebanz, N., H. Bekkering, and G. Knoblich. 2006. Joint Action: Bodies and Minds Moving Together. *Trends Cogn. Sci.* **10**:70–76. [7, 8]

Sebanz, N., and G. Knoblich. 2009. Prediction in Joint Action: What, When, and Where. *Top. Cogn. Sci.* **1**:353–367. [10]

Sebanz, N., G. Knoblich, and W. Prinz. 2003. Representing Others' Actions: Just Like One's Own? *Cognition* **88**:B11–B21. [8]

Seligman, M. E. P., and J. L. Hager. 1972. Biological Boundaries of Learning. New York: Appleton-Century-Crofts. [14]

Shah, F., M. Evens, J. Michael, and A. Rovick. 2002. Classifying Student Initiatives and Tutor Responses in Human Keyboard-to-Keyboard Tutoring Sessions. *Discourse Process.* **33**:23–52. [12]

Shanahan, T. 1998. On the Effectiveness and Limitations of Tutoring in Reading. *Rev. Res. Educ.* **23**:217–234. [12]

She, L., and J. Y. Chai. 2017. Interactive Learning of Grounded Verb Semantics Towards Human-Robot Communication. In: Proc. 55th Annual Meeting of the Association for Computational Linguistics, pp. 1634–1644. Vancouver: ACL. [9]

She, L., S. Yang, Y. Cheng, et al. 2014. Back to the Blocks World: Learning New Actions through Situated Human-Robot Dialogue. In: Proc. SIGDIAL 2014, pp. 89–97. Philadelphia: ACL. [9]

Shuell, T. 1990. Teaching and Learning as Problem Solving. *Theory. Pract.* **29**:102–108. [13]

Shute, V. J. 2008. Focus on formative feedback. *Rev. Educ. Res.* **78**:153–189. [12]

Sibert, C., and W. D. Gray. 2017. The Tortoise Only Wins When the Race Is Long: How the Task Environment Changes the Behavior of Tetris Models. Poster Presented at the 39th Annual Conference of the Cognitive Science Society. http://homepages. rpi.edu/~grayw/pubs/papers/2017/sibert17csc.paper.poster.pdf (accessed Aug. 23, 2017). [10]

———. 2018. The Tortoise and the Hare: Understanding the Influence of Sequencelength and Variability on Decision-Making in Skilled Performance. *Comput. Brain Behav,* **1**:215–227. [10]

Sibert, C., W. D. Gray, and J. K. Lindstedt. 2017. Interrogating Feature Learning Models to Discover Insights into the Development of Human Expertise in a Real-Time, Dynamic Decision-Making Task. *Top. Cogn. Sci.* **9**:374–394. [10]

Siler, S., and K. VanLehn. 2015. Investigating Micro-Adaptation in One-to-One Tutoring. *J. Experiment. Edu.* **83**:344–367. [11, 12]

Simon, H. A. 1975. The Functional Equivalence of Problem Solving Skills. *Cogn. Psychol.* **7**:268–288. [10]

———. 1989. The Scientist as Problem Solver. In: Complex Information Processing: The Impact of Herbert A. Simon, ed. D. Klahr and K. Kotovsky, pp. 375–398. Hillsdale: Erlbaum Associates. [10]

———. 1992. What Is an "Explanation" of Behavior? *Psychol. Sci.* **3**:150–161. [10]

———. 1996. The Sciences of the Artificial, 3rd edition. Cambridge, MA: MIT Press. [3, 13]

Simon, H. A., and W. G. Chase. 1973. Skill in Chess. *American Scientist* **61**:394–403. [10]

Simon, H. A., and K. Gilmartin. 1973. A Simulation of Memory for Chess Positions. *Cogn. Psychol.* **5**:29–46. [10]

Simonton, D. K. 2003. Scientific Creativity as Constrained Stochastic Behavior: The Integration of Product, Person, and Process Perspectives. *Psychol. Bull.* **129**:475. [14]

Sleeman, D., A. E. Kelly, R. Martinak, R. D. Ward, and J. L. Moore. 1989. Studies of Diagnosis and Remediation with High School Algebra Students. *Cogn. Sci.* **13**:551–568. [11, 12]

Sloman, A., and M. Scheutz. 2002. A Framework for Comparing Agent Architectures. In: Proc. UKCI'02: UK Workshop on Computational Intelligence. Birmingham: Univ. of Birmingham. [3]

Smith, N. V. 1982. Mutual Knowledge. London: Academic Press. [2]

Spangenberg, M., and D. Henrich. 2015. Grounding of Actions Based on Verbalized Physical Effects and Manipulation Primitives. In: 2015 IEEE/RSJ Intl. Conf. on Intelligent Robots and Systems S pp. 844–851. Hamburg: IEEE. [9]

Spelke, E. S. 1990. Principles of Object Perception. *Cogn. Sci.* **14**:29–56. [16]

Spelke, E. S., K. Breinlinger, J. Macomber, and K. Jacobson. 1992. Origins of Knowledge. *Psychol. Rev.* **99**:605–632. [16]

Spelke, E. S., R. Kestenbaum, D. J. Simons, and D. Wein. 1995. Spatiotemporal Continuity, Smoothness of Motion and Object Identity in Infancy. *Br. J. Dev. Psychol.* **13**:113–142. [16]

Steels, L. 1990. Components of Expertise. *AI Mag.* **11**:30–49. [13]

———. 2017. Basics of Fluid Construction Grammar. In: Verb Phrase and Fluid Construction Grammar. Constructions and Frames, ed. L. Steels and K. Beuls, vol. 9, pp. 178–225. Amsterdam: John Benjamins. [13]

Steels, L., and M. Hild, eds. 2012. Language Grounding in Robots. New York: Springer. [13]

Steels, L., and M. Tokoro, eds. 2003. The Future of Learning. Amsterdam: IOS Press. [13]

Stevens, A., and A. Collins. 1977. The Goal Structure of a Socratic Tutor. In: Proc. Natl. ACM Conf., pp. 256–263. New York: ACM. [12]

Stevens, A., A. Collins, and S. E. Goldin. 1979. Misconceptions in Student's Understanding. *Int. J. Man Mach. Stud.* **11**:145–156. [12]

Stivers, T., N. J. Enfield, P. Brown, et al. 2009. Universals and Cultural Variation in Turn-Taking in Conversation. *PNAS* **106**:10587–10592. [7, 8]

Stocco, A., C. Lebiere, and J. R. Anderson. 2010. Conditional Routing of Information to the Cortex: A Model of the Basal Ganglia's Role in Cognitive Coordination. *Psychol. Rev.* **117**:541–574. [6]

Stocco, A., C. Lebiere, R. C. O'Reilly, and J. R. Anderson. 2012. Distinct Contributions of the Caudate Nucleus, Rostral Prefrontal Cortex, and Parietal Cortex to the Execution of Instructed Tasks. *Cogn. Affect. Behav. Neurosci.* **12**:611–628. [15]

Suchanek, F. M., G. Kasneci, and G. Weikum. 2007. YAGO: A Core of Semantic Knowledge. In: Proc. 16th Intl. Conf. on World Wide Web, pp. 697–706. Banff: ACM. [15]

Taatgen, N. A. 2013. The Nature and Transfer of Cognitive Skills. *Psychol. Rev.* **120**:439–471. [6]

Taatgen, N. A., D. Huss, and J. R. Anderson. 2006. How Cognitive Models Can Inform the Design of Instructions. In: Proc. 7th Intl. Conf. on Cognitive Modeling, pp. 304–309. Trieste, Italy: Edizioni Goliardiche. [4]

Taatgen, N. A., D. Huss, D. Dickison, and J. R. Anderson. 2008. The Acquisition of Robust and Flexible Cognitive Skills. *J. Exp. Psychol. Gen.* **137**:548. [3]

Taatgen, N. A., and F. J. Lee. 2003. Production Compilation: A Simple Mechanism to Model Complex Skill Acquisition. *Human Factors* **45**:61–76. [3]

Tellex, S., T. Kollar, S. Dickerson, et al. 2011. Understanding Natural Language Commands for Robotic Navigation and Mobile Manipulation. In: Proc. 25th AAAI Conf. on Artificial Intelligence, pp. 1507–1514. San Francisco: AAAI Press. [9]

Tellex, S., P. Thaker, J. Joseph, and N. Roy. 2014. Learning Perceptually Grounded Word Meanings from Unaligned Parallel Data. *Mach. Learn.* **94**:151–167. [9]

Tennie, C., J. Call, and M. Tomasello. 2009. Ratcheting up the Ratchet: on the Evolution of Cumulative Culture. *Phil. Trans. R. Soc. B* **364**:2405–2415. [7]

Tenorth, M., and M. Beetz. 2009. KNOWROB—Knowledge Processing for Autonomous Personal Robots. In: 2009 IEEE/RSJ Intl. Conf. on Intelligent Robots and Systems pp. 4261–4266. St. Louis: IEEE. [9]

Thomason, J., S. Zhang, R. Mooney, and P. Stone. 2015. Learning to Interpret Natural Language Commands through Human-Robot Dialog. In: Proc. 24th Intl. Joint Conference on Artificial Intelligence (IJCAI), pp. 1923– 1929. Palo Alto: AAAI Press/IJCAI. [9]

Thomaz, A. L., and C. Breazeal. 2006. Transparency and Socially Guided Machine Learning. *IEEE Trans. Auton. Ment. Dev.* **2**:108–111. [9]

———. 2008. Teachable Robots: Understanding Human Teaching Behavior to Build More Effective Robot Learners. *Artif. Intell.* **172**:716–737. [17]

Thomson, R., C. Lebiere, J. R. Anderson, and J. Staszewski. 2015. A General Instance-Based Learning Framework for Studying Intuitive Decision-Making in a Cognitive Architecture. *J. Appl. Res. Mem. Cogn.* **4**:180–190. [3, 4]

Thorndike, E. L. 1898. Animal Intelligence: An Experimental Study of the Associative Processes in Animals. *Psychol. Rev.* **8**:109. [14]

———. 1901. The Evolution of the Human Intellect. *Pop. Sci. Monthly* **60**:58–65. [14]

———. 1913. Educational Psychology Vol II: The Psychology of Learning. New York: Teachers College, Columbia Univ. [10, 14]

Tiles, M. 1984. Mathematics: The Language of Science? *Monist* **67**:3–17. [14]

Tomasello, M. 1990. Cultural Transmission in the Tool Use and Communicatory Signaling of Chimpanzees? In: "Language" and Intelligence in Monkeys and Apes: Comparative Developmental Perspectives, ed. S. Parker and K. Gibson. Cambridge: Cambridge Univ. Press. [7]

———. 2003. Constructing a Language: A Usage-Based Approach to Child Language Acquisition. Cambridge, MA: Harvard Univ. Press. [7]

———. 2008. Origins of Human Communication. Cambridge, MA: MIT Press. [7]

Torreira, F., and E. Valtersson. 2015. Phonetic and Visual Cues to Questionhood in French Conversation. *Phonetica* **72**:20–42. [8]

Torrey, L., and M. Taylor. 2013. Teaching on a Budget: Agents Advising Agents in Reinforcement Learning. In: Proc. of the Intl. Conf. on Autonomous Agents and Multi-Agent Systems, pp. 1053–1060. St. Paul: IFAAMAS. [5]

Truit, E. R. 2015. Medieval Robots. Philadelphia: Univ. Pennsylvania Press. [8]

Van Eecke, P., and K. Beuls. 2017. Meta-Layer Problem Solving for Computational Construction Grammar. In: Proc. of AAAI Spring Symp. on Computational Construction Grammar and Natural Language Understanding, Tech. Report SS-17-02, ed. L. Steels and J. Feldman. Palo Alto: AAAI Press. [13]

VanLehn, K. 1988. Student Modeling. In: Foundations of Intelligent Tutoring Systems, ed. M. Polson and J. Richardson, pp. 55–78. Hillsdale: Erlbaum. [12, 13]

———. 1996. Cognitive Skill Acquisition. *Annu. Rev. Psychol.* **47**:513–539. [12]

———. 1999. Rule Learning Events in the Acquisition of a Complex Skill: An Evaluation of Cascade. *J. Learn. Sciences* **8**:179–221. [12]

———. 2008. Intelligent Tutoring Systems for Continuous, Embedded Assessment. In: The Future of Assessment: Shaping Teaching and Learning, ed. C. A. Dwyer, pp. 113–138. New York: Erlbaum. [12]

———. 2011. The Relative Effectiveness of Human Tutoring, Intelligent Tutoring Systems and Other Tutoring Systems. *Educ. Psychol.* **46**:197–221. [12]

———. 2016. Regulative Loops, Step Loops and Task Loops. *Int. J. Artific. Intel. Educ.* **26**:107–112. [11, 12]

VanLehn, K., S. Siler, C. Murray, T. Yamauchi, and W. B. Baggett. 2003. Human Tutoring: Why Do Only Some Events Cause Learning? *Cogn. Instr.* **21**:209–249. [12]

Van Rooij, I., J. Kwisthout, M. Blokpoel, et al. 2011. Intentional Communication: Computationally Easy or Difficult? *Front. Hum. Neurosci.* **5**:1–18. [8]

Veenman, M. V. J., H. A. M. Van Hout-Wolters, and P. Afflerbach. 2006. Metacognition and Learning: Conceptual and Methodological Considerations. *Metacogn. Learn.* **1**:3–14. [13]

Vesper, C., E. Abramova, J. Bütepage, et al. 2016. Joint Action: Mental Representations, Shared Information and General Mechanisms for Coordinating with Others. *Front. Psychol.* **7**:2039. [8]

Vesper, C., L. Schmitz, N. Sebanz, and G. Knoblich. 2009. Joint Action Coordination through Strategic Reduction of Variability. In: Proc. 31st Annual Meeting of the Cognitive Science Society, ed. N. A. Taatgen and H. v. Rijn, pp. 1522–1527. Austin: Cognitive Science Society. [10]

Vollmer, A.-L., K. S. Lohan, K. Fischer, et al. 2009. People Modify Their Tutoring Behavior in Robot-Directed Interaction for Action Learning. In: IEEE 8th Intl. Conf. on Development and Learning, pp. 1–6. Shanghai: IEEE. [11]

Vollmer, A.-L., M. Mühlig, J. J. Steil, et al. 2014. Robots Show Us How to Teach Them: Feedback from Robots Shapes Tutoring Behavior during Action Learning. *PLoS One* **9**:e91349. [11]

Vygotsky, L. S. 1967. Play and Its Role in the Mental Development of the Child. *Soviet Psychol.* **5**:6–18. [14]

Waldersee, R., and F. Luthans. 1994. The impact of positive and corrective feedback on customer service performance. *J. Org. Behav.* **15**:83–95. [12]

Wallach, D., and C. Lebiere. 2003. Conscious and Unconscious Knowledge: Mapping to the Symbolic and Subsymbolic Levels of a Hybrid Architecture. In: Attention and Implicit Learning, ed. L. Jimenez, pp. 112–143. Amsterdam: John Benjamins. [4]

Walsh, M. W., K. A. Gluck, G. Gunzelmann, T. Jastrzembski, and M. Krusmark. 2018. Evaluating the Theoretical Adequacy and Applied Potential of Computational Models of the Spacing Effect. *Cogn. Sci.* **42**:644–691. [11]

Wang, Y., and M. Kosinski. 2017. Deep Neural Networks Can Detect Sexual Orientation from Faces. *J. Personality Soc. Psychol.* **114**:246–257. [18]

Warneken, F., and M. Tomasello. 2006. Altruistic Helping in Human Infants and Young Chimpanzees. *Science* **311**:1301–1303. [16]

Wasik, B. A. 1998. Volunteer Tutoring Programs in Reading: A Review. *Read. Res. Q.* **33**:266–291. [12]

Wasik, B. A., and R. E. Slavin. 1993. Preventing Early Reading Failure with One-to-One Tutoring: A Review of Five Programs. *Read. Res. Q.* **28**:178–200. [12]

West, R. L., and C. Lebiere. 2001. Simple Games as Dynamic, Coupled Systems: Randomness and Other Emergent Properties. *Cogn. Syst. Res.* **1**:221–239. [4]

Whitehead, A. N. 1979. Process and Reality, 2nd edition. New York: The Free Press. [14]

Whitney, D., M. Eldon, J. Oberlin, and S. Tellex. 2016. Interpreting Multimodal Referring Expressions in Real Time. In: 2016 IEEE Intl. Conf. on Robotics and Automation (ICRA), pp. 3331–3338. Stockholm: IEEE. [9]

Wierzbicka, A. 1985. Different Cultures, Different Languages, Different Speech Acts. *J. Pragmatics* **9**:145–178. [18]

Williams, A. M., P. R. Ford, D. W. Eccles, and P. Ward. 2011. Perceptual-Cognitive Expertise in Sport and Its Acquisition: Implications for Applied Cognitive Psychology. *Appl. Cogn. Psychol.* **25**:432–442. [10]

Williams, T., P. Briggs, and M. Scheutz. 2015. Covert Robot-Robot Communication: Human Perceptions and Implications for HRI. *J. Hum. Robot. Interact.* **4**:23–49. [18]

Wilson, J., E. Krause, M. Scheutz, and M. Rivers. 2016. Analogical Generalization of Actions from Single Exemplars in a Robotic Architecture. In: Proc. 15th Intl. Conf. on Autonomous Agents and Multiagent Systems. Singapore: AAMAS. [3]

Wimmer, H., and J. Perner. 1983. Beliefs About Beliefs: Representation and Constraining Function of Wrong Beliefs in Young Children's Understanding of Deception. *Cognition* **13**:103–128. [15]

Wittwer, J., M. Nuckles, N. Landmann, and A. Renkl. 2010. Can Tutors Be Supported in Giving Effective Explanations? *J. Educ. Psychol.* **102**:74–89. [12]

Wood, D. J., J. S. Bruner, and G. Ross. 1976. The Role of Tutoring in Problem Solving. *J. Child Psychol. Psychiatry* **17**:89–100. [12]

Woodward, A. 1998. Infants Selectively Encode the Goal Object of an Actor's Reach. *Cognition* **69**:1–34. [16]

Wray, R. E., S. Lisse, and J. Beard. 2004. Investigating Ontology Infrastructures for Execution-Oriented Autonomous Agents. In: Knowledge Representation and Ontology for Autonomous Systems ed. C. Schlenoff and M. Uschold. AAAI Spring Symposium, Technical Report SS-04-04. Palo Alto: AAAI Press. [4]

Wray, R. E., and A. Woods. 2013. A Cognitive Systems Approach to Tailoring Learner Practice. In: Proc. 2nd Annual Conf. on Advances in Cognitive Systems, ed. J. E. Laird and M. Klenk, pp. 21–38. Baltimore: ACS. [3]

Xu, F., and S. Carey. 1996. Infants' Metaphysics: The Case of Numerical Identity. *Cogn. Psychol.* **30**:111–153. [16]

Yang, S., Q. Gao, C. Liu, et al. 2016. Grounded Semantic Role Labeling. In: Proc. 15th Annual Conf. of the North American Chapter of the Association for Computational Linguistics: Human Language Technologies (NAACL-HLT), ed. R. Mihalcea et al., pp. 149–159. Denver: ACL. [9]

Yechiam, E., I. Erev, and A. Parush. 2004. Easy First Steps and Their Implication to the Use of a Mouse-Based and a Script-Based Strategy. *J. Exp. Psychol. Appl.* **10**:89–96. [10]

Yechiam, E., I. Erev, V. Yehene, and D. Gopher. 2003. Melioration and the Transition from Touch-Typing Training to Everyday Use. *Human Factors* **45**:671–684. [10]

Yee-King, M., M. Krivenski, H. Brenton, A. Grimalt, and M. d'Inverno. 2014. Designing Educational Social Machines for Effective Feedback. In: Internaional Conference on E-Learning, pp. 239–248. Intl. Assoc. for Development of the Information Society. [11]

Yost, G. R. 1992. TAQL: A Problem Space Tool for Expert System Development. PhD thesis, School of Computer Science, Carnegie Mellon Univ., Pittsburgh. [3]

Young, S., M. Gasic, B. Thomson, and J. Williams. 2013. Pomdp-Based Statistical Spoken Dialogue Systems: A Review. *Proc. IEEE* **101**:1160–1179. [9]

Yu, H., and J. M. Siskind. 2013. Grounded Language Learning from Video Described with Sentences. In: Proc. 51st Annual Meeting of the Association for Computational Linguistics, pp. 53–63. Sofia: ACL. [9]

Zacharias, G. L., J. MacMillan, and S. B. Van Hemel, eds. 2008. Behavioral Modeling and Simulation: From Individuals to Societies. Washington, D.C.: National Academies Press. [17]

Zimmerman, B. 2008. Investigating self-regulation and motivation: Historical background, methodological developments and future prospects. *Am. Educ. Res. J.* **45**:166–183. [12]

———. 2010. Becoming a Self-Regulated Learner: An Overview. *Theory. Pract.* **41**:64–70. [13]

Subject Index

Strüngmann Forum Report Series

Agrobiodiversity: Integrating Knowledge for a Sustainable Future
Edited by Karl S. Zimmerer and Stef de Haan, ISBN: 9780262038683

Rethinking Environmentalism: Linking Justice, Sustainability, and Diversity
Edited by Sharachchandra Lele, Eduardo S. Brondizio, John Byrne,
Georgina M. Mace and Joan Martinez-Alier, ISBN: 9780262038966

Emergent Brain Dynamics: Prebirth to Adolescence
Edited by April A. Benasich and Urs Ribary, ISBN: 9780262038638

The Cultural Nature of Attachment: Contextualizing Relationships and Development
Edited by Heidi Keller and Kim A. Bard
Hardcover: ISBN: 9780262036900, ebook: ISBN: 9780262342865
Winner of the Ursula Gielen Global Psychology Book Award

Investors and Exploiters in Ecology and Economics: Principles and Applications
edited by Luc-Alain Giraldeau, Philipp Heeb and Michael Kosfeld
Hardcover: ISBN: 9780262036122, eBook: ISBN: 9780262339797

Computational Psychiatry: New Perspectives on Mental Illness
edited by A. David Redish and Joshua A. Gordon, ISBN: 9780262035422

Complexity and Evolution: Toward a New Synthesis for Economics
edited by David S. Wilson and Alan Kirman, ISBN: 9780262035385

The Pragmatic Turn: Toward Action-Oriented Views in Cognitive Science
edited by Andreas K. Engel, Karl J. Friston and Danica Kragic
ISBN: 978-0-262-03432-6

Translational Neuroscience: Toward New Therapies
edited by Karoly Nikolich and Steven E. Hyman, ISBN: 9780262029865

Trace Metals and Infectious Diseases
edited by Jerome O. Nriagu and Eric P. Skaar, ISBN 978-0-262-02919-3

Rethinking Global Land Use in an Urban Era
edited by Karen C. Seto and Anette Reenberg, ISBN 978-0-262-02690-1

Schizophrenia: Evolution and Synthesis
edited by Steven M. Silverstein, Bita Moghaddam and Til Wykes,
ISBN 978-0-262-01962-0

Cultural Evolution: Society, Technology, Language, and Religion
edited by Peter J. Richerson and Morten H. Christiansen,
ISBN 978-0-262-01975-0

Language, Music, and the Brain: A Mysterious Relationship
edited by Michael A. Arbib, ISBN 978-0-262-01962-0

Available at https://mitpress.mit.edu/books/series/strungmann-forum-reports

Evolution and the Mechanisms of Decision Making
edited by Peter Hammerstein and Jeffrey R. Stevens, ISBN 978-0-262-01808-1

Cognitive Search: Evolution, Algorithms, and the Brain
edited by Peter M. Todd, Thomas T. Hills and Trevor W. Robbins,
ISBN 978-0-262-01809-8

Animal Thinking: Contemporary Issues in Comparative Cognition
edited by Randolf Menzel and Julia Fischer, ISBN 978-0-262-01663-6

Disease Eradication in the 21st Century: Implications for Global Health
edited by Stephen L. Cochi and Walter R. Dowdle, ISBN 978-0-262-01673-5

Dynamic Coordination in the Brain: From Neurons to Mind
edited by Christoph von der Malsburg, William A. Phillips and Wolf Singer,
ISBN 978-0-262-01471-7

Linkages of Sustainability
edited by Thomas E. Graedel and Ester van der Voet, ISBN 978-0-262-01358-1

Biological Foundations and Origin of Syntax
edited by Derek Bickerton and Eörs Szathmáry, ISBN 978-0-262-01356-7

*Clouds in the Perturbed Climate System: Their Relationship to Energy Balance,
Atmospheric Dynamics, and Precipitation*
edited by Jost Heintzenberg and Robert J. Charlson, ISBN 978-0-262-01287-4
Winner of the Atmospheric Science Librarians International Choice Award

*Better Than Conscious? Decision Making, the Human Mind, and Implications
For Institutions*
edited by Christoph Engel and Wolf Singer, ISBN 978-0-262-19580-5